AUTOMATIC TECHNOLOGY IN FOOD INDUSTRY

AUTOMATIC TECHNOLOGY IN FOOD INDUSTRY

Cristina García Jaime

ARCLER PRESS

www.arclerpress.com

Automatic Technology in Food Industry

Cristina García Jaime

Arcler Press

2010 Winston Park Drive,

2nd Floor

Oakville, ON L6H 5R7

Canada

www.arclerpress.com

Tel: 001-289-291-7705
001-905-616-2116
Fax: 001-289-291-7601
Email: orders@arclereducation.com

ISBN: 978-1-77361-365-9 (Hardcover)

ABOUT THE AUTHOR

Cristina Garcia Jaime is a Spanish food specialist who studied Food Quality and Technology in the University of Granada and Human Nutrition and Dietetics in Pablo de Olavide University (Seville). She has worked in different food areas: scientific, industry and clinic. Currently, she has founded a company established in Europe and called "Foodise" dedicated to help small business and companies to manage all issues related to food.

TABLE OF CONTENTS

LIST OF FIGURES

LIST OF TABLES

PREFACE

Due to growing completion and rapidly changing environment, all food processing industries are going through an era of transition where they are changing their way of processing food products. In comparison to other manufacturing industries, the adaptation of automatic technologies in food industries is slow. Process controls on the basis of computers is an essential tool in automatic food processing. Enhancement of the final food product and overall process can be simply attained by implementing the recently developed technology called "Automatic Technology in Food Industry." Automation can be employed into the food industry from the raw material to final processed food product. For next coming several years, automation will continue to rise at a higher pace.

The changes in technology have been influenced by a variety of factors: significant increases in the expenditures of both energy and labor, by public pressure and legislation to reduce negative ecological effects of processing, especially regard to air or water pollution and energy consumption. Food processing equipment has now reached an innovative stage where all levels of control are computerized to lessen processing costs, enable rapid change-overs between diminutive production runs, to enhance product quality and to provide improved records for management resolutions. Microprocessors are now almost generally used to control food-processing equipment. The automation of entire processes, from the reception of materials, through processing and packaging to warehousing, has become a certainty. This requires a higher capital investment by industrialists but, together with value-added quality assurance, reduces production costs and wastage. It increases production efficiency, uses less energy and fewer operators, and creates better revenue and market share from goods that have higher quality.

This book will trace how the term automatic technology was coined and its evolution from a theoretical concept to a practical one. There is a plethora of information out there on the development of process controls in food production but each situation demands a customized approach. The existing food processing practices can at best give clues on what improvements can be made, but it is the thorough study of individual process patterns that can give rise to efficient strategies that can work in real life. As it is defined in this book, automation generally refers to the adaptation of science and technology to carry out the process which maximum accuracy and efficiency. It combines several factors that affect the processing of a food product. At the same time, the book will offer very clear insights on the perceptions that can be worked upon for change which will eventually drive the industry's overall efficiency and productivity.

The subject matter of this book starts from establishing a clear picture of how each automation plays out to the dynamics of the food processing. The education of this approach will contribute to widen the understanding on improving existing food technology. This would be supported by real-life case studies. This section would also present the existing areas of improvement and challenges included under the various segments aimed at improving the utilization of resources. Towards the end, a comprehensive detail of the existing challenges would be covered.

CHAPTER 1

AUTOMATIC PROCESS CONTROL FOR THE FOOD INDUSTRY

"In the 21st century our tastes buds, our brain chemistry, our biochemistry, our hormones and our kitchens have been hijacked by the food industry."

—**Mark Hyman**

CONTENTS

As a way to make sure food security in food-manufacturing operations, automated management methods is looked-for. With the operation of automated procedure-control systems the deviation of the controlled variables from standards can be consistently monitored, adjusted, and minimized to improve the manner operations on an everyday foundation. Proportional integral-derivative (PID) control has been widely used inside the food industry. Version-based control has been developed to enhance the performance of manage structures inside the food enterprise. This chapter overviews the ideas, strategies, and systems of automated technique control for the food industry, and initiatives the destiny of computerized manner manage.

1.1. INTRODUCTION

The food industry has usually fall behind other industries in embracing new innovation, and plant mechanization is no special case. In any case, quick advances in PC innovation and uplifted desires of purchasers and regulatory agencies for enhanced food quality and well-being have constrained the food industry to think about computerization of most assembling forms. In spite of the fact that food industry presents numerous exceptional difficulties to finish computerization, the industry has been effective in instituting numerous programmed forms.

The following huge advancement will be to incorporate these "islands of mechanization" into a general arrangement of plant computerization, from getting crude materials to transportation completed items. New innovative instruments, for example PC vision, master frameworks, PC incorporated assembling, adaptable assembling frameworks, frameworks building, and so on., have empowered coordination of numerous batch operations into a general assembling framework configuration to give on-line and persistent control capacity. This pattern will proceed at a significantly quicker pace in the following quite a while.

The computerization of assembling plants has been effectively sought after for over 50 years. What's more, it will keep on being in this way, considerably more forcefully, amid the following 50 years. The expanded enthusiasm in modern robotization is predominantly because of the unstable development in PC equipment and programming innovation. As PCs attack relatively every part of our day-by-day lives, general society everywhere has generally expected an abnormal state of computerization in each feature of the assembling forms.

The degree of modern robotization depends on incredible arrangement on the sort of industry. The automobile and semiconductor businesses signify the most developed nature in receiving plant mechanization standards with about the sum total of what forms having been robotized and genuinely very much incorporated. At the opposite end of the range is may be the food business, signifies to bring down levels of robotization, which has generally lingered behind in embracing innovative advances. The present level of computerization in the sustenance business has been portrayed as "islands of mechanization." In any case, the food business now positions among the quickest developing portions for plant robotization. For instance, the food industry is among the main ten in utilizing machine vision innovation, a key part in plant mechanization. In any case, most frameworks are disengaged, bunch compose activities that objective a particular undertaking. With the end goal for robotization to be effective, it must be coordinated into the general assembling framework outline and give on-line, ceaseless control capacity.

1.1.1. Automation

Automation generally refers to the science and technology to carry out the process which maximum accuracy and efficiency. Automation aims to take full advantage of value-cycle opportunities by amalgamation of knowledge and automation technologies. This will extend the company's capabilities and promote global collaboration by providing continuous access to information. To achieve this, it's necessary to pay special attention to the career development of automation professionals, more leverage of existing automated facilities and partners, and development of best practices in delivery and maintenance.

The Oxford English Dictionary defines automation in terms of manufacturing as: "Automatic control of the manufacture of a product through a number of successive stages; the application of automatic control to any branch of industry or science; by extension, the use of electronic or mechanical devices to replace human labor (2006). Automation is a technology concerned with the application of mechanical, electronic, and computer based systems to operate and control production. This technology includes:

- automatic machines tools to process parts.
- automatic assembly machines.

- industrial robots.
- automatic material handling and storage systems.
- automatic inspection systems for quality control.
- feedback control and computer process control.
- computer systems for planning, data collection, and decision making to support manufacturing facilities.

1.1.2. Manufacturing Support Systems

To operate the production facilities efficiently, a company must organize itself to design the processes and equipment, plan and control the production orders, and satisfy product quality requirement. These functions are accomplished by manufacturing support systems —people and procedures by which a company manages its production operations. Most of these support systems do not directly contact the product, but they plan and control its progress through the factory. Manufacturing support involves a cycle of information-processing activities. The information-processing cycle, represented by the outer ring, can be described as consisting of four functions: (1) business functions, (2) product design, (3) manufacturing planning, and (4) manufacturing control.

1.1.2.1. Business Functions

The business functions are the principal means of communicating with the customer. They are therefore, the beginning and the end of the information-processing cycle. Included in this category are sales and marketing, sales forecasting, order entry, cost accounting, and customer billing. The order to produce a product typically originates from the customer and proceeds into the company through the sales and marketing department of the firm. The production order will be in one of the following forms:

- an order to manufacture an item to the customer's specifications.
- a customer order to buy one or more of the manufacturer's proprietary products.
- an internal company order based on a forecast of future demand for a proprietary product.

1.1.2.2. Product Design

If the product is to be manufactured to customer design, the design will have been provided by the customer. The manufacturer's product design department will not be involved. If the product is to be produced to customer specifications, the manufacturer's product design department maybe contracted to do the design work for the product as well as to manufacture it. If the product is proprietary, the manufacturing firm is responsible for its development and design. The cycle of events that initiates a new product design often originates in the sales and marketing department. The departments of the firm that are organized to accomplish product design might include research and development, design engineering, drafting, and perhaps a prototype shop.

1.1.2.3. Manufacturing Planning

The information and documentation that constitute the product design flows into the manufacturing planning function. The information-processing activities in manufacturing planning include process planning, master scheduling, requirements planning, and capacity planning. Process planning consists of determining the sequence of individual processing and assembly operations needed to produce the part. The manufacturing engineering and industrial engineering departments are responsible for planning the processes and related technical details. Manufacturing planning includes logistics issues, commonly known as production planning.

The authorization to produce the product must be translated into the master production schedule. The master production schedule is a listing of the products to be made, when they are to be delivered and in what quantities. Months are traditionally used to specify deliveries in the master schedule. Based on this schedule, the individual components and subassemblies that make up each product must be planned. Raw materials must be purchased or requisitioned from storage, purchased parts must be ordered from suppliers and all of these items must be planned so that they are available when needed. This whole task is called material requirements planning. A function called capacity planning is concerned with planning the manpower and machine resources of the firm.

1.1.2.4. Manufacturing Control

Activities include production control, product realization, Inspection and Testing, packaging and labeling, final verification and release, production change control.

1.1.3. Automated Manufacturing Systems

Automated manufacturing systems operate in the factory on the physical product. They perform operations such as processing, assembly, inspection, or material handling, in some cases accomplishing more than one of these operations in the same system. They are called automated because they perform their operations with a reduced level of human participation compared with the corresponding manual process. In some highly automated systems, there is virtually no human participation. Examples of automated manufacturing systems include:

(a) Automated machine tools that process parts.

(b) Transfer lines that perform a series of machining operations.

(c) Automated assembly systems.

(d) Manufacturing systems that use industrial robots to perform processing or assembly operations.

(e) Automatic material handling and storage systems to integrate manufacturing operations.

(f) Automatic inspection systems for quality control.

Automated manufacturing systems can be classified into three basic types. In the following sections, we shall explore the topic of automation in great detail. These are fixed, programmable and flexible automation.

Fixed Automation

It is a system in which the sequence of processing (or assembly) operations is fixed by the equipment configuration. The operations in the sequence are usually simple. It is the integration and coordination of many such operations into one piece of equipment that makes the system complex. The typical features of fixed automation are:

(i) high initial investment for custom–engineered equipment.

(ii) high production rates.

(iii) Relatively inflexible in accommodating product changes.

The economic justification for fixed automation is found in products with very high demand rates and volumes. The high initial cost of the equipment can be spread over a very large number of units, thus making the unit cost attractive compared to alternative methods of production. Examples of fixed automation include mechanized assembly and machining transfer lines.

Programmable Automation

Programmable automation is a form of automation for producing products in batches. The products are made in batch quantities ranging from several dozen to several thousand units at a time. For each new batch, the production equipment must be reprogrammed and changed over to accommodate the new product style. This reprogramming and changeover take time to accomplish, and there is a period of nonproductive time followed by a production run for each new batch. Production rates in programmable automation are generally lower than in fixed automation, because the equipment is designed to facilitate product changeover rather than for product specialization. A numerical-control machine tool is a good example of programmable automation. The program is coded in computer memory for each different product style, and the machine tool is controlled by the computer program. Industrial robots are another example.

Flexible Automation

The features of flexible automation can be summarized as follows:

(a) High investment for a custom–engineered system.

(b) Continuous production of variable mixtures of products.

(c) Medium production rates.

(d) Flexibility to deal with product design variations.

The essential features that distinguish flexible automation from programmable automation are:

(a) The capacity to change part programs with no lost production time.

(b) The capability to changeover the physical setup, again with no lost production time.

These features allow the automated production system to continue production without the downtime between batches that is characteristic of programmable automation. Changing the part programs is generally accomplished by preparing the programs offline on a computer system and electronically transmitting the programs to the automated production system. Therefore, the time required to do the programming for the next job does not interrupt production on the current job. Advances in computer systems technology are largely responsible for this programming capability in flexible automation.

Changing the physical setup between parts is accomplished by making the changeover offline and then moving it into place simultaneously as the next part comes into position for processing. The use of pallet fixtures that hold the parts and transfer into position at the workplace is one way of implementing this approach. For these approaches to be successful; the variety of parts that can be made on a flexible automated production system is usually more limited than a system controlled by programmable automation.

1.1.4. Production System Facilities

The facilities in the production system are the factory, production machines and tooling, material handling equipment, inspection equipment, and the computer systems that control the manufacturing operations. Facilities also include the plant layout, which is the way the equipment is physically arranged in the factory. The equipment is usually arranged into logical groupings, and we refer to these equipment arrangements and the workers who operate them as the manufacturing systems in the factory. Manufacturing systems can be individual work cells, consisting of a single production machine and worker assigned to that machine. We more commonly think of manufacturing systems as groups of machines and workers, for example, a production line. The manufacturing systems come in direct physical contact with the parts and/or assemblies being made. A manufacturing company attempts to organize its facilities in the most efficient way to serve the particular mission of that plant. Over the years, certain types of production facilities have come to be recognized as the most appropriate way to organize for a given type of manufacturing. Of course, one of the most important factors that determine the type of manufacturing is the type of products that are made.

Production quantity refers to the number of units of a given part or product produced annually by the plant. The annual part or product quantities produced in a given factory can be classified into three ranges:

(a) Low production: Quantities in the range of 1 to 100 units per year.

(b) Medium production: Quantities in the range of 100 to 10,000 units annually.

(c) High production: Production quantities are 10,000 to millions of units.

The boundaries between the three ranges are somewhat subjective. Depending on the types of products we are dealing with, these boundaries may shift by an order of magnitude or so. Some plants produce a variety of different product types, each type being made in low or medium quantities. Other plants specialize in high production of only one product type. It is instructive to identify product variety as a parameter distinct from production quantity.

Product variety refers to the different product designs or types that are produced in a plant. Different products have different shapes and sizes and styles; they perform different functions; they are sometimes intended for different markets; some have more components than others; and so forth. The number of different product types made each year can be counted. When the number of product types made in a factory is high, this indicates high product variety. There is an inverse correlation between product variety and production quantity in terms of factory operations. When product variety is high, production quantity tends to be low; and vice versa. We can use the three production quantity ranges to identify three basic categories of production plants.

Low Quantity Production

The type of production facility usually associated with the quantity range of 1 to 100 units per year is the job shop, which makes low quantities of specialized and customized products. The products are typically complex, such as space capsules, aircraft, and special machinery. Job shop production can also include fabricating the component parts for the products. Customer orders for these kinds of items are often special, and repeat orders may never occur. Equipment in a job shop is general purpose and the labor force is highly skilled. The distinguishing feature of job shop production is low volume.

The manufacturing lot sizes are small, often one of a kind. Job shop production is commonly used to meet specific customer orders, and there is a great variety in the type of work the plant must do. Therefore, the production equipment must have multipurpose or be flexible to allow for this variety of work. Also, the skill level of job shop workers must be relatively high so that they can perform a range of different work assignments. Examples of products manufactured in a job shop include space vehicles, aircraft, machine tools, special tools and equipment, and prototypes of future products.

A job shop must be designed for maximum flexibility to deal with the wide part and product variations encountered (hard product variety). If the product is large and heavy, and therefore difficult to move in the factory, it typically remains in a single location, at least during its final assembly. Workers and processing equipment are brought to the product, rather than moving the product to the equipment. This type of layout is referred to as a fixed-position layout. In the pure situation, the product remains in a single location during its entire fabrication. Examples of such products include ships, aircraft, railway locomotives, and heavy machinery. In actual practice, these items are usually built in large modules at single locations, and then the completed modules are brought together for final assembly using large-capacity cranes.

The individual parts that comprise these large products are often made in factories that have a process layout, in which the equipment is arranged according to function or type. The lathes are in one department, the milling machines are in another department.

Medium Quantity Production

In the medium quantity range (100–10,000 units annually), we distinguish between two different types of facility, depending on product variety. When product variety is hard, the traditional approach is batch production, in which a batch of one product is made, after which the facility is changed over to produce a batch of the next product, and so on. Orders for each product are frequently repeated. The production rate of the equipment is greater than the demand rate for any single product type, and so the same equipment can be shared among multiple products. The changeover between production runs takes time. This is the Setup time or changeover time and it is the time to change tooling and to set up and reprogram the machinery. This is lost production time, which is a disadvantage of batch manufacturing. Batch production is commonly used in make-to-stock conditions, in which items are manufactured to replenish inventory that has been gradually depleted by demand.

The equipment is usually arranged in a process layout. This category involves the manufacture of medium-sized lots of the same item or product. The lots may be produced only once, or they may be produced at regular intervals. The purpose of batch production is often to satisfy continuous customer demand for an item. However, the plant is capable of a production rate that exceeds the demand rate. Therefore, the shop produces to build up an inventory of the item. Then it changes over to other orders. When the stock of the first item becomes depleted, production is repeated to build up the inventory again. The manufacturing equipment used in batch production is general-purpose but designed for higher rates of production.

An alternative approach to medium range production is possible if product variety is soft. In this case, extensive changeovers between one product style and the next may not be required. It is often possible to configure the equipment so that groups of similar parts or products can be made on the same equipment without significant lost time for changeovers. The processing or assembly of different parts or products is accomplished in cells consisting of several workstations or machines. The term cellular manufacturing is often associated with this type of production. Each cell is designed to produce a limited variety of part configurations; that is, the cell specializes in the production of a given set of similar parts or products, according to the principles of group technology. The layout is called a cellular layout.

High Production

The high quantity range (10,000 to millions of units per year) is often referred to as mass production. The situation is characterized by a high demand rate for the product, and the production facility is dedicated to the manufacture of that product. Two categories of mass production can be distinguished as:

(a) Quantity production.

(b) Flow line production.

Quantity production involves the mass production of single parts on single pieces of equipment. The method of production typically involves standard machines (such as stamping presses) equipped with special tooling (e.g., dies and material handling devices), in effect dedicating the equipment to the production of one part type. The typical layout used in quantity production is the process layout. Mass production is characterized by very high production rates, equipment that is completely dedicated to the manufacture of a particular product, and very high demand rates for the product.

Not only is the equipment dedicated to one product, but also the entire plant is often designed for the exclusive purpose of producing the particular product. The equipment has a singular purpose and is not for general usage. The investment in machines and specialized tooling is high. In a sense, the production skill has been transferred from the operator to the machine. Consequently, the skill level of labor in a mass production plant tends to be lower than in a batch plant or job shop. Flow line production involves multiple workstations arranged in sequence, and the parts or assemblies are physically moved through the sequence to complete the product.

The workstations consist of production machines and/or workers equipped with specialized tools. The collection of stations is designed specifically for the product to maximize efficiency. The layout is called a product layout, and the workstations are arranged into one long line or into a series of connected line segments. The work is usually moved between stations by powered conveyor. At each station, a small amount of the total work is completed on each unit of product.

1.1.5. Computer Integrated Manufacturing (CIM)

The term computer integrated manufacturing (CIM) denotes the pervasive use of computer systems to design the products, plan the production, control the operations, and perform the various business—related functions needed in a manufacturing firm. True CIM involves integrating all of these functions in one system that operates throughout the enterprise. Other terms are used to identify specific elements of the CIM system. A number of activities and operations viz. designing, analyzing, testing, manufacturing, packaging, quality control, etc. are involved in the life cycle of a product or a system. Application of principles of automation to each of these activities enhances the productivity only at the individual level. These are termed as 'islands of automation.' Integrating all these islands of automation into a single system enhances the overall productivity. Such a system is called as "Computer Integrated Manufacturing (CIM)."

The Society of Manufacturing Engineers (SME) defined CIM as 'CIM is the integration of the total manufacturing enterprise through the use of integrated systems and data communications coupled with new managerial philosophies that improve organizational and personal efficiency.'

Today CIM has moved a step ahead by including and integrating the business improvement activities such as customer satisfaction, total quality and continuous improvement. These activities are now managed by computers. Business and marketing teams continuously feed the customer feedback to the design and production teams by using the networking systems. Based on the customer requirements, design and manufacturing teams can immediately improve the existing product design or can develop an entirely new product. Thus, the use of computers and automation technologies made the manufacturing industry capable to provide rapid response to the changing needs of customers.

Computer Integrated Manufacturing (CIM) encompasses the entire range of product development and manufacturing activities with all the functions being carried out with the help of dedicated software packages. The data required for various functions are passed from one application software to another in a seamless manner. For example, the product data is created during design. This data has to be transferred from the modeling software to manufacturing software without any loss of data. CIM uses a common database wherever feasible and communication technologies to integrate design, manufacturing and associated business functions that combine the automated segments of a factory or a manufacturing facility. CIM reduces the human component of manufacturing and thereby relieves the process of its slow, expensive and error-prone component. CIM stands for a holistic and methodological approach to the activities of the manufacturing enterprise in order to achieve vast improvement in its performance.

Role of CIM

(i) Marketing: The need for a product is identified by the marketing division. The specifications of the product, the projection of manufacturing quantities and the strategy for marketing the product are also decided by the marketing department. Marketing also works out the manufacturing costs to assess the economic viability of the product.

(ii) Product Design: The design department of the company establishes the initial database for production of a proposed product. In a CIM system this is accomplished through activities such as geometric modeling and computer aided design while considering the product requirements and concepts generated by the creativity of the design engineer. Configuration management is an important activity in many designs. Complex designs are usually carried out by several teams working simultaneously, located often in different parts of the world. The design process is constrained by the costs that will be incurred in actual production and by the capabilities of the available production equipment and processes. The design process creates the database required to manufacture the part.

(iii) Planning: The planning department takes the database established by the design department and enriches it with production data and information to produce a plan for the production of the product.

Planning involves several subsystems dealing with materials, facility, process, tools, manpower, capacity, scheduling, outsourcing, assembly, inspection, logistics etc. In a CIM system, this planning process should be constrained by the production costs and by the production equipment and process capability, in order to generate an optimized plan.

(iv) Purchase: The purchase departments is responsible for placing the purchase orders and follow up, ensure quality in the production process of the vendor, receive the items, arrange for inspection and supply the items to the stores or arrange timely delivery depending on the production schedule for eventual supply to manufacture and assembly.

(v) Manufacturing Engineering: Manufacturing Engineering is the activity of carrying out the production of the product, involving further enrichment of the database with performance data and information about the production equipment and processes. In CIM, this requires activities like CNC programming, simulation and computer aided scheduling of the production activity. This should include online dynamic scheduling and control based on the real time performance of the equipment and processes to assure continuous production activity. Often, the need to meet fluctuating market demand requires the manufacturing system flexible and agile.

(vi) Factory Automation Hardware: Factory automation equipment further enriches the database with equipment and process data, resident either in the operator or the equipment to carry out the production process. In CIM system this consists of computer controlled process machinery such as CNC machine tools, flexible manufacturing systems (FMS), Computer controlled robots, material handling systems, computer controlled assembly systems, flexibly automated inspection systems and so on.

(vii) Warehousing: Warehousing is the function involving storage and retrieval of raw materials, components, finished goods as well as shipment of items. In today's complex outsourcing scenario and the need for just-in-time supply of components and subsystems, logistics and supply chain management assume great importance.

(viii) Finance: Finance deals with the resources pertaining to money. Planning of investment, working capital, and cash flow control, realization of receipts, accounting and allocation of funds are the major tasks of the finance departments. ix. Information Management: Information Management is perhaps one of the crucial tasks in CIM. This involves master production scheduling, database management communication, manufacturing systems integration and management information systems.

1.2. WHY AUTOMATE?

The need to computerize mechanical procedures is driven by a few key prerequisites for focused achievement and, in a few enterprises, suitability of the assembling plants. They can be recorded as those expecting to enhance efficiency, item quality, and benefit. This is portrayed schematically in Figure 1.1.

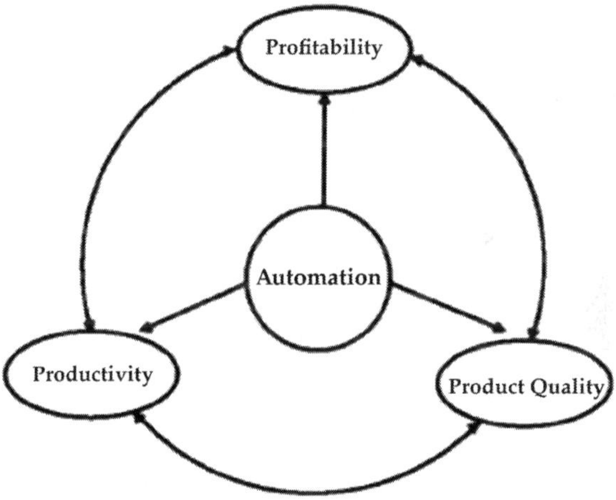

Figure 1.1: Plant automation can improve productivity, product quality, and profitability, (*Source: S. Gunasekaran, Automation of Food Processing, Food Engineering–Vol. 4*).

1.2.1. Improved Productivity

Plant efficiency might be characterized as the amount of final results produced per unit of working parameters–plant estimate, number of laborers, time of task, and so forth. In this way, profitability is straightforwardly identified with how productively the input resources are used in making an interpretation of them into attractive finished results. This is conceivable in light of the fact that mechanization takes into consideration productive planning of work process and work utilize. The capacity to maintain great records and data about past procedures can plainly feature zones that can be focused for a more productive distribution of assets. One plant revealed a 30% expansion in plant efficiency by utilizing three discrete microchip based controllers intended to play out every single nonstop circle including intricate, incorporated calculations, valve interlocking, and some sequencing. Comparable controls can likewise be utilized to improve details, generation planning, and process demonstrating.

1.2.2. Improved Product Quality

Quality affirmation is a standout amongst the most imperative objectives of any industry. The capacity to make astounding items reliably is the reason for achievement in the exceedingly aggressive food industry. Excellent items empower client unwaveringness and brings about a growing piece of the overall industry. Quality confirmation techniques utilized as a part of the food industry have generally included human visual assessment. Such techniques are dreary, arduous, tedious, and conflicting. As plant profitability expanded and quality resilience fixed, it ended up fundamental for the food industry to utilize programmed techniques for quality confirmation and quality control. Indeed, this part of food manufacture is one of the territories that has gotten the most consideration as far as mechanization. Because of advances in PC vision innovation, considerable changes have been executed in nourishment plants to encourage programmed sustenance quality assessment.

1.2.3. Improved Profitability

Expanded or improved profitability is maybe most critical from the viewpoint of administration. Enhanced productivity adds to investor esteem as well as enables administration to put deliberately in growing plant tasks, expanding product offerings, additionally enhancing item quality, and so on. As talked about already, robotization enhances efficiency and item quality. Both of these contribute specifically to enhanced benefit.

Another imperative factor that makes robotization to a great degree basic for the food industry is the need to consent to sustenance security and ecological administrative organizations. PC controlled plant tasks give basically boundless chances to keep up records of all occasions in plant activity. Besides, the capacity to gather, store, recover, and process information enables plants to distinguish territories of concern. This data would then be able to promptly be utilized for enhanced efficiency, item quality, and benefit. For instance, producing fixing utilization reports helps in dynamic stock control. Such reports can be produced for day-by-day, week after week, month to month, and yearly use to give a quantitative picture of correlations fundamental for future arranging. Shrewd frameworks can likewise screen and record occasional and transient varieties in item factors. An administrator can utilize these records to screen ongoing, modify set focuses, change framework setups, and perform testing, and so on.

1.3. UNIQUENESS OF THE FOOD INDUSTRY

A standout amongst the most imperative explanations behind expanded enthusiasm for mechanizing the food industry is its cost structure. Food processing is very work concentrated, with work costs at anything up to 50% of the item cost. Enhancing efficiency and diminishing work expenses will in this way significantly affect benefit. A great part of the manual work in food processing requires quick, dull, and tedious development and, thus, low levels of inspiration are frequently found. This prompts low quality control and a high frequency of mechanical mishaps. The tedious idea of the work has brought about a significant therapeutic cost to the business. Mechanizing dull errands will enhance quality control and productivity and lessen the abnormal state of mischances.

One of the most imperative snags in the mechanization of food manufacturing is the organic variety in size, shape, and homogeneity of the crude materials. A few materials (e.g., dairy) loan themselves promptly to programmed preparing on the grounds that the crude material (drain) can be taken care of in mass. As needs be, the dairy business is among the most robotized. In any case, materials, for example, organic products, vegetables, meat, and so on, should be taken care of on a more individual unit premise. This has hampered robotization colossally. Consequently, nourishment industry computerization requires a level of adaptability unprecedented to other develop enterprises.

Extra issues are because of the absence of finish physical and compound portrayal of foods. Notwithstanding when finish data is accessible, the crude material or the final result can change. Changes in the crude material emerge from the presentation of new assortments or potentially varieties in agronomic conditions. The finished result can change because of nonstop reformulation of product offerings to pick up piece of the pie. Use of PC vision innovation is generously changing the quality assessment errands in the food business.

Notwithstanding an items physical attributes, factors, for example, microbiological and biochemical concerns put extra confinements on taking care of and preparing techniques utilized. The mechanical, warm, and tactile properties of food materials additionally require particular breaking points on the nature and degree of preparing steps. These requirements muddle process computerization.

Materials that are not very much characterized fit as a fiddle are frequently exhibited in an irregular, unconstrained introduction. They should regularly be taken care of painstakingly to avoid harm and subsequently challenge the abilities of current innovation.

Subsequently, there are couple of cases of "hard" robotization–types that give practically no stipend for changeability. For instance, programmed hardware for peeling natural product still depends vigorously on standard shapes and sizes. Some amazing cases of mechanization incorporate handling and bundling of crisp eggs and dairy items. Albeit most heating is done in programmed stoves, prepared items are still physically reviewed and bundled. In the poultry business, robotization is conceivable as long as the feathered creatures are evaluated into various weight classes. Advancement in this industry can offer programmed butchering, culling, and washing, beheading, and gutting of poultry cadavers at a genuinely high rate. Another business activity is programmed angle handling. A mechanized butchering framework isolates the eatable loin segments of transverse fish cuts. Each cut is checked by a PC vision framework, and control signals are transmitted to a cutting arm that at that point isolates the palatable segment. Vision frameworks have additionally been effectively utilized as a part of robotizing sweet corn preparing.

Food plant floors are somewhat unfriendly to gadgets and PCs. Temperature extremes, vibration, clean and, particularly wash down, would all be able to meddle with the activity of PC equipment. Lower levels of equipment, for example, controllers are not by and large influenced by these issues since they are particularly intended for plant utilize. In any case, PCs are more inclined to harm. Processors needing PCs on plant floors ought to either purchase units particularly adjusted for roughness, which cost a few times that of general units, or else find all registering hardware in a mechanical nook far from the plant floor. Clearly, cost and comfort ought to be measured precisely to dodge costly PC downtime when legitimate safety measures are not taken.

1.4. TOOLS OF AUTOMATION

1.4.1. Computer Vision Systems

PC vision is the science that builds up the hypothetical and algorithmic premise by which helpful data around a question or scene can be consequently extricated and investigated from a watched picture, picture set, or picture arrangement. A picture can be characterized as a spatial portrayal of a question or scene. A computerized monochrome picture is a two-dimensional (2-D) light power work signified by I(x,y), where the esteem or sufficiency of force I at spatial directions x and y is commonly relative to the brilliant vitality got in the electromagnetic band, to which the sensor or indicator (camera) is delicate in a little zone around the point (x,y). To the extent the PC is concerned, the picture is a lattice x,y of numeric esteems, each speaking to a quantized picture power esteem. Every lattice passage is known as a pixel (short for picture component). The aggregate number of pixels in a picture is dictated by the span of the 2-D exhibit utilized as a part of the camera. The force of the monochrome picture is known as the dim level. The impact of a protest's size, shape, position, introduction, and different traits from the spatial dissemination of dark levels requires the ability to deduce which pixels have a place with the question and which don't. At that point, from the pixels having a place with the question, it requires the ability to recognize those protest highlights of intrigue. Numerous calculations and preparing strategies have been produced to interpret the dim levels of a pixel in a way that highlights wanted highlights.

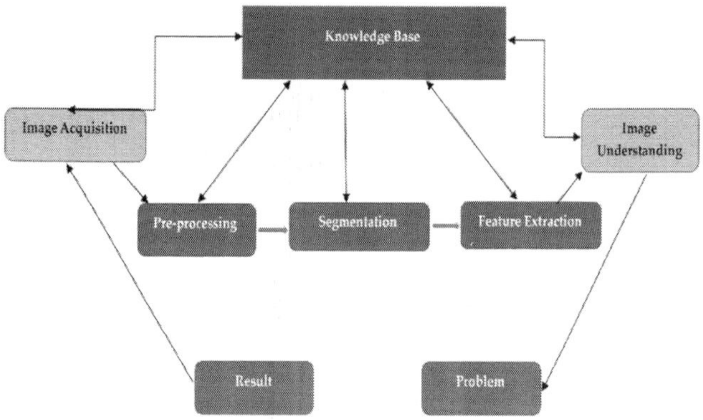

Figure 1.2: Essential steps in computer vision systems application, (Source: S. Gunasekaran, *Automation of Food Processing, Food Engineering–Vol. IV)*.

By and large, a machine vision framework is utilized to gauge some part of the consequences of the assembling procedure (e.g., shape, measure, surface, area) that is demonstrative of the precision, proficiency, or nature of the procedure. The deliberate parameters would then be able to be utilized as input in a constant control circle that streamlines the assembling procedure through varieties in process parameters (speed, temperature, stream rate, and so on.). The basic strides in such a procedure are picture securing, picture preparing, and picture understanding (Figure 1.2). The picture handling step includes a few key activities, for example, pre-preparing, division, and highlight extraction. The picture preparing step depends vigorously on the information of the item being assessed and the idea of the imperfection or quality characteristics of intrigue. Together, they encourage picture understanding and empower official choices to be made.

PC vision applications right now run from basic review to vision-guided automated get together. Most down to earth applications can be gathered into six general classifications:

- Gauging–performing exact dimensional estimations.
- Verification–subjectively guaranteeing that at least one wanted highlights are available or potentially undesired highlights are truant.
- Flow recognition–finding and separating undesirable highlights of obscure size, area, and shape
- Identification–deciding the personality of a protest from images, including alphanumeric characters.
- Recognition–deciding the character of a protest from watched highlights.
- Locating–deciding the area and introduction of a protest.

Some key prerequisites in choosing suitable PC vision take after:

- Speed of task–the framework must work progressively, i.e., at creation line speed.
- Robustness–the capacity to work legitimately in a nourishment plant condition.
- Tolerance–the capacity to endure satisfactory varieties in the item.
- Accuracy–the capacity to distinguish required highlights with an abnormal state of precision.

- Flexibility–the capacity to consider changes in set focuses, working calculations, controls and so forth.
- Reliability–the capacity to play out an assortment of examination undertakings reliably and more than once finished a drawn out stretch of time.

Because of advances in hardware and PC innovation, vision frameworks can be introduced in all food plants for a financially savvy quality assessment/ control activity. Be that as it may, vision frameworks must be precisely planned around the specific qualities of items being assessed with the goal that they perform dependably under plant conditions. Noteworthy research and investigation is important to decide the best technique for upgrading and recognizing item absconds. This implies pre-testing countless items with required deformities or highlights.

Recent improvements in vision frameworks incorporate shading picture handling and three dimensional (3-D) picture preparing. These advancements offer extra advantages, particularly for all the more difficult investigation undertakings. PC vision frameworks for dissecting on-line or moving scenes are additionally being intended for various applications. Contribution to a dynamic or moving scene investigation is as an arrangement of picture outlines taken from an evolving world. The camera used to get the picture arrangement may likewise be in movement. Each casing speaks to a picture of the scene at a specific moment. Changes in a scene might be because of movement of the camera or the items, enlightenment changes, or changes in a protest's structure, size, or shape. It is typically expected that adjustments in scene are because of camera and additionally question movement. A framework must identify changes, decide movement attributes of the onlooker and items, describe the movement utilizing abnormal state reflection, recoup the structure of the articles, and perceive moving articles.

Contingent upon the plan of the imaging framework, distinctive picture preparing methods are required. In the food industry, the most well-known plan is that of a stationary camera and moving items. On the off chance that an edge grouping is procured at a rate with the end goal that no sensational change happens between two sequential casings, at that point no unexpected change in movement can be watched for most physical articles. This is the reason for almost all on-line applications right now accessible in the nourishment business. The vital factor, at that point, is to set the picture securing rate sufficiently quick to limit picture obscure so an examination of picture information can happen outline by-outline. Ongoing picture preparing loads up and constant processors are accessible to aid on-line continuous PC vision applications. Figure 1.3 portrays an on-line item quality assessment and control framework utilizing a PC vision framework. For a nonstop stream of material streaming along a transport line, a PC vision framework can be outlined utilizing a line-filter camera for picture procurement. A line-filter camera contains a one-dimensional exhibit of photosensitive destinations. The line-check camera is appropriate for decently quickly moving article scenes. Notwithstanding higher paces, line scan cameras offer high determination and the capacity to deal with vastly long picture scenes. Another type of cameras, known as time-delay and coordinated cameras, are line-filter cameras, which utilize charge couple gadget picture sensor innovation to pick up speed or increment affectability up to 100-crease (that of ordinary cameras) while giving extraordinary spatial determination.

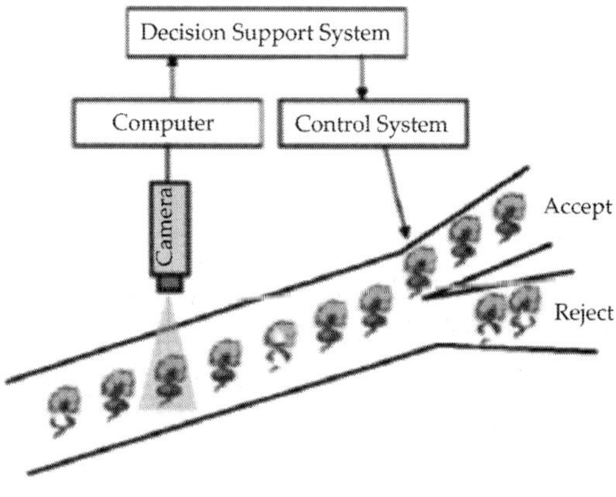

Figure 1.3: Schematic representation of on-line food quality evaluation using a computer vision system, (*Source: S. Gunasekaran, Automation of Food Processing, Food Engineering–Vol IV*).

General prerequisites for on-line applications are throughput, speed, precision, consistency, solidness, enhancement, adaptability, and versatility. Contemplations of these conditions and imperatives must be considered at all phases of framework plan and advancement. Speed of assessment is maybe the most striking necessity. It has been assessed that an on-line apple evaluating framework may need to look at no less than 3600 focuses/min. A few business frameworks are being utilized to look at 3.5 million bits of organic product (apples, oranges, and so on.) in an 8-h day. Another on-line fill-tallness examination framework has been accounted for to deal with rates of up to 1400 containers/min. PC vision innovation is turning into a necessary piece of the food industry turn towards mechanization. The nearness of a machine vision framework on a generation line has come to speak to an unmistakable exhibit of an industry's responsibility regarding quality.

1.5. SENSORS FOR FOOD PROCESSING

Sensors are the fundamental interface between the control framework and the procedure. In the food processing industry, sensors are required for the accompanying: approaching material examination; material waste control; process quality control; bundling review; hardware upkeep/disappointment expectation; natural control. To satisfy the developing interest for minimal effort, on-line, solid sensors from the food business, sensor advancement programs intended to meet the particular needs of this industry are being set up. Figure 1.4 is an incomplete rundown of the sensors regularly required for nourishment handling. A decent portrayal of particular sensors for the sustenance business is given in the writing. Of genuine significance sooner rather than later for mechanization in the food business will be the utilization of sensors for in-line quality observing and control. These include: detecting of subjective properties, for example, delicateness, season, notice, taste, surface and so on and the improvement of new biosensors, substance sensors and immune sensors, for example, sensors for meat freshness appraisal, sugar substance of products of the soil, particular particle sensors and sensors for evaluating deterioration. A depiction and continuous utilizations of these sensors will be the subject of future papers.

Today	Pressure
	Temperature
	Flow
	Level
	Position
	Vibration
	Object recognition
	Colour
	Humidity
	Viscosity
	Turbidity
	Mass flow
	Specific ions
	Gas compositions
Tommorrow	Subjective properties
	-smell, tastes, testure, flavour
	Bionsensors
	- meat freshness
	Chemical sensors
	- acidity, sugar content

Figure 1.4: Sensors for the food industry, (*Source: Dahm M., Mathur A., Automation in the food industry, Food Control–January 1990*).

1.6. FOOD INDUSTRY: CONTROL SYSTEMS

The level of mechanization in the food industry shifts impressively between organizations, and in addition between plants inside a similar organization. Most food preparing plants have developed from little tasks with a fairly preservationist, steady way to deal with mechanical changes. In the food business, computerization started with the use of programmable controllers and single board PCs on assembling hardware, and extremely straightforward control frameworks. These gadgets, especially programmable controllers, were generally acknowledged by food processors since they were easy to work and keep up. They were constrained in what they could do: ordinarily supplant transfers, clocks and counters.

Because of expanded rivalry, be that as it may, the requirement for generation adaptability, visit changes in the assembling procedure and little designing staffs, food processors can profit by utilizing adaptable, effortlessly arranged control and process administration frameworks. Today food processors have a decision of overhauling their programmable rationale controllers (PLCs) or introducing conveyed control frameworks (DCS). They can expand upon their current PLCs to accomplish completely incorporated control frameworks performing different capacities, for example, process control, crude material administration, deals arrange preparing, monetary administration and revealing. DCS-based engineering has played out these capacities for huge persistent procedures. The DCS with every one of these advantages is currently accessible little nonstop and clump activities at an aggressive cost. The rest of this paper portrays the conveyed control framework and its focal points.

1.7. DISTRIBUTION CONTROL SYSTEM (DCS) FOR THE FOOD INDUSTRY

Distributed control systems have evolved significantly in the last ten years. From large systems tailored to the needs of continuous industries, such as refineries, DCSs can now respond to the requirements of small batch-oriented processes and can address a variety of automation projects.

Many elements make up the subsystems of the DCS. Figure 1.5 shows a hardware configuration of the DCS.

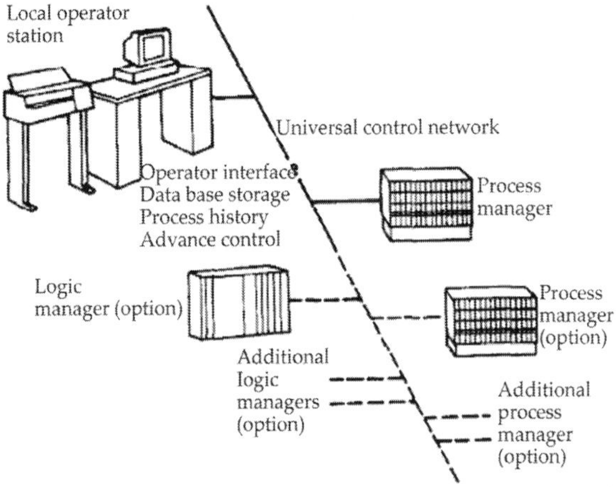

Figure 1.5: Distribution control system hardware, (*Source: Dahm M., Mathur A., Automation in the food industry, Food Control–January 1990*).

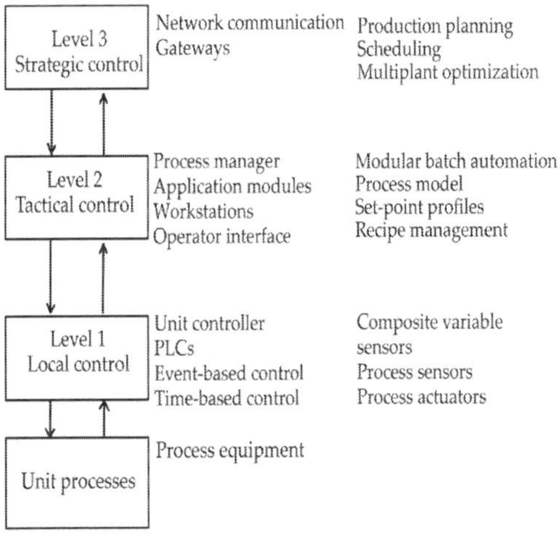

Figure 1.6: Control system hierarchy, (*Source: Dahm M., Mathur A., Automation in the food industry, Food Control–January 1990*).

The rationality of disseminated control is to separate an extensive application into littler subsystems and convey the level of control down to the unit level when suitable to diminish general framework reaction time. This makes it conceivable to trade data between the distinctive control units and takes into consideration incorporated basic leadership at the product offering or plant level.

The control of a procedure happens at a wide range of levels of sensible and theoretical refinement. There are numerous approaches to classify the control capacities into various progressive levels. For motivations behind talk, a calculated three-level structure is appeared in Figure 1.6.

At the base of the framework progressive system, the information/ yield gadgets, sensors, and actuators straightforwardly collaborate with the procedure. Administrative control of process factors, for example, weights and temperatures utilizing relative, essential and subordinate (PID) control and legitimate tasks, for example, time-and occasion based control activities are completed at this level. Singular controllers work at the unit control level to control gear, for example, blenders.

To enhance control at this level, a few choices are as of now accessible to the clients. For instance, savvy transmitters with some measure of information handling accommodate nearer observing of the procedure factors, programmed adjustment, linearization; auto tuners accommodate programmed tuning of the PID controller constants as the procedure conditions change. Sooner rather than later, prescient controllers will be accessible as other options to PID controllers. Predictive controllers will make the activity of controller tuning simple and natural. For instance, rather than specifying picks up for the PID segments of the circle controller – amounts with no instinctive elucidation – the clients would indicate the settling time for the procedure or the coveted rate of progress of the controlled variable; the control activities important to accomplish this will be processed by the prescient controllers. With these controllers, improvement at the local level will likewise be conceivable.

The following level of chain of command, named as the strategic level, enhances control by incorporating the control of autonomous process parameters. For instance, if the item quality is out of particular, at that point set-point profiles or the formula may must be adjusted on-line. To perform on-line change of set focuses or formulas, a model of the procedure is required. The model can be an arrangement of heuristic guidelines, a scientific portrayal of the procedure, or a blend of the two. DCS and other robotization frameworks give a client programming office to such canny control of the process. For cluster forms, sellers are at present offering preconfigured programming modules. Clients characterize the group computerization program by arranging and connecting modules instead of by programming. This component is vital in light of the fact that a procedure architect or physicist can design or alter the product without the help of a PC master. At the following larger amount, correspondence systems for remote and neighborhood data trade empower combination and coordination of the diverse subsystems. The dispersed control framework bolsters entryways to corporate processing frameworks, permitting administration a continuous window into plant task.

1.7.1. Benefits of the DCS

The appropriate control frameworks give the accompanying advantages. Right off the bat, there are littler sensible blocks including incremental programming and checkout, together with simple blame separation and upkeep. An immediate preferred standpoint of dispersed control is to permit the product program to be separated into littler legitimate pieces, autonomous of framework equipment. This influences framework to outline more productive and easy to change. Programming programs empower proficient formula administration, exact chronicle and dissecting of creation information, and factual process and quality control capacities.

Furthermore, there is an effortless corruption from disappointment. The disseminated idea of the general framework gives enough independence to the subparts with the goal that huge disappointment of the framework is improbable. This is rather than a direct advanced control technique in view of a solitary PC controlling the whole procedure.

A third advantage is that islands of computerization are expelled through a coordinated framework. The disseminated control frameworks empower correspondence over the diverse subunits of the assembling procedure, giving simple coordination of general creation instead of separated islands of mechanization in view of single controllers.

A further advantage is from open design. Simplicity of coordinating existing gadgets into the framework, for example, PLCs officially exhibit on the industrial facility floor, is basic and requires that the chose DCS have the capacity to speak with gadgets from different merchants.

At last, there is the advantage from worked in data dealing with ability; including data handling, information documenting, restart, and graphing/inclining.

Dispersed control frameworks offer the adaptability and expandability essential in an industry experiencing fast changes, as is as of now the case in the food industry. Through its simplicity of execution and task, the dispersed control framework contributes fundamentally to the general generation improvement process, opening the ways to complex instruments for creation arranging and planning, quality control, multiline advancement, and multiplane coordination, concentrated and robotized recordkeeping.

1.7.2. Examples of DCS Implementation in Food Industry

One case is of a dessert maker hoping to update a preparing plant from the current level of computerization in view of secluded PLCs. The DCS arrangement gave the adaptability to run a few items on a similar line. The DCS likewise gave adaptable sequencing of activities and diminished changeover time, in this way lessening general generation cost. Better checking of the procedure factors and factual examination of process patterns brought about enhanced quality control. Intuitive illustrations and presentations gave plant work force a superior comprehension of the procedure status in an auspicious manner. The capacity to secure the framework utilizing passwords included the advantage of creation wellbeing, which is basic in nourishment fabricating. At last, the expandability of the framework gave the maker certainty that the chose mechanization system would stay pertinent to creation needs as the prerequisites for more complex controls expanded.

The second illustration is the situation of a dry bundled nibble food plant expecting to overhaul its current frameworks to coordinate its procedure control and data administration capacities. DCS arrangement gave better observing and control of crude material taking care of, cluster handling, framing, cooking, and bundling and warehousing activities were the primary drivers for plant modernization. Despite the fact that the underlying expense of a DCS arrangement was more prominent than a PLC-based framework, the distinction was paid back in enhanced usefulness and profitability inside four to a half year. The arrangement additionally kept on diminishing assembling costs and enhance efficiency at a speedier rate than would the PLC-based arrangement.

1.8. PROCESS CONTROL SYSTEMS AND STRUCTURE IN THE FOOD INDUSTRY

In a procedure control framework, PCs are critical to driving and dealing with the activity of the framework. A PC for process control incorporates equipment, for example, the PC itself, peripherals, instrumentation, input-yield gear, and framework and application programming. Process control engineers, food researchers, and specialists are in charge of building up the application programming based control calculations.

Two sorts of process control frameworks exist: open-loop and closed-loop. In open-loop frameworks, the framework yield is controlled specifically just by the information motion, with the yield having no impact on the framework input. The framework can't adjust for any surprising conditions in the framework yield. Closed loop frameworks screen the framework yield, and feed the yield estimation back to the control PC, which constantly limits the distinction between the deliberate yield and wanted yield, by changing the controller input. Criticism on how the framework is really performing enables the controller to progressively make up for unsettling influences to the framework. Contrasted with the closed loop control frameworks, open-loop control frameworks are less regularly utilized in light of the fact that they are less precise. With the advantage of criticism, the closed loop control frameworks have been broadly utilized as a part of the food industry. The estimation of the framework yield from food producing task is sustained back to be contrasted with the coveted value(s) of the variable and to change the framework yield to limit the mistake between the deliberate and wanted yields.

1.8.1. Process Control Methods in the Food Industry

In the food industry, different control plans have been outlined and utilized. Traditional PID controllers have been connected to various procedures. Be that as it may, PID controllers can't function admirably reliably, on the grounds that they are for the most part appropriate for forms with low-arrange straight flow. So as to enhance the execution of process control frameworks for forms that have high-arrange non-direct progression with time-variation parameters, further developed control plans have been composed and connected.

Demonstrate construct controllers are composed situated in light of the forecast of process models (expository and experimental) and process control necessities. Artificial Neural Networks (ANNs) and fuzzy logic (FL) are two of the most widely recognized ways to deal with setting up the connection between the info and yield of the framework to assess the procedure yield as opposed to estimating it specifically. Neuro-fuzzy controllers take the benefits of ANNs and FL to build up the connection between the information and yield of the framework to induce the procedure yield that can't be estimated straightforwardly. FL and ANNs are two delicate registering procedures. They have been utilized as a part of food science and innovation. Other delicate figuring strategies, for example, hereditary calculations (GAs) and bolster vector machines (SVMs), likewise can possibly be utilized as a part of planning process controllers.

1.8.2. Proportional-Integral-Derivative Controller

PID controllers have been produced for the exemplary closed-loop criticism control conspire. This improvement depends on the investigation of the progression of second-arrange straight frameworks. So as to guide the framework to execute as wanted a PID controller can be added at the contribution to shape an input shut circle. A PID controller can be communicated by the accompanying condition (Figure 1.7):

$$u(t) = K_p e(t) + K_j f e(\pi) d\pi + K_d d e(t)$$

dt

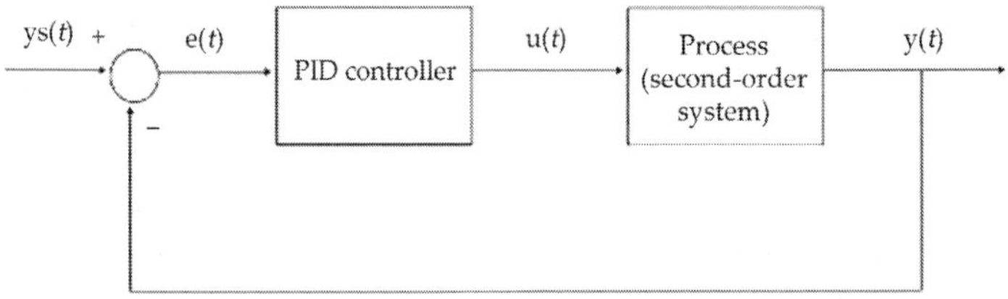

Figure 1.7: A feedback closed-loop control scheme with PID controller, (*Source: G. Caldwell Darwin (2013). Robotics and Automation in the Food Industry: Current and Future Technologies. Woodhead Publishing*).

where u(t) is the yield of the PID controller and the contribution to the procedure at time moment t, $e(t) = y^s(t) - y(t)$ is the distinction between the framework yield y(t) and the coveted esteem y^s (t) at time moment t, K_p is the corresponding addition steady, K_i is the essential increase consistent, and K_d is the subsidiary increase consistent.

Figure 1.8 demonstrates the meaning of controller tuning parameters in light of a stage input. The relative control mode essentially has the impact of decreasing the ascent time, and lessens however never disposes of the enduring state blunder. The basic control mode has the impact of dispensing with the unfaltering state blunder, yet it might build the overshoot and time to exacerbate the transient reaction. The subsidiary control demonstrate has the impact of expanding security of the framework, decreasing the overshoot, and enhancing the transient reaction. Regularly, PID controllers are utilized with all the three modes together utilizing K_p, K_t and K_d to measure the relative significance of the corresponding, vital, and subsidiary controls. A few applications may just require utilization of maybe a couple modes to give the fitting framework control. This is accomplished by setting maybe a couple of K_p, K_j, and K_d to zero. A PID controller will be known as a PI, PD, P, I or D controller without other control activities.

PID controllers have been utilized as a part of different regions of process control, including the food industry. Be that as it may, the outcomes shift. On occasion they can perform attractively, even with no tuning, yet on different events they can perform ineffectively, paying little respect to how the controllers are tuned. Specifically, framework non-direct qualities contribute contrarily to the execution of an ordinary PID controller amid transient conditions. These non-linearity can bring about expanded motions, overshoots, and long settling circumstances. To decrease these negative impacts, further developed non-straight control is expected to decidedly upgrade the control execution of the closed-loop framework. Keeping in mind the end goal to build up a further developed controller, a procedure show is expected to accomplish better execution in process control (Figure 1.8).

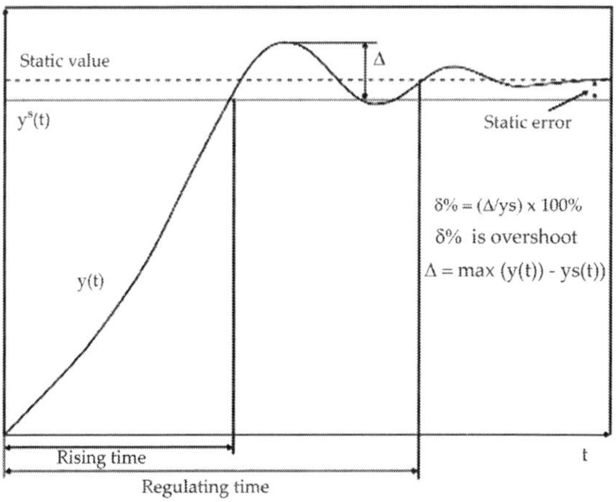

Figure 1.8: Typical output response of control system under step input, (*Source: G. Caldwell Darwin, Robotics and Automation in the Food Industry: Current and Future Technologies, Woodhead Publishing*).

1.8.3. Advanced Process Controller

PID control is a controller that does not utilize a procedure display for controller outline. Keeping in mind the end goal to accomplish better execution in process control, particularly when PID control performs ineffectively, a procedure display is helpful. This is additionally called display based control. In display-based control, process models are required. The models can be scientifically in view of first standards or exactly in view of exploratory information. Logical models are worked by considering the specific procedure as a white box with all essential data accessible. In food process control, if all the physical and concoction forms are known in a unit activity, the scientific model can be worked for investigation and control of task. Ozilgen (1998) gave cases of expository displaying in food preparing, for example, transport wonders demonstrating and motor demonstrating. Be that as it may, the physical and synthetic data on different procedures in obscure. In this way, it is hard to build up a scientific model for some procedures. In these cases, exact models can be fabricated utilizing exploratory information. Experimental models are back-boxed models. A discovery procedure has no earlier data accessible. With no information of its inward working systems, a discovery procedure can be thinks about exclusively as far as its information, yield and exchange qualities. Framework distinguishing proof gives a numerical way to deal with discovery process displaying. In demonstrate based control, the experimental procedure models can be fabricated web based utilizing framework distinguishing proof. The online distinguished models can be utilized to tune the parameters of the controllers.

1.8.3.1. Process Modeling

Nonexclusively an assembling procedure can be thought to be administered by the accompanying discrete-time auto-backward moving normal with exogenous info (ARMAX) relationship:

$$y(k) = f\left(y(k-1).....y(k-p), u(k-1).....u(k-q), \in(k-1)..... \in(k-r), O\right) \quad [1.1]$$

where f is the obscure utilitarian connection between process info and yield, y(k) is the deliberate procedure yield at moment k, u(k) is the procedure contribution at moment k, ϵ(k) is the clamor in the deliberate yield at moment k, which can be expected as Gaussuan repetitive sound known change σ2ϵ, p is the request of past yield, q is the request of past information, r is the request of past yield commotion, and O is the arrangement of parameters in the practical relationship.

By and by, a great deal of dynamic procedures can be satisfactorily portrayed by the direct ARMAX display as long as the varieties in the working conditions are sufficiently little. Nonetheless, much of the time, the genuine connection between process info and yield over the scope of intrigue is non-direct. The straight model could be utilized to gauge the non-direct process flow piecewise by expecting that the varieties of the working conditions are sufficiently little locally. The above condition could be utilized straightforwardly to delineate y(k) and u(k) with ϵ(k).

In demonstrating direct procedures, the ARMAX model can show deterministic and educational parts of the framework freely. Be that as it may, the technique for direct relapse can't be utilized for display parameters estimation due to the moving normal piece of the model with ϵ(k). Along these lines, the straight ARMAX model can be disentangled into the direct autoregressive with exogenous info (ARX) demonstrate. Essentially, when demonstrating non-direct procedures, the bland non-straight ARMAX display as depicted in above condition can be disentangled into a non-straight ARX show by and large:

$$y(k)=f\left(y(k-1)\ldots.y(k-p),u(k-1)\ldots..u(k-q),O\right)+\in(k) \quad [1.2]$$

This model has been produced for snack-food frying procedure control. In this procedure, the sensors of the item quality files, for example, item shading and dampness, are situated toward the finish of the generation line. They are far from the actuators that control the components, for example, gulf oil temperature and submerged introduction time that have impacts on the item quality records. In this manner, the model in the above condition must be altered for the procedure to manage the time slack between the procedure information and yield:

$$y(k)=f\left(y(k-1)\ldots..y(k-p),u(k-d-1)\ldots..u(k-d-q),O\right)+\in(k) \quad [1.3]$$

where *d* is the time lag between process input and output.

1.8.3.2. Model-Based Control

Model-based control fundamentally utilizes the forecast of the procedure model to transform the procedure model to outline the controller. There are two kinds of expectation from the procedure demonstrate, one-advance ahead forecast and multi-venture ahead. One-advance ahead expectation predicts the procedure yield from the procedure show one stage in front of the present moment k. Multi-venture ahead forecast predicts the procedure yield from the procedure display l-ventures ahead (l> 1) of the present moment k.

From the non-linear ARX model, one-step-ahead prediction is as follows:

y(k+1) = f(y(k)….y(k-p+1),u(k)…..u(k-q+1), O) [1.4]

In the case of linear modeling, the one-advance-ahead forecast is a blend of the present and past y and u.

For the plan of a non-linear controller, in view of ARX one advance ahead forecast as appeared in Equation 1.4, if there is a special answer for u(k) from the inverse off, at the point of control law can be spoken to as represented as

$$\hat{U}(k) = g\left(y(k), y(k+1), y(k)......y(k-p+1), u(k-1).....u(k-q+1), O\right) \quad [1.5]$$

where g is the inverse of f for u(k)

For linear models, the opposite is clear. Be that as it may, for non-linear models, the control law is acquired by altering the procedure demonstrate at each inspecting point numerically:

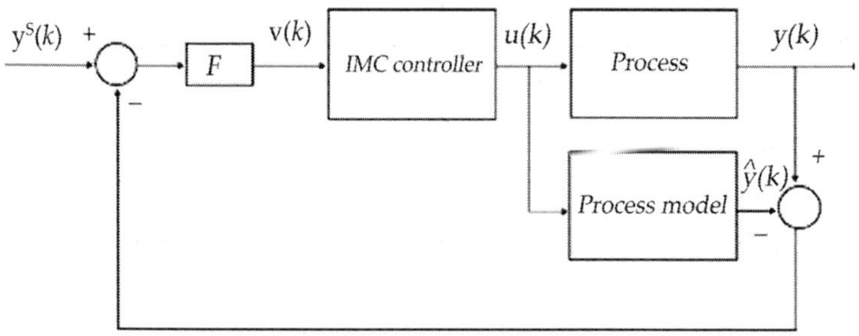

Figure 1.9: IMC Control loop where F is a filter to improve controller robustness, (*Source: G. Caldwell Darwin, Robotics and Automation in the Food Industry: Current and Future Technologies, Woodhead Publishing*).

This control approach builds a procedure show based internal model control (IMC) scheme (Figure 1.9). IMC is a control technique that is utilized to make a model-based controller by joining the procedure demonstrates with the controller framework execution detail. IMC was first proposed for direct frameworks. It later was reached out to general non-straight frameworks. IMC has the properties of double solidness, consummate control and zero balance.

In IMC, a function can be defined to be used to compute the control action for a unique solution of u(k):

$$U\left(u\left(k\right)\right)=v\left(k\right)-y\left(k+1\right) \tag{6}$$

where v(k) is the tracking signal of the process model prediction.

In the IMC loop, in the frequency domain, the relationship between v and e can be derived as:

$$v(z) = F(z)\,(y^s(z) - y(z) + y(z))$$
$$= F(z)\,(e(z) + y(z)) \qquad [1.7]$$

If the process model is perfect and u can be solved from the process model in the form of one-step ahead prediction, then

$$v(z) = F(z)y(z) \qquad [1.8]$$

The combination of Equations (7) and (8) produces the relationship as

$$v\left(z\right)=\frac{F\left(z\right)}{1-z^{-1}}e\left(z\right) \qquad [1.9]$$

Since e is the mistake flag of the great input control circle, and v is the blunder flag of the IMC control circle, Equation (9) outlined that an IMC controller is identical to the criticism controller based on the procedure demonstrate expectation.

The other method to plan a model-based controller is to advance online an assigned target work in view of multi-venture ahead expectation from the procedure model to accomplish the control activity, u(k) at each time instant.

1.8.3.3. Neural Network-Based Control

For complex food forms, ANNs could give process demonstrate expectations to controller plan. A run of the mill procedure for ANN application is the nibble sustenance broiling process. This procedure is one of the unit activities in the nibble sustenance fabricating industry. The manufacturing of snack food includes various unit tasks in nibble nourishment fabricating plants, for example, expulsion, broiling, preparing, and drying. Broiling is a procedure in which a bite is cooked by skimming or being drenched in hot oil. There are two sorts of snack foods browning: group and consistent. Consistent fricasseeing is utilized everywhere scale tasks, for example, snack food plants. In the consistent broiling process, constant contribution of expelled snack food material happens toward one side of the fryer, pushed through by a submerger with oil stream, and afterward separated at the opposite end of the fryer. The essential worry in fricasseeing is the ease creation of nutritious snack food at reliable quality and with least waste. With a specific end goal to guarantee consistency in snack item consistency, programmed control is wanted. The snack food broiling process is a mind-boggling process. It has entangled collaborations between the item quality sign factors, for example, snack food dampness, shading and oil content, and those components that can be changed autonomously to influence the item quality sign factors, for example, broiling bay oil temperature and snack food material living arrangement time in the fryer, which is identified with the paces of the sub merger transport and the takeout transport (Figure 1.10). To control this mind-boggling process, it is important to demonstrate it first. Because of the intricacy, an expository model is hard to create. A multi-input-multi-yield (MIMO) model can be assembled utilizing the strategy for framework ID. In the snack food searing procedure, the item quality list sensors are situated toward the finish of the generation line. These sensors are at a separation from the actuators directing bay oil temperature and transport speed. In this manner, there exist huge time-slacks between the item quality files (process yield) and their effect factors (process input), which result in worldwide non-linearity of process yield in light of process input. A multivariate non-straight model is expected to deal with the non-linearity caused when slacks. ANNs can outline nonexclusive non-straight connection between the procedure information and yield with no earlier presumptions on the model shape. ANNs are given model forecasts to controller plan for controlling the snack food searing procedure.

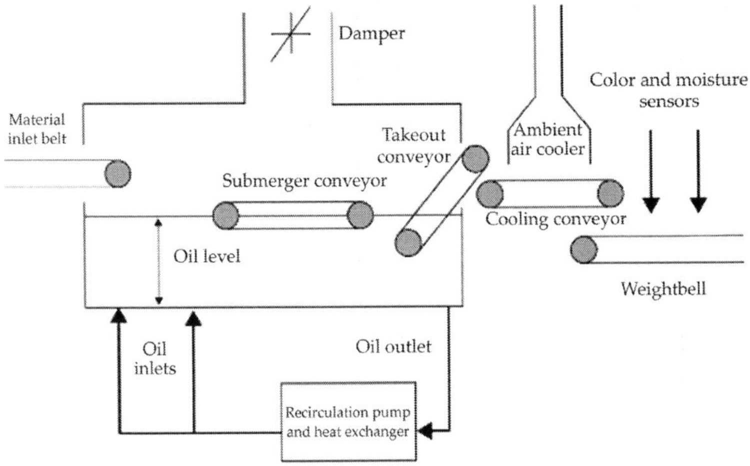

Figure 1.10: Schematic diagram of continuous snack-food frying process, (*Source: Huang et al., 2001, p.9*).

1.9. MATERIALS HANDLING

Materials handling is simply the loading, moving and unloading of materials. To do it safely and economically, different types of tackles, gadgets and equipment are used, when the materials handling is referred to as mechanical handling of materials. Since primitive men discovered the use of wheels and levers, they have been moving materials mechanically. Any human activity involving materials need materials handling. However, in the field of engineering and technology, the term materials handling is used with reference to industrial activity. In any industry, be it big or small, involving manufacturing or construction type work, materials have to be handled as raw materials, intermediate goods or finished products from the point of receipt and storage of raw materials, through production processes and up to finished goods storage and dispatch points.

Materials' handling, as such, is not a production process and hence does not add to the value of the product. It also costs money; therefore it should be eliminated or at least reduced as much as possible. However, the important point in favor of materials handling is that it helps production. Depending on the weight, volume and throughput of materials, mechanical handling of materials may become unavoidable. In many cases, mechanical handling reduces the cost of manual handling of materials, where such material handling is highly desirable. All these facts indicate that the type and extent of use of materials handling should be carefully designed to suit the application and which becomes cost effective.

A well designed materials handling system attempts to achieve the following:

- Improve efficiency of a production system by ensuring the right quantity of materials delivered at the right place at the right time most economically.
- Cut down indirect labor cost.
- Reduce damage of materials during storage and movement.
- Maximize space utilization by proper storage of materials and thereby reduce storage and handling cost.
- Minimize accident during materials handling.
- Reduce overall cost by improving materials handling.
- Improve customer services by supplying materials in a manner convenient for handlings.

- Increase efficiency and salability of plant and equipment with integral materials handling features.

1.10. AUTOMATION ON DOMESTIC FOOD PROCESSING

Food is the one of the most basic physiological needs essential for the survival of a living being. Some of the living beings have the capacity to prepare their own food (like most plants) and therefore are designated as primary food producers. Those who depend on these primary food producers for food form the primary consumers' class (herbivores). Some of the organisms relying on these primary consumers are the secondary food consumers (carnivores). There is a third category of consumers called tertiary food consumers/ apex food consumers that feed on both the primary food producers and the primary and secondary food consumers. Humans form an essential part of the apex predators and are generally at the top of the food chain. The early stages of the development of the modern human can be traced back to the times when the fire was discovered. This discovery also paved the way for the earliest form of food processing. In the present days, the old notion of eating raw/brute food is long gone and food processing has become very trenchant in the lives of the modern human. The food habits of the modern human i.e. Homo sapiens, reveals that they depend on other individuals for preparing their own food.

There is an increase in dependence on these "processing individuals" for preparing the food before the modern human can consume it. It has created a further shift between the primitive and the modern human in the food chain hierarchy. The processing of food has a direct impact on the economy of the individual (consumer). Also, most of the individuals depend on other processing individuals for the preparation of food.

Automation: the food industries are in a state of change driven by cost of operations. This change is driving the need for automation solutions and IT equipment that can enable the industry to become leaner and more agile. Increasing safety and liability issues are a major concern for food companies and they are affecting their investment decisions. The need for more automated traceability has become a basic requirement. Although Programmable Logic Controller (PLC)–based architecture has been commonly used in the food industry, they were not designed for making secure electronic records, nor were they designed with version control and software security in mind. It is becoming increasingly difficult to make PLCs compliant with Food and Drug Administration (FDA) regulations. The FDA and USDA (U.S. Department of Agriculture) will be looking for evidence of control system validation for regulated food processes. There are about 28,000 establishments manufacturing food in the US. Processed food sales worldwide are in the order of trillions of dollars and over tens of millions of people work in the industry.

Implementation of advanced automation and control strategy in the food packaging domain plays a vital role. Compared to traditional methods, advanced control methods have several advantages. The traditional methods use discrete electronics, PLCs, Industrial Computers and even high-computing systems such as the DSP (Digital Signal Processor) platform. These automation platforms have issues such as reliability, flexibility, scalability. For example, PC- or PLC-based systems are considered to be redundant because they require maintenance and additional expertise. They have restricted interoperability due to incompatibility in host-platform requirements. Similarly, DSP systems require customization and expertise in addition to being very costly.

Conversely, the modern food industry entails sophisticated control architecture. The food processing and packaging control system embeds a generic set of essential requirements with regard to timing, error diagnostics, coordination, periphery-control acknowledgement, and synchronization. Moreover, its imperative error recovery and fault corrections are needed during run time. Furthermore, the control systems should satisfy other fault-tolerant features such as interlock checking, lockout checking, and user friendly, intuitive status display with error messaging. Applying modern control network technology to the food industry can markedly improve food productivity, manufacturing, preservation, and handling. Currently, however, the industry's automation systems are predominantly centralized. Though there is some modular approach, the overall system is implemented with static functionality. The conventional PC-, PLC-, or DSP-based centralized automation strategy characterizes a central processing unit with peripheral devices such as sensors, actuators, and drives. One of the important requirements is the integration of drive systems, I/O interfacings, instrumentation hardware, supports for diagnostics and prognostics, data collection and operator interfaces. The architecture needed to be flexible in terms of extendibility, scalability and configurability to enable features to be developed independently and added, modified or removed, along with the addition of new features required for the future. In order to be able to know about the appropriate control strategy for the food industry, the author conducted a survey.

It was found that the industry preferred to implement the Distributed Control System (DCS) instead of a more centralized one, because of its advantageous flexibility over the whole operating range. Other benefits include low implementation cost, easy maintenance, configurability, scalability, and modularity. The potential benefits of adopting decentralized automation have been widely recognized. However, fully integrated, distributed, agile and open control systems still need to be addressed. DCS interconnects devices with a single serial link and it is evidently a four-layer automation service. The bottom layer, referred to as the device level, includes the physical components such as micro-controller and sensors, actuators and switches. The interface layer is similar to the MAC sub-layer of the link layer protocol. The process layer includes application layer features. Since the control system does not transfer raw data through a physical media, an application layer should exist in the automation architecture.

The application layer defines the variables, which are responsible for transferring data from one place to another when they are connected logically. Advances in distributed control principles indeed influenced many kinds of automation and control applications. Although a dedicated networking system for the food industry does not yet exist, the available proprietary standards are somehow adequate because they provide essential features in order to implement both DCS (Distributed Control Systems) and SCADA (Supervisory Control and Data Acquisition) architectures. Fieldbus technology is preferred for DCS implementation.

The technology includes a protocol, which can transmit and receive digital data through multiple media and channels. Recently, many fieldbuses have been made available in the technology marketplace. For industrial control applications, the type of bus used is described as a sensor-bus or device-bus, and it usually deals with devices such as sensors, motors, drives, switches and valves providing process variable information. In summary, the fieldbus provides a good basis for distributed real-time control implementation which is essential in food processing and packaging areas. It has been successfully tried and tested in many applications, including a wide range of general industrial applications from food processing, to conveyor system automation and packaging plant automation.

1.11. REASONS FOR AUTOMATION

(a) Increased productivity: Automation of manufacturing operations holds the promise of increasing the productivity of labor. This means greater output per hour of labor input. Higher production rates (output per hour) are achieved with automation than with the corresponding manual operations.

(b) High cost of labor: The trend in the industrialized societies of the world has been toward ever-increasing labor costs. As a result, higher investment in automated equipment has become economically justifiable to replace manual operations. The high cost of labor is forcing business leaders to substitute machines for human labor. Because machines can produce at higher rates of output, the use of automation results in a lower cost per unit of product.

(c) Labor shortages: In many advanced nations there has been a general shortage of labor. Labor shortages stimulate the development of automation as a substitute for labor.

(d) Trend of labor toward the service sector: This trend has been especially prevalent in India. There are also social and institutional forces that are responsible for the trend. There has been a tendency for people to view factory work as tedious, demeaning, and dirty. This view has caused them to seek employment in the service sector of the economy government, insurance, personal services, legal, sales, etc. Hence, the proportion of the work force employed in manufacturing is reducing.

(e) Safety: By automating the operation and transferring the operator from an active participation to a supervisory role, work is made safer.

(f) High cost of raw materials: The high cost of raw materials in manufacturing results in the need for greater efficiency in using these materials. The reduction of scrap is one of the benefits of automation.

(g) Improved product quality: Automated operations not only produce parts at faster rates but they produce parts with greater consistency and conformity to quality specifications.

(h) Reduced manufacturing lead time: With reduced manufacturing lead time automation allows the manufacturer a competitive advantage in promoting good customer service.

(i) Reduction of in-process inventory: Holding large inventories of work-in-process represents a significant cost to the manufacturer because it ties up capital. In-process inventory is of no value. It serves none of the purposes of raw materials stock or finished product inventory. Automation tends to accomplish this goal by reducing the time a workpart spends in the factory.

(j) High cost of manual production: A significant competitive advantage is gained by automating a manufacturing plant. The benefits of automation show up in intangible and unexpected ways, such as, improved quality, higher sales, better labor relations, and better company image. All of these factors act together to make production automation a feasible and attractive alternative to manual methods of manufacture.

1.12. THE PRINCIPLES OF FOOD PROCESSING

1.12.1. What are Foods?

Foods are materials, raw, processed, or formulated, that are consumed orally by humans or animals for growth, health, satisfaction, pleasure, and satisfying social needs. It is known that excessive amounts could be lethal, for example, salt, fats, and sugar. Chemically, foods are mainly composed of water, lipids, fat, and carbohydrate with small proportions of minerals and organic compounds. Minerals include salts and organic substances include vitamins, emulsifiers, acids, antioxidants, pigments, polyphenols, and flavor-producing compounds. The different classes of foods are perishable, nonperishable, harvested, fresh, minimally processed, preserved, manufactured, formulated, primary, secondary derivatives, synthetic, functional, and medical foods.

The fundamental necessity for food is to sustain life. The main reason for food processing is to make it microbiologically safe to eat. Processing foods can transform unpalatable or unacceptable raw materials into desirable and attractive products. Typical processing operations may include raw material handling, ingredient formulation, heating and cooling, cooking, freezing, shaping, and packaging (Heldman and Hartel, 1998). These could broadly be categorized into primary and secondary processing. Primary processing is the processing of food that occurs after harvesting or slaughter to make food ready for consumption or use in other food products. Primary processing ensures that foods are easily transported and are ready to be sold, eaten or processed into other products (e.g., after the primary processing of peeling and slicing, an apple can be eaten fresh or baked into a pie). Secondary processing turns the primary-processed food or ingredient into other food products. It ensures that foods can be used for a number of purposes, are healthy and wholesome to eat, and are available all year (e.g., seasonal foods).

1.13. ENGINEERING PROPERTIES OF FOOD

Knowledge of various engineering (physical, thermal, and thermodynamic) properties of food, biological and packaging material is critical for successful product development, quality control, and optimization of food processing operations. For example, data on density of food material are significant to separation, size reduction or mixing processes. Knowledge of thermal properties of food (thermal conductivity, specific heat, thermal diffusivity) is useful in identifying the extent of process uniformity during thermal processes such as pasteurization and sterilization. For liquid foods, knowledge of rheological characteristics, including viscosity, helps in the design of pumping systems for different continuous flow operations. Different food process operations (heating, cooling, and concentration) can alter product viscosity during processing, and this needs to be considered during design. Phase and glass transition characteristics of food materials govern many food processing operations such as freezing, dehydration, evaporation, and distillation. For example, the density of water decreases when the food material is frozen and as a result increases product volume. This should be considered when designing freezing operations. Thus, food scientists and process engineers need to adequately characterize or gather information about relevant thermo physical properties of food materials being processed.

1.13.1. Density

The density of liquids is fairly straightforward, but solids in particulate form, such as peas or powders, have a bulk density as well as a solid density to consider. The bulk density of a material depends on the solids density and the geometry, size and surface properties of the individual particles. Gases and vapors, unlike solids and liquids, are considered to be compressible and some foods such as whipped cream and ice-cream have air incorporated into them during their preparation, the amount of air being measured by the overrun. The density of a substance is equal to the mass of the substance divided by the volume that it occupies (Table 1.1):

$$density = \frac{mass}{volume}$$

Density has dimensions of [ML^{-3}]. In the SI system of units it is measured in kilograms per cubic meter (kg m^{-3}). In the Imperial system, density is measured in pounds per cubic foot (lb ft^{-3}), where 1 lb ft^{-3} = 16.02 kg m^{-3}.

The density of liquids can be expressed as specific gravity (SG), a dimensionless number which is found by dividing the mass (or density) of a liquid by the mass (or density) of an equal volume of pure water at the same temperature:

$$SG = \frac{mass\ of\ liquid}{mass\ of\ water} \qquad SG = \frac{density\ of\ liquid}{density\ of\ water}$$

Table 1.1: Densities of Some Common Solid Materials

Materials	**Density (kg/m³)**	**Temperature (°C)**
Aluminum	2640	0
Cast iron	7210	0
Copper	8900	0
Mild steel	7840	18
Stainless steel	7950	20
Brick	1760	20
Concrete	2000	20
Glass soda	2240	20
Wood	200	30

Source: Lewis, M. (1987). Physical Properties of Foods and Food Processing Systems

1.13.2. Viscosity

Viscosity is an important property of fluid foods. It is defined as the internal friction of a liquid or its ability to resist flow. The internal friction in a fluid can be easily demonstrated by observing a liquid that has been vigorously stirred to create a vortex. Once the stirring has stopped the speed of the vortex is gradually reduced and rotation of the liquid eventually stops. This happens as a result of the frictional force within the liquid and this force has to be overcome in order for the liquid to flow. Viscosity is also a characteristic of the texture of food. This means that the viscosity of a product must be controlled and measured in production so that each batch is consistent from day to day.

Liquid food systems can be highly shear-sensitive and the structure can break down with the application of shear, as is the case during mixing or when pumped through pipes. There are a number of characteristic behaviors that can be observed. Many liquid foods are non-Newtonian, including emulsions and suspensions, and concentrated solutions that contain starches, pectins, gums and proteins.

Newtonian behavior is displayed by simple liquids consisting of small molecules that do not interact or form any connected structure. However, it must be pointed out that long chain polymers at low concentration can also show Newtonian behavior. An easy way to demonstrate Newtonian behavior is to double the shear stress during a viscosity test and this should result in doubling of the shear rate. If this is not observed then the liquid is non-Newtonian (Figure 1.11).

Non-Newtonian can be categorized into;

(i) Dilatant fluid: In this fluid, there is an increase in viscosity with respect to increase in shear rate. It is also referred to as "shear thickening" flow behavior.

(ii) Pseudoplastic fluid: This type of fluid displays a decreasing viscosity with respect to an increase in shear rate. Examples are starch paste, emulsions and suspensions.

(iii) Bingham or casson plastic fluids (tomato ketchup).

(iv) Thixotropic fluid: This fluid undergoes a decrease in viscosity with time while it is subjected to constant shear. Examples are creams, paint, cosmetics, asphalt, and glue.

(v) Rheopetic fluid: This is the opposite of thixotropic behavior, in that the fluid's viscosity increases with time as it is sheared at a constant rate. Examples are whipping cream and Gypsum paste.

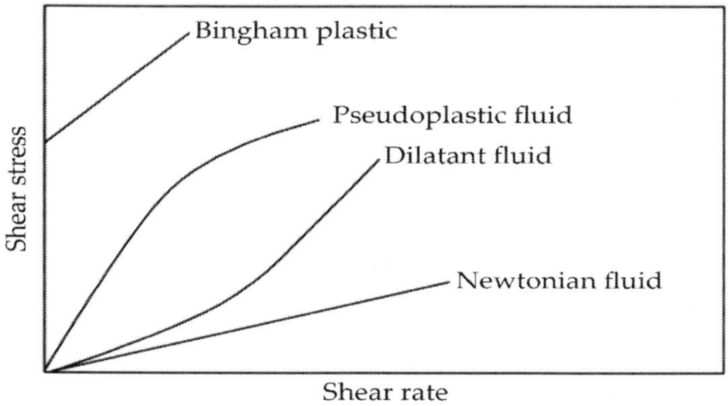

Figure 1.11: Classification of non-Newtonian fluids. (*Source:https://www.researchgate.net/publication/258387572_Computational_Fluid_Dynamics_Analysis_of_Pulsatile_Blood_Flow_Behavior_in_Modelled_Stenosed_Vessels_with_Different_Severities/figures?lo=1*).

1.13.3. Surface Activity

Many foods are a mixture of multi-phase materials (i.e., a mix of solid, liquid, and gas). The boundary between these phases, the interface, plays a significant role in the properties of the food. Examples include foam, a gas in a liquid, whipped creams and toppings. An emulsion—a liquid in a liquid—example is mayonnaise. A solid foam example is the meringue on pies. Both oil and water, which do not readily mix, exist in many foods. Milk and cream (oil-in-water) and butter and margarine (water-in-oil) are examples of emulsions. Uniform distribution of oil and water throughout a food material incorporates the use of surface properties. Surface properties are important in unit operations such as spray drying, aeration, hydrogenation, and fermentation. An understanding of surface properties is also important in cleaning operations. Emulsifying agents and detergents clean because they reduce the surface tension to more readily remove solids (dirt). These liquids wet solid surfaces and dissolve water-soluble and fat-soluble components. Detergents reduce surface tension to break apart large particles and promote spreading over the surface. Detergents allow mechanical force such as sprays or bulk turbulent flow in cleaned-in-place systems to be more effective. The forces acting on the surface of a liquid tending to minimize its area are known as surface tension.

The surface tension of water is twice as high as for cooking oils. Thus, water tends to form spherical droplets on clean surfaces. Materials when mixed with water to reduce its surface tension are referred to as surface active, i.e., surfactants. The addition of low concentrations of surfactants decreases the surface tension of water significantly. Wettability and solubility of powders affect their mixing with liquids. The liquid's surface properties control the rate of wetting and dissolution of the powder's particles. Wetting properties of milk powder during spray drying are affected by particle size, surface tension, presence of wetting agents, and the form and location of fat molecules. Although data on the surface properties of food materials are not quantified as extensively as other properties, it is no less important; and empirical, qualitative data are often used as the only information available.

1.13.4. Rheology and Texture

The science of rheology has many applications in the fields of food acceptability, food processing, and food handling. Rheology concerns the flow and deformation of substances and, in particular, to their behavior in the transient area between solids and fluids. Moreover, rheology attempts to define a relationship between the stress acting on a given material and the resulting deformation and/or flow that takes place. Rheological properties are determined by measuring force and deformation as a function of time.

Some of the most popular foods, such as gelatin desserts, cooked egg whites, frankfurters, surimi based seafood analogs, and fruit jellies, can be considered gels. A gel is a solid-in-liquid colloid in which the solid phase forms a network structure that immobilizes the liquid and produces solid-like properties. A gel can also be described as a substantially diluted system that exhibits no steady state flow. The initial state can be a solution, dispersion, or suspension. Some food gels are formed irreversibly by cooking, while others like gelatin form reversible gels. Gelation arises either from chemical cross-linking by way of covalent reactions or from physical cross-linking through polymer-polymer interactions. The macromolecular substances responsible for network formation in food systems are primarily polysaccharides and proteins. The unifying property among these foods is that they are mostly fluids but respond as viscoelastic solids with a high degree of elasticity.

They usually fracture rather than flow when deformed. Fracture is inherent in sensory biting and mastication of foods so it is important to relate the fracture character of gels to their sensory texture. Fracture properties are determined by deforming a sample to the point of abrupt mechanical yield (sometimes referred to as failure or ultimate properties). Deforming forces can be applied to gels during shear, compression, or tension. Shear changes the shape of a specimen but does not change volume. Tension tends to increase volume and compression tends to decrease volume. When compressive or tensile forces on a specimen are limited to a single direction they are termed uniaxial, differentiating them from forces causing bulk changes such as hydrostatic pressure (equal force per unit area applied perpendicularly to the specimen surf face at all points on the specimen surface), which changes the volume but not the shape of a specimen. Uniaxial tension or uniaxial compression can change both shape and volume. There are methods for determining fracture properties of foods based on uniaxial compression, uniaxial tension, or torsion.

Of these methods, uniaxial compression is the most commonly used method because it does not require attachment of the specimen to the testing machine and several suitable commercial machines are available to perform the test. Most machines capable of uniaxial compression testing can also be used to deform samples in uniaxial tension; however, tension requires a strong attachment of the specimen to the machine, which can be difficult with food samples. Torsional testing of fracture properties also requires a strong attachment of the material to the machine and, until recently, no commercially produced machine has been available specifically designed for torsion testing of food samples. Commercial availability of testing machines and minimal sample preparation have made uniaxial compression testing methods dominant in determining fracture properties. However, the limitations of uniaxial compression testing may make it advantageous in some situations to utilize methods that deform samples in tension or torsion.

Texture of food materials plays a key role in consumer acceptance and market value. Texture characteristics are important factors for raw products and for processing, preparation, and consumption. Texture features are also important considerations in quality assurance, including hazard analysis critical control program (HACCP) and food safety issues. Additionally, flavor, juiciness, color, and other appearance characteristics are important factors in food selection and contribute to texture. Some of the descriptive terms utilized to characterize food properties include toughness vs. tenderness, hard vs. soft, dry vs. moist, brittle vs. elastic, roughness vs. smoothness, crispness vs. sogginess, firmness, stickiness, ripeness, springiness, blandness, flakiness, grittiness, aroma, etc. Many of the methods utilized by food scientists and engineers to classify characteristics of food products represent attempts to develop techniques that will reasonably predict how consumers will perceive the item and subsequently purchase (Figure 1.12).

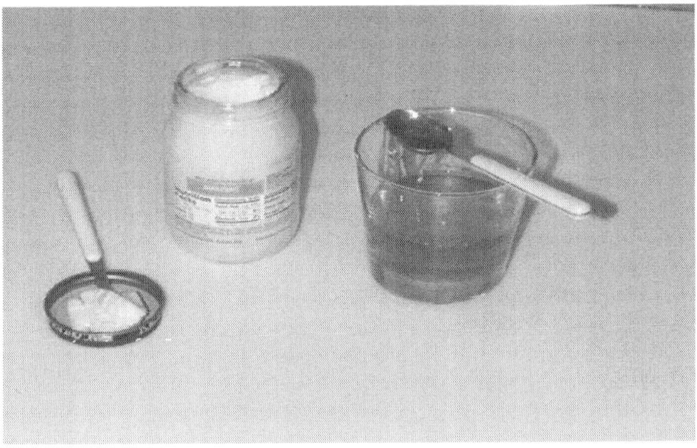

Figure 1.12: Differences in rheological properties of mayonnaise and honey, (*Source: http://www.rheology.org/sor/publications/rheology_b/jan04/default.htm*).

1.13.5. Moisture in Foods

Foods are composed of larger amount or percentage of water. Fresh fruit, vegetables, and milk contain over 80% water. The amount of water in a food is denoted by its moisture content. A food's storability is directly related to moisture content, along with temperature and oxygen availability. High amounts of available moisture lead to mold growth and microbial activity. Fruit must be dried below 30% and agricultural grains below 12% for good long-term storage. Water is a major constituent of most food products. The approximate, expected moisture content of a food can affect the choice of the method of measurement. It can also guide the analyst in determining the practical level of accuracy required when measuring moisture content, relative to other food constituents. The ease of water removal from foods depends on how it exists in the food product. The three states of water in food products are:

(a) Free water: This water retains its physical properties and thus acts as the dispersing agent for colloids and the solvent for salts.

(b) Adsorbed water: This water is held tightly or is occluded in cell walls or protoplasm and is held tightly to proteins.

(c) Water of hydration: This water is bound chemically, for example, lactose monohydrate; also some salts such as $Na_2SO_4.10H_2O$.

1.13.6. Mass Transfer

Mass transfer can be defined as the migration of a substance through a mixture under the influence of a concentration gradient in order to reach chemical equilibrium. A group of unit operations for separating the components of mixtures is based on the transfer of material from one homogenous phase to another. The driving force for transfer is a concentration difference or a concentration gradient; much like a temperature difference or a temperature gradient provides the driving force for heat transfer. Biochemical and chemical engineering operations, such as absorption, humidification, distillation, crystallization and aeration involve mass transfer principles. In food processing, mass transfer phenomena are present in freeze-drying, osmotic dehydration, salting or desalting, curing and pickling, extraction, smoking, baking, frying, drying, membrane separation and the transmission of water vapor, gases or contaminants across a packaging film. Food stability and the preservation of its quality are also affected by mass transfer of environmental components that can affect the rate of reactions. Among the components involved in these mass transfer processes are water, sugars, salt, oils, proteins, acids, flavor and aroma substances, oxygen, carbon dioxide, residual monomers or polymer additives, and toxins or carcinogens produced by microorganisms. Furthermore, mass transfer phenomena are important in the scale-up of processes to pilot or commercial scale plants, and in the control and optimization of the processes such as drying, extraction, distillation, absorption, adsorption, crystallization from solution, and Ion exchange.

1.14. HEAT TRANSFER

Heat transfer is one of the fundamental processing principles applied in the food industry and has applications in various unit operations, thermal processing, evaporation (concentration) and drying, freezing and thawing, baking, and cooking. Heating is used to destroy microorganisms to provide a healthy food, prolong shelf life through the destruction of certain enzymes, and promote a product with acceptable taste, odor, and appearance. Heat transfer is governed by heat exchange between a product and its surrounding medium. The extent of heat transfer generally increases with increasing temperature difference between the product and its surrounding.

Food materials often experience the movement of heat around and through them in various forms and degrees during growth, harvest, handling, processing, transport, storage, and preparation for consumption. Only a few foods such as fresh fruit and some vegetables go from field to the table without any thermal processing. Most foods are thermally processed to extend their shelf life and maintain high quality. Foods are exposed to heat transfer numerous times during heating, cooling, freezing, frying, and/or baking. Thermal operations are also used to insure safe food products for extended periods. A classic example is the pasteurization of milk, a heating process that eliminates bacteria. Pasteurization, followed by proper cooling, provides a safe milk product for weeks. Canned foods, which are good for years, are sterilized with heat processes. Juice, having been heated and aseptically packaged in boxes, is shelf stable without refrigeration for months.

Heat transfer occurs by conduction, convection, and radiation. These mechanisms can occur individually or simultaneously. In food processing, heat transfer is usually a combination of conduction and convection. Conduction is principally involved during heat transfer within solid-like materials, i.e., solids or static liquids. Convection is the transfer between solids (walls of pipes, vats, rooms) and fluids (food materials). Heat is transferred in fluid foods by bulk movement of fluids as a result of a temperature gradient, and this process is referred to as convective heat transfer. Convective heat transfer can be further classified as natural convection and forced convection. Natural convection is a physical phenomenon wherein a thermal gradient due to density difference in a heated product causes bulk fluid movement and heat transfer. Movement of liquids inside canned foods during thermal sterilization is an example of natural convection. If the movement and heat transfer are facilitated by mechanical agitation (such as use of mixers), this is called forced convection.

Radiant heat transfer is less frequently used but is becoming more common in microwave and infrared heating. Having an understanding of the mechanisms of heat transfer allows the food engineer to better design equipment and processes. The movement of heat to and through a food material depends primarily on the existence of a temperature difference and on physical characteristics such as size, shape, density, and thermal properties. The thermal properties are unique for each food. Each mechanism of heat transfer has an associated thermal property.

Specific heat (C_p) or heat capacity is the heat required to increase the temperature of one unit of mass by one degree. The subscript (p) is included because specific heat of solids and liquids is determined at constant pressure. In steady state conduction through a solid-like material the important thermal property is thermal conductivity (k). Thermal conductivity is a measure of the ease with which heat flows through a material. Heat is conducted quickly through a metal like copper; hence its thermal conductivity value is high. Heat flows more slowly through materials like wood or fiberglass insulation; their thermal conductivity is low. The thermal conductivity of most food materials is in a relatively narrow range between 0.2 and 0.5 W m^{-1} K^{-1}.

1.14.1. Heat Exchangers

Heat exchangers are devices for the exchange of heat between two fluids separated by a heat conducting partition. Heat exchangers are extensively used in the food industry for heating (e.g., pasteurizers), cooling (chilled water generators) and heat induced phase change (freezing, evaporation). Each one of the two fluids may be confined or unconfined (free), stagnant or flowing. The partition is a heat conducting solid wall, usually made of metal. The design of a heat exchanger usually involves two main domains, namely thermal analysis and hydraulic calculations. This section will deal only with the thermal performance of heat exchangers.

1.14.1.1. Tubular Heat Exchangers

If one or both of the materials that are exchanging heat are fluids, flowing continuously through the equipment and acquiring/giving up heat, the process is very efficient, and these equipment are called "continuous flow heat exchangers." This equipment are often employed to pasteurize milk or other beverages, and the heat exchanging fluid is almost exclusively water or water steam (depending on process temperature). One of the fluids is usually passed through pipes or tubes, and the other fluid is passed round or across these. At any point in the equipment, the local temperature differences and the heat transfer coefficients control the rate of heat exchange. The fluids can flow in the same direction through the equipment (parallel flow) or in opposite directions (counter flow), or they can also flow at right angles to each other (cross flow). Various combinations of these directions of flow can occur in different parts of the exchanger; in fact, most heat exchangers of this type have a mixed flow pattern.

In parallel flow, the maximal temperature difference between the coldest and the hottest stream is at the entry to the heat exchanger, but at the exit the two streams approach each other's temperature. In a counter flow exchanger, leaving streams can approach the temperatures of the entering stream of the other component and so counter flow exchangers are often preferred. To further improve heat exchange efficiency, the surface of the tubes can be "corrugated," to extend the available surface and also to provoke dynamic turbulence in the fluids, improving thermal exchange up to 90% (especially in the case of low viscosity fluids).These equipment can in principle be used also to cool down beverages, e.g., using a sodium chlorine brine as exchanger fluid, but other approaches are more often used for this purpose.

1.14.1.2. Shell and Tube Exchanger

Shell and Tube Exchanger consists of a number of tubes mounted inside a cylindrical shell. Two fluids can exchange heat, one fluid flows over the outside of the tubes while the second fluid flows through the tubes. The fluids can be single or two phase and can flow in a parallel or a cross/counter flow arrangement. The shell and tube exchanger consists of four major parts:

(a) Front end–this is where the fluid enters the tube side of the exchanger.

(b) Rear end–this is where the tube side fluid leaves the exchanger or where it is returned to the front header in exchangers with multiple tube side passes.

(c) Tube bundle–this comprises of the tubes, tube sheets, baffles and tie rods etc. to hold the bundle together.

(d) Shell-this contains the tube bundle. The three most common types of shell-and-tube exchangers are fixed tubesheet design, U-tube design, and floating-head type. In all three types, the front-end head is stationary while the rear-end head can be either stationary or floating depending on the thermal stresses in the shell, tube, or tubesheet, due to temperature differences as a result of heat transfer.

The popularity of shell and tube exchangers has resulted in a standard being developed for their designation and use. This is the Tubular Exchanger Manufactures Association (TEMA) Standard. Tubular heat exchangers are particularly suitable for heating or cooling highly viscous products and where relatively high pressures must be applied. They are therefore utilized for the bulk in-flow sterilization of products containing solid particles or for the heat treatment of cooling of tomato paste prior to aseptic packaging. Tubular heat exchangers are also the heat transfer component in tubular evaporators.

1.14.1.3. Plate Heat Exchangers

A popular heat exchanger for fluids of low viscosity, such as milk, is the plate heat exchanger, where heating and cooling fluids flow through alternate tortuous passages between vertical plates as illustrated. (see figure below). The plates are clamped together, separated by spacing gaskets, and the heating and cooling fluids are arranged so that they flow between alternate plates. Suitable gaskets and channels control the flow and allow parallel or counter current flow in any desired number of passes. A substantial advantage of this type of heat exchanger is that it offers a large transfer surface that is readily accessible for cleaning. The banks of plates are arranged so that they may be taken apart easily (Figure 1.13).

Figure 1.13: A Plate Heat Exchanger, (*Source: http://www.thornhillheatexchangers.co.uk/Plate-Heat-Exchangers/).*

1.14.1.4. Jacketed Pans

In this kind of heat exchanger, the fluid (liquid food up to paste consistence food can be treated with this equipment) to be heated is contained in a vessel, which may also be provided with an agitator to keep the fluid moving across the heat transfer surface, to assure its homogeneous heating. Where there is no agitation, heat transfer coefficients are lower or even halved. The source of heat is commonly steam condensing in the vessel jacket: there must be the minimum of air within the steam in the jacket, because air hinders heat exchange. The pan itself can be made of cast iron, stainless steel, or copper. Heat transfer coefficients are not very high: depending on the pan material and on the viscosity of the fluid to be heated can range from 300 J m^{-2} s^{-1} $°C^{-1}$ (for pastes heated in stainless steel pans) to 1800 J m^{-2} s^{-1} $°C^{-1}$ (for thin liquids heated in copper pans). Furthermore, usually this equipment only allows discontinuous food processing.

1.14.1.5. Scraped Surface Heat Exchangers

One type of heat exchanger, that finds considerable use in the food processing industry particularly for products of higher viscosity, consists of a jacketed cylinder with an internal cylinder concentric to the first and fitted with scraper blades (see figure below). The blades rotate, causing the fluid to flow through the annular space between the cylinders with the outer heat transfer surface constantly scraped. Coefficients of heat transfer vary with speeds of rotation, which is about 600–700 rpm, but they are of the order of 900–4000 Jm^{-2} s^{-1} $°C^{-1}$. These machines are used in the freezing of ice cream and in the cooling of fats during margarine manufacture. The scraped surface exchanger is an expensive piece of equipment, both in price and in operating cost (moving parts) (Figures 1.14 and 1.15).

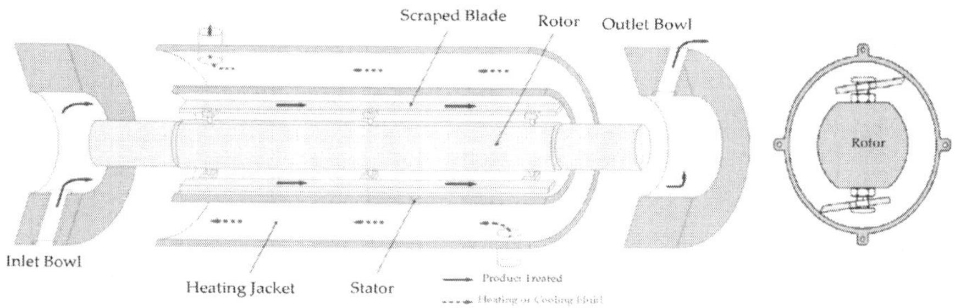

Figure 1.14: How scraped surface heat exchangers work, (*Source: https://www.sciencedirect.com/science/article/pii/S1359431116335748*).

Figure 1.15: Scraped Surface Heat Exchangers, (*Source: http://apienergy.co.uk/product/api-energy-scraped-surface-heat-exchangers/*).

1.15. FLUID FLOW

Fluid flow involves transporting liquid food through pipes during processing. Powders and small-particulate foods are handled by pneumatic conveying, whereas fluids are transported by gravity flow or through the use of pumps. The centrifugal pump and the positive displacement pump are two pumps commonly used for fluid flow. The study of fluids is of great significance in food processing. It is divided into fluid statics (stationary fluids) and fluid dynamics (moving fluids). A property of static liquids is the pressure that they exert on the containing vessel. The pressure is related to the density of the liquid and the depth or the mass of liquid in the vessel. Liquids at the base of a vessel are at a higher pressure than at the surface, owing to the weight of liquid above (the hydrostatic head). This is important in the design of holding tanks and processing vessel, to ensure that the vessel is constructed using materials of adequate strength. A large hydrostatic head also affects the boiling point of liquids, which is important in the design of some types of evaporation equipment. Computational Fluid Dynamics finds application in several processing operations in the food industry including baking, drying, sterilization, pasteurization, heat exchangers, mixing, refrigeration, etc.

Fluid flow is characterized by a dimensionless group named the *Reynolds number (Re).*

The Reynolds number for fluid flowing through a non-circular channel is calculated by substituting the hydraulic diameter of the flow path for the pipe diameter.

$$R_e = \frac{\rho V D_h}{\mu}$$

From a host of experimental measurements on fluid flow in pipes, it has been found that the flow remains calm or "streamline" for values of the Reynolds number up to about 2100. For values above 4000 the flow has been found to be turbulent. Between approximately 2100 and 4000 the flow pattern is unstable; any slight disturbance tends to upset the pattern but if there is no disturbance, streamline flow can be maintained in this region. The expression is given below as;

- Laminar flow: $Re < 2000$
- Transitional flow: $2000 < Re < 4000$
- Turbulent flow: $Re > 4000$

1.16. MIXING

Mixing is a common unit operation used to evenly distribute each ingredient during manufacturing of a food product. Mixing increases the homogeneity of a system by reducing non-uniformity or gradients in composition, properties or temperature. Secondary objectives of mixing include control of rates of heat and mass transfer, reactions and structural changes. In food processing applications, additional mixing challenges include sanitary design, complex rheology, desire for continuous processing and the effects of mixing on final product texture and sensory profiles. Mixing ensures delivery of a product with constant properties. For example, consumers expect all containers of soups, breakfast cereals, fruit mixes, etc. to contain the same amount of each ingredient. If mixing fails to achieve the required product yield, quality, organoleptic or functional attributes; production costs may increase significantly.

Dough and pastes are mixed in machines that have, of necessity, to be heavy and powerful. Because of the large power requirements, it is particularly desirable that these machines mix with reasonable efficiency, as the power is dissipated in the form of heat, which may cause substantial heating of the product. Such machines may require jacketing of the mixer to remove as much heat as possible with cooling water.

1.17. SIZE ADJUSTMENT

In size adjustment, the food is reduced mostly into smaller pieces during processing, as the raw material may not be at a desired size. This may involve slicing, dicing, cutting, grinding, etc. However, increasing a product size is also possible. For example, aggregation, agglomeration (instant coffee), and gelation are examples of size adjustment that lead to an increase in size. In the case of liquid foods, size reduction is often achieved by homogenization. During milk processing, fats are broken into emulsions via homogenization for further separation. Size reduction and emulsification have little or no preservative effect. They are used to improve the eating quality or suitability of foods for further processing and to increase the range of products available.

1.18. SEPARATION

This aspect of food processing involves separation and recovery of targeted food components from a complex mixture of compounds. This may involve separating a solid from a solid (e.g., peeling of potatoes or shelling of nuts), separating a solid from a liquid (e.g., filtration, extraction) or separating liquid from liquid (Fellows, 2009). Industrial examples of separation include crystallization and distillation, sieving, and osmotic concentration. Separation is often used as an intermediate processing step, and is not intended to preserve the food.

Separation from solids or liquids involves the transfer of selected components across the boundary of the food. In many processes another stream or phase is involved, for example in extraction processes. However, this is not always so, for example expression, centrifugation or filtration. In expression, fruit juice or oil is squeezed from the food by application of pressure. In centrifugation, fat can be separated from water due to their density differences, by the application of a centrifugal force. In filtration there is a physical barrier to the transfer of certain components and the liquid is forced through the barrier by pressure, whilst the solids are retained. Separation processes may be batch or continuous. A single separation process, for example a batch extraction, involves the contact of the solvent with the food. Initially concentration gradients are high and the rate of extraction is also high. The extraction rate falls exponentially and eventually an equilibrium state is achieved when the rate becomes zero. The extraction process may be accelerated by size reduction, inducing turbulence and increasing the extraction temperature.

Continuous processes may be single- or multiple-stage processes. The stages themselves may be discrete entities, for example a series of stirred tank reactors, or there may be many stages built into one unit of equipment, for example a distillation column or a screw extractor. The flow of the two streams can either be co-current or counter-current; although counter-current is normally favored as it tends to give a more uniform driving force over the length of the reactor as well as a higher average driving force over the reactor. In some instances a combination of co-current and counter-current may be used; for example in hot air drying, the initial process is co-current to take advantage of the high initial driving rates, whereas the final drying is counter-current to permit drying to a lower moisture content.

1.19. MECHANICAL PROCESSES

1.19.1. Handling of Raw Materials

Raw material handling is the very first step in the food processing. Raw material handling includes postharvest transportation (farm to plant), sorting, cleaning or sanitizing before loading into equipment in the plant. These could also be considered as part of primary processing of the food materials. Microorganisms could attach to inert non-porous surfaces in raw foods and it has been demonstrated that these cells transfer from one surface to another to another when contact occurs (Zottola and Sasahara, 1994). Appropriate raw material selection and handling affect microbial safety and final product quality. Future food preservation studies need to consider the impact of raw material (including postharvest handling prior to preservation) on the final processed product.

1.19.1.1. Cleaning

Cleaning and sanitation of raw food material could be considered the first step in controlling any contamination of foreign materials or microorganisms during food processing. Cleaning removes foreign materials (i.e. soil, dirt, animal contaminants) and prevents the accumulation of biological residues that may support the growth of harmful microbes, leading to disease and/ or the production of toxins. Sanitization is the use of any chemical or other effective method to reduce the initial bacterial load on the surface of raw materials or food processing equipment. Efficient sanitization includes both the outside and the inside of the plant such as specific floor plan, approved materials used in construction, adequate light, air ventilation, direction of air flow, separation of processing areas for raw and finished products, sufficient space for operation and movement, approved plumbing, water supply, sewage disposal system, waste treatment facilities, drainage, soil conditions, and the surrounding environment. Cleaning is the unit operation in which contaminating materials are removed from the food and separated to leave the surface of the food in a suitable condition for further processing.

Peeling fruits and vegetables, skinning meat or descaling fish may also be considered as cleaning operations. In vegetable processing, blanching also helps to clean the product. The presence of contaminants (or foreign bodies) in processed foods is the main cause of prosecution of food companies. Cleaning should take place at the earliest opportunity in a food process both to prevent damage to subsequent processing equipment by stones, bone or metals, and to prevent time and money from being spent on processing contaminants which are then discarded. In addition, the early removal of small quantities of food contaminated by microorganisms prevents the subsequent loss of the remaining bulk by microbial growth during storage or delays before processing. Cleaning is thus an effective method of reducing food wastage, improving the economics of processing and protecting the consumer. Equipment for cleaning is categorized into wet procedures (for example soaking, spraying, flotation washing and ultrasonic cleaning) and dry procedures (separation by air, magnetism or physical methods). The selection of a cleaning procedure is determined by the nature of the product to be cleaned and by the types of contaminant to be removed. In general, more than one type of cleaning procedure is required to remove the variety of contaminants found on most foods (Figure 1.16).

A classification of the type of contaminants found on raw foods is shown in the table 1.2 below.

Table 1.2: Contaminants in Raw Foods

Type of contaminant	**Examples**
Metals	Ferrous and non-ferrous metals, bolts, filings
Mineral	Soil, engine oil, grease, stones
Plant	Leaves, twigs, weed seeds, pods and skins
Animal	Hair, bone, excreta, blood, insects, larvae

Chemical	Fertilizer, pesticides, herbicides
Microbial cells	Soft rots, fungal growth, yeasts
Microbial products	Colors, flavors, toxins

Source: http://www.webpal.org/SAFE/aaarecovery/2_food_storage/Food%20 Processing%20Technology.pdf, Adapted from Brennan et al. (1990).

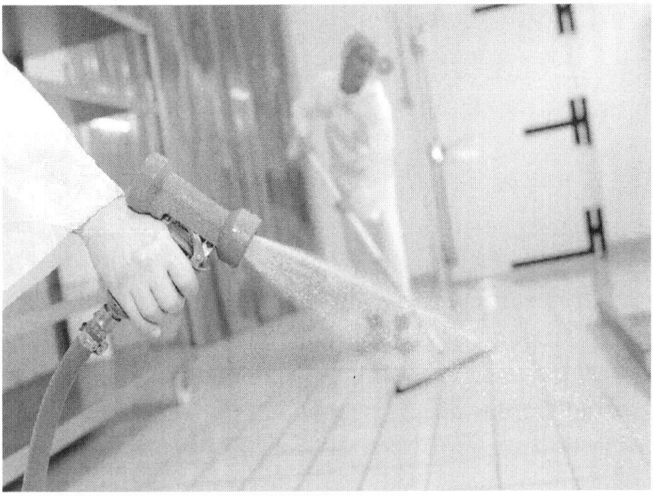

Figure 1.16: Cleaning process in a food company, (*Source: http://www.myprocessexpo.com/blog/category/industry-perspectives/page/5/*).

1.19.1.1.1. Wet Cleaning

The main cleaning method used in most food processing facilities is wet cleaning. This involves using a liquid (most often water) and some form of agitation (scrubbing or scraping) to remove soil. Tools such as brushes, high pressure pumps, air or steam are used in wet cleaning.

Wet cleaning is more effective than dry methods for removing soil from root crops or dust and pesticide residues from soft fruits or vegetables. It is also dustless and causes less damage to foods than dry methods. Different combinations of detergents and sterilants at different temperatures allow flexibility in operation. However, the use of warm cleaning water may accelerate chemical and microbiological spoilage unless careful control is exercised over washing times and subsequent delays before processing.

Furthermore wet procedures produce large volumes of effluent, often with high concentrations of dissolved and suspended solids (measured as biological oxidation demand (BOD) or chemical oxidation demand (COD)). There is then a requirement to purchase clean water and to either pay for high effluent disposal charges or build in-factory water treatment facilities. To reduce costs, recirculated, filtered and chlorinated water is used whenever possible. Examples of wet-cleaning equipment include spray washers, brush washers, drum or rod washers, ultrasonic cleaners and flotation tanks.

1.19.1.1.2. Dry Cleaning

Not all operations can be wet cleaned. In bakeries, flourmills, dryblending facilities and similar operations, microorganisms are of less concern than molds, insects, rodents and foreign objects. In these facilities, clean-up crews use brooms, brushes, shovels and vacuum systems to remove waste and spill. Dry cleaning methods are used for products that are smaller, have greater mechanical strength and possess lower moisture content. The main advantages of dry cleaning methods are that, these methods are generally inexpensive and involve cheaper equipment than wet cleaning methods and produce a concentrated dry effluent which may be disposed of more cheaply but it suffers from various disadvantages such as, it is prone to production of dust, which can be a source of product recontamination and in some cases, a fire and explosion hazard. The main groups of equipment used for dry cleaning are:

(i) Aspirators.
(ii) Magnetic separators.
(iii) Separators based on screening of foods.

Forced air separators or aspirators use large quantities of low-pressure air and consume considerable energy. Control of dust is necessary for safety and health reasons and to prevent the spread of contaminants. This method should not be used for products, which are sensitive to oxidation. Magnetic cleaning is generally used to remove magnetic contaminants that can damage preparatory equipment from the raw materials. Rotating or stationary magnetic drums, magnetic belts, magnets located over belts carrying the food, through which the food is passed, are used involving permanent and electromagnets. The contaminants that adhere to the magnets should be frequently removed to prevent a buildup that can be swept into the product stream, which results in a gross localized recontamination. Screens remove contaminants that have sizes different from the product. A screen in a frame is the simplest form used which has now been largely replaced by continuous types, such as centrifugal screens and the flat bed screens.

Cleaning by abrasion between food particles or between the food and moving parts of cleaning machinery is used to loosen and to remove adhering contaminants. Trommels, tumbler vibrators, abrasive discs and rotating brushes are used for this purpose. Attention should be paid to dust removal to limit recontamination, protect operators and prevent dust explosions.

1.19.1.2. Sorting

1.19.1.2.1. Reverse Sorting

Reverse sorting is a software-driven capability included on select laser, laser/camera, and hyperspectral sorters that enables the user to quickly switch the definitions of what is accepted and what is rejected. It is ideal for rework and when incoming defect loads are higher than 50%. Typically, sorters are programmed to reject FM and/or EVM but when running in reverse-sort mode, they are programmed to target good product. This approach uses less compressed air and, more importantly, it improves the results with a cleaner end-product when defect loads are high. Many sorter suppliers claim a reverse sort capability but often, the adjustment requires a labor intensive recalibration, which may take more time than it is worth. On select sorters, the switch from a forward sort to a reverse sort is achieved in seconds via the touchscreen control panel, with no recalibration or mechanical adjustments required.

1.19.1.2.2. Shape and Size Sorting

For nuts, green beans, and other select products, shape sorting can be an extremely important capability. With nuts, broken products sell for less money. With green beans, shape sorting differentiates beans from same-color stems. Advanced shape sorting can be accomplished with monochromatic or color cameras, coupled with powerful software algorithms. A sorter can be dedicated to shape sorting or configured to achieve shape sorting in addition to FM, EVM, and color sorting. Effective shape sorters must avoid shade effects while creating the images to maximize the contrast of the shape.

Sorting into size categories requires some sort of screen. The main categories of screens are fixed-aperture and variable-aperture designs. Flatbed and rotary screens are the main geometries of fixed bed screen and a number of screens may be used in series or in parallel to sort units into several size categories simultaneously. The problem with fixed screens is usually contacting the feed material with the screen which may become blocked or overloaded. Fixed screens are often used with smaller particulate foods such as nuts or peas. Variable-aperture screens have either a continuous diverging or stepwise diverging apertures. These are much gentler and are commonly used with larger, more delicate items such as fruit.

1.19.1.2.3. Color Sorting

Manual sorting by color is still widely used but is increasingly expensive in labor costs, operator training and the space required for sorting tables. There has therefore been considerable development of machine vision sorting systems which are said to have lower operating costs and greater accuracy than manual methods. These include monochrome (black and white), bichrome (4100 shades of red and green) and trichromatic or full color (262000 shades of red, green and blue, with optional infrared).Each is controlled by a programmable logic controller which has pre-set programs for different products that are easily changeable by operators using a video display. They are used for example, to sort potatoes for defects and blemishes by identifying dark areas on the potato surface. Light sensitive cells in the camera (termed 'pixels') produce a voltage that is proportional to the intensity of light received.

An electronic circuit that receives a lower voltage than the pre-set value can thus detect darker objects or areas which reflect less light than normal. The voltage produced in the electronic circuit can be adjusted to alter the sensitivity of detection. Up to 10 tons of product per hour pass beneath the cameras on conveyors operating at 150–180 m per min. Defective items are removed by electronically controlled air jets that can operate for 20 milliseconds, thus covering 50 mm of the belt length in a single blast. In another system, vegetables in free-fall are scanned 1000 times per second, as they leave a conveyor belt, using concentrated helium-neon or laser light beams and a high-speed rotating mirror. The machine detects differences in reflectivity between good product and unwanted material.

1.19.1.2.4. Weight Sorting

The weight of a food product is proportional to the cube of its characteristic dimension and hence, weights sorting is more precise compared to dimensional sorting. Fruits such as apples, pears and citrus fruits, vegetables such as potatoes, carrots and onions are sorted by weight. Weight is usually the most precise method of sorting, as it is not dependent on the geometry of the products. Eggs, fruit, or vegetables may be separated into weight categories using spring-loaded, strain gauge, or electronic weighing devices incorporated into conveying systems. Using a series of tipping or compressed air blowing mechanisms set to trigger at progressively lesser weights, the heavier items are removed first followed by the next weight category, and so on. These systems are computer controlled and can additionally provide data on quantities and size distributions from different growers. An unconventional system is to use the "catapult" principle where units are thrown into different collecting chutes, depending on their weight, by spring-loaded catapult arms. A disadvantage of weight sorting is the relatively long time required per unit and other methods are more appropriate with smaller items such as legumes or cereals, or if faster quantity is required.

1.20. AN INDUSTRIAL PROCESS

An industrial process is a manufacturing unit in which a series of continuous or regularly occurring actions take place in a predetermined or planned manner to produce the desired product. The product changes from the raw material stage to the final product stage within a certain span of time. During this transition, there might be more than one form of physical or chemical change of property of the material. On the other hand, there can be addition or removal of materials in quantity. The physical and chemical changes that take place in the materials or components in an industrial process are mainly responsible for the quality of the final product. The causes of the physical and chemical changes are the variations of some parameters that get reflected in some other parameters. These parameters are called process parameters or process variables.

It is evident that the process action mainly depends on the process variables, so careful attention to the process variables is very much essential to knowing the state of the process action. The process variables that indicate the correctness of the process action are the process outputs or system outputs. The process variables that can change the process action are the process inputs or system inputs. Process control is an integral part of modern processing industries; and the food processing industry is no exception. The fundamental justification for adopting process control is to improve the economics of the process by achieving, amongst others, the following objectives:

(a) Reduce variation in the product quality, achieve more consistent production and maximize yield.

(b) Ensure process and product safety.

(c) Reduce manpower and enhance operator productivity.

(d) Maintaining certain variables of a process (such as temperatures, pressures, concentrations, flow rates etc.) within specified limits or changing said variables according to a preset program.

1.21. MANUAL CONTROL

In manual control, an operator periodically reads the process parameter which requires to be controlled and, when its value changes from the set value, initiates the control action necessary to drive the parameter towards the set value. An operator is constantly monitoring the temperature in the pipe. As the temperature changes from the set point on the thermometer, the operator adjusts the steam valve, i.e. either increases or decreases its flow, in order to get the temperature in the pipe back to the set point.

Further action may be required if the temperature does not return to the set point within a reasonable time. It is clear that the success of manual control operation depends on the skill of individual operators in knowing when and how much adjustment to make. Therefore, manual control may be used in those applications where changes in the manipulated parameter cause the process to change slowly and by a small amount. This is possible in plants where there are few processing steps with infrequent process upsets and the operator has sufficient time to correct before the process parameter overshoots acceptable tolerance. Otherwise, this approach can prove to be very costly in terms of labor, product inconsistencies and product loss.

1.22. COMPUTER-BASED SYSTEMS

The increasingly rapid use of microprocessor-based process controllers over the years is as a result of flexibility in operation, their ability to record data for subsequent calculations and their significant reduction in their cost. Computers can only be programmed to read data from sensors and send signals to process control devices, but they can also store and analyze data and be connected to printers, computers and controllers and communication devices through a plant. They can also be easily reprogrammed by operators to accommodate new process or product changes.

1.22.1. Programmable Logic Control

A programmable logic controller (PLC) is a simple microprocessor-based, specialized computer that carries out logical control functions of many types, with the general structure shown. A PLC usually operates on ordinary house current but may control circuits of large amperage and voltage of 440V and higher. A PLC typically has a detachable programmer module, used to construct a new program or modify an existing one, which can be taken easily from one PLC to another. The program instructions are typed into the memory by the user via a keyboard. The central processing unit (CPU) is the heart of the PLC, which has three parts: the processor, memory, and power supply. In the input-output (I/O) modules, the input module has terminals into which the user enters outside process electrical signals and the output module has another set of terminals that send action signals to the processor. A remote electronic system for connecting the I/O modules to long-distance locations can be added, so that the actual operating process under the PLC control can be some miles away from the CPU and its I/O modules. Optional units include racks, chassis, printer, program recorder/player, etc.

1.22.2. Types of Control Systems

The different combinations in which PLCs and larger computers can be linked together in an integrated control system can be described in three categories:

(1) Dedicated systems.
(2) Centralized systems.
(3) Distributed systems.

1.22.2.1. Dedicated Control Systems

These use local equipment controllers (PLCs) which are an integral part of the process plant, dedicated to the control of a single unit operation. Examples include controlling the temperature of a heat exchanger or blending different ingredients into a formulated batch. They do not communicate information to other computers but simply receive on/off instructions from a central control panel. They are relatively unsophisticated computers that are able to receive data from sensors and send signals to actuators. They may also have the capability for data-logging, report generation, and automatic set-point adjustment. They have remained popular because they are easy to fit to existing processes without requiring significant changes to the control system; they are easy to use and low cost. However, they are usually designed for a particular application and the present operating sequence is difficult to change (Figure 1.17).

Figure 1.17: Programmable Logic Controller, (*Source: http://www.equipment-direct.net/automation/programmable-logic-controllers/*).

1.22.2.2. Centralized Control Systems

In these systems, a mainframe computer or large mini-computer, located in a centralized control room, monitors and controls different on-line controllers that in turn control the process in specified zones. Each on-line controller may have a printer, data logger and graphics display to inform operators of the status of the process. They can be reprogrammed relatively easily to accommodate process changes and can have facilities for report generation and communication with other computers.

Each process area has a mimic panel to indicate continuously the status of the process variables. Closed-circuit television cameras view the plant and relay the information to monitors located in the central control room. The central computer checks the positions of valves, fluid levels, pressures, flow rates, densities and temperatures in the processing equipment, at a rate of 998 inputs every 7 s. When a fault occurs, the computer sounds an alarm and produces a printout for the operators in the control room of the faulty equipment, its location and the nature of the fault. The computer also prints detailed production data and the status of stocks at any time. Larger systems are able to control 200 plant actuators and monitor 5000 inputs at a rate of 2000 per second. Although centralized computer control systems have been used in large companies for several years, their major disadvantage is that any failure in the central computer could cause a total plant shutdown, unless an expensive standby computer of equal capacity is available to take over. For this reason, distributed control systems which do not have this disadvantage, are now more commonly installed.

1.22.2.3. Distributed Control Systems (DCS)

These are an integrated system of control in which each area of a process is independently controlled by a PLC, and the PLCs are both linked together (process interlocking), and linked to a central computer via a communications network (Lasslett, 1988). Each PLC controller is located close to the equipment that it controls to reduce the cost of wiring, and each has an operator's console with a graphics display and control inputs. Although the capital cost and programming costs are higher than the other systems, distributed control systems are highly flexible in being able to change processing conditions and do not have the vulnerability to total plant shutdown if one component fails (Persson, 1987; Dahm and Mathur, 1990).

Distributed control systems have evolved significantly in the last ten years. From large systems tailored to the needs of continuous industries, such as refineries, DCSs can now respond to the requirements of small batch-oriented processes and can address a variety of automation projects. The philosophy of distributed control is to break down a large application into smaller subsystems and bring the level of control down to the unit level when appropriate to decrease overall system response time.

This makes it possible to exchange information between the different control units and allows for integrated decision making at the product line or plant level. One example is of an ice cream manufacturer looking to upgrade a processing plant from the existing level of automation based on isolated PLCs. The DCS solution provided the flexibility to run several products on the same line. The DCS also provided flexible sequencing of operations and reduced changeover time, thereby reducing overall production cost. Better monitoring of the process variables and statistical analysis of process trends resulted in improved quality control. Interactive graphics and displays gave plant personnel a better understanding of the process status in a timely fashion. The ability to secure the system using passwords added the benefit of production safety, which is essential in food manufacturing. Finally, the expandability of the system gave the manufacturer confidence that the selected automation strategy would remain applicable to production needs as the requirements for more sophisticated controls increased.

Another example is the case of a dry packaged snack foods plant needing to upgrade its existing systems to integrate its process control and information management functions. DCS solution provided better monitoring and control of raw material handling, batch processing, forming, cooking, and packaging and warehousing operations were the main drivers for plant modernization. Even though the initial cost of a DCS solution was greater than a PLC-based system, the difference was paid back in improved functionality and productivity within four to six months. The solution also continued to reduce manufacturing costs and improve productivity at a faster rate than would the PLC-based solution.

1.23. AUTOMATIC CONTROL

In automatic control, the process parameters measured by various sensors and instrumentation may be controlled by using control loops. A typical control loop consists of three basic components:

(i) Sensor: the sensor senses or measures process parameters and generates a measurement signal acceptable to the controller.

(ii) Controller: the controller compares the measurement signal with the set value and produces a control signal to counteract any difference between the two signals.

(iii) Final control element: the final control element receives the control signal produced by the controller and adjusts or alters the process by bringing the measured process property to return to the set point, e.g., liquid flow can be controlled by changing the valve setting or the pump speed. An automatic control system can be classified into five main types:

(1) On/off (two position) controller.

(2) Proportional controller (P-controller).

(3) Integral controller (I-controller).

(3) Proportional integral controller (PI controller).

(4) Proportional integral derivative controller (PID controller). We shall discuss the following components of an automatic control system.

1.23.1. Control Algorithm

1.23.1.1. On–Off Control Action

An on–off controller is the simplest and most inexpensive of controllers. An on–off controller results in an output action, in only two states, depending on the level of the manipulated variable. Describing the control action with reference to a valve on a flow pipe, the control actions are: "the valve is fully open" and "the valve is fully closed." One state of the output is used when the manipulated variable is above the desired level (set point) and the other state is used when the manipulated variable is below the setpoint level.

Consider an on–off control action for a temperature-controlled batch retort used for closing and heat sterilization operation of cans to achieve microbiological stability in canning. As soon as the temperature crosses the set level, the steam entry valve is turned off. This causes the temperature to come down to the set level because the valve is still in the off position.

Cycling of temperature and switching of the valve, on and off, continues simultaneously. In this mode of control, the actuator can have only one of two positions: either on or off (open or closed, all or nothing, 1 or 0). There are no intermediate positions such as partially open. This is the simplest and the least costly of the controllers. On-off control is very common in household devices, laboratory instruments and in some cases of industrial control systems (air compressors, refrigeration engines, limit switches and alarms etc.). In on-off control, the controlled variable must be allowed to fluctuate about the set point, within a range known as the differential band (or dead zone). For example, in a thermostat set to 25 °C, the switch turns the heater on at 24 °C and off at 26 °C. Between these two values, the heater would be either on or off, depending on the direction of change.

Without a differential band, the controller would cycle between two contradictory positions at very high frequency (chatter), causing damage to contractors and valves or causing the actuator to 'freeze' altogether. The differential band or 'hysteresis' is either stored in the controller as an analog or digital datum or built-in in the hardware as the physical lag between the controller and the actuator. In practice, the width of the differential band is usually set between 0.5% and 2% of the control range.

A narrow band results in more precise control but high frequency of change and vice versa. On/Off control action is shown in Figure 1.18. The heating power is either fully On when the temperature is below setpoint or fully Off when it is above. As a result, the temperature oscillates about the setpoint. The amplitude and time period of the oscillation is a function of the thermal lag between the heating source and the temperature sensor. To prevent the output 'chattering' as the measured temperature crosses the setpoint, the controller does not turn On and Off at precisely the same point. Instead, a small differential known as the 'hysteresis' is applied. A typical value is 1°C On/Off control is satisfactory for non-critical heating applications where some oscillation in the temperature is permissible.

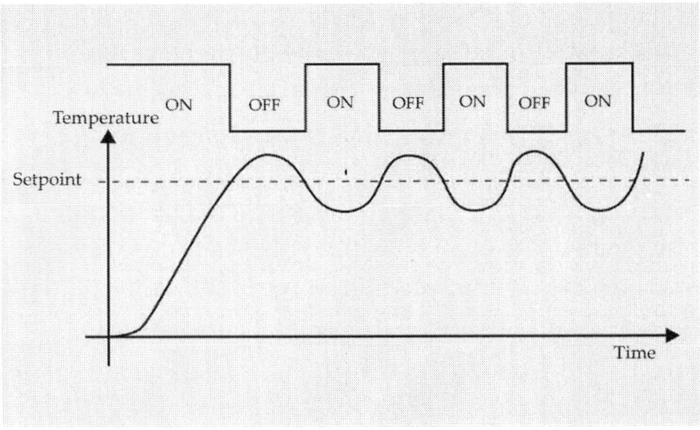

Figure 1.18: On/Off control action, (*Source: https://www.eurotherm.com/principles-of-pid-control-and-tuning*).

1.23.1.2. Proportional Controller

In proportional (P) control, the magnitude of the correction signal m is proportional to the error e.

$$m = Ke + M \qquad (10)$$

The proportionality factor K is called proportional gain. The constant M is known as the controller bias because it represents the magnitude of the correction signal when no correction is needed (e 0). K is dimensionless. M, e, and m are usually expressed as percentage. Proportional control serves to eliminate the oscillation associated with on-off controllers.

The magnitude of the corrective action is reduced as the controlled variable approaches the setpoint. In control systems of the proportional type, the actuator can assume intermediate positions between two extremes, depending on the amplitude of the correction signal from the controller. In some simpler systems, the actuator works in on-off fashion and the proportionality is achieved by regulating the ratio of on-time/off-time. At the set point, that ratio is equal to 1. The proportional gain K is usually a fixed property of the controller but, in some proportional controllers, K is manually adjustable. If K is increased, the sensitivity of the controller to error is increased but the stability is impaired. The system approaches the behavior of on-off controlled systems and its response becomes oscillatory.

The bias M is usually adjustable. As a rule, it is customary to adjust M so as to stabilize the system at a state slightly different from the set point. The difference between the measurement c and the set point r at steady (stable) state is called the offset. It is possible to eliminate the offset by adjusting the bias. This action is called reset. However, adjusting the controller to zero offset under a given set of process conditions would require re-adjusting every time the load or any other process condition changes. In automatic control, this would be highly problematic. For fine-tuning of a controller, the offset must be removed. This can be done manually or automatically. Manual reset is a traditional method in which a potentiometer is used to nullify the offset electrically. The amount of proportional band shifting is manually done, step-by-step, until the manipulated input to controller (process heat demand) conforms to the setpoint level of temperature (Figure 1.19).

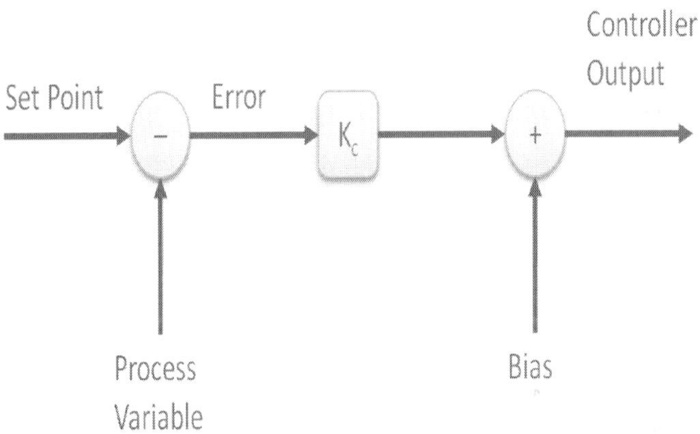

Figure 1.19: Proportional-only controller algorithm, (*Source: http://blog.opti-controls.com/archives/344*).

1.23.1.3. Integral Controller

The need for manual reset in the previous subsection led to the development of automatic reset or the Integral Control Mode, as we know it today. An automatic reset proportional controller performs a reset operation with the help of an electronic integrator. The error is integrated and added with the error signal. Due to the integral term in the control law, a sudden change in deviation is tracked continuously, at a certain rate, determined by the reset rate (inverse of the integral time). As long as there is an error present (process variable not at set point), the integral control mode will continuously cause an increment or reduction in the controller's output to reduce the error.

Given enough time, integral action will drive the controller output far enough to reduce the error to zero. Performance of a proportional controller is expressed by two terms: proportional gain and integral time constant. During integral time, equal parts of the output signal due to integral action advance as that of the output signal due to proportional action. Removal of the offset is the main advantage of I control. However, the response of I-controlled systems is slow and may be highly oscillatory. The response of such systems to a step change in the controlled variable is a ramp, with a slope proportional to the proportionality constant R (Figure 1.20).

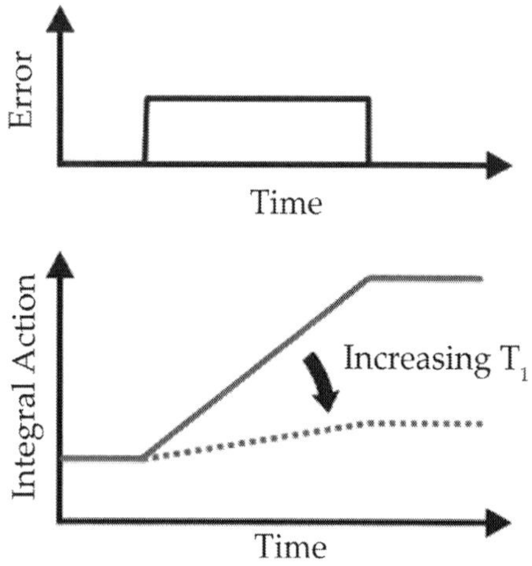

Figure 1.20: Integral control action, (*Source: http://blog.opticontrols.com/archives/344*).

1.23.1.4. PI Controller

A variation of Proportional Integral Derivative (PID) control is to use only the proportional and integral terms as PI control. The PI controller is the most popular variation, even more than full PID controllers.

$$e(t) = SP - PV$$

$$u(t) = u_{bias} + K_c e(t) + \frac{K_c}{\tau_1}\int_0^t e(t)dt \qquad (11)$$

The u_{bias} term is a constant that is typically set to the value of u(t) when the controller is first switched from manual to automatic mode. This gives "bumpless" transfer if the error is zero when the controller is turned on. The two tuning values for a PI controller are the controller gain, K_c and the integral time constant τ_I. The value of K_c is a multiplier on the proportional error and integral term and a higher value makes the controller more aggressive at responding to errors away from the setpoint. The setpoint *(*SP*)* is the target value and process variable (PV) is the measured value that may deviate from the desired value. The error from the setpoint is the difference between the SP and PV and is defined as e(t) = SP−PV (Figure 1.21).

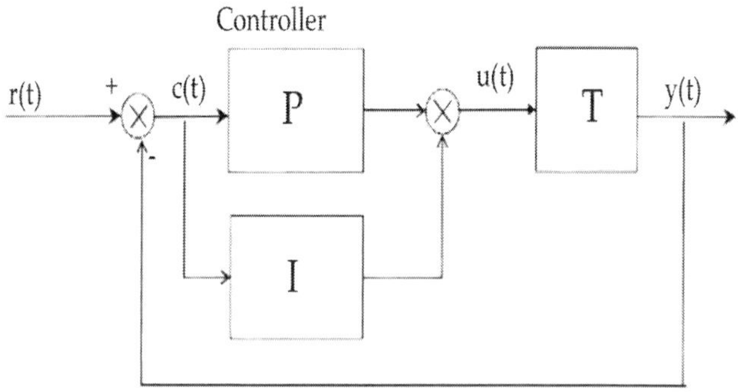

Figure 1.21: PI controller, (*Source: http://cs4hs.cs.pub.ro/wiki/roboticsisfun/chapter5/ch5_1_optimizations*).

1.23.1.5. Proportional Integral Derivative Controller

The PID algorithm is extensively used in industrial process control. For optimal performance, the controller must be 'tuned' by adjusting the parameters of the three terms: K, R and T. In line with the PI controller, the PID controller produces an output signal to the final control element which can be expressed as:

$$M = K\left(e + R\int e\partial t + T\frac{\partial e}{\partial t}\right) = M \tag{12}$$

In the differential mode, the correction signal is proportional not to the instantaneous error but to the rate of change of the error. If the error increases rapidly, the correction signal is larger. It can be said that the differential action is predictive. The response of differential-controlled systems is fast but highly unstable. The differential mode is, therefore, not applied alone but in combination with other control types. The PID mode provides proportional control with the automatic reset feature of integral control and the rapidity and predictive action of differential control.

REFERENCES

1. Abdelrahan, M., (1988). Sensors and AI; the Sixth Sense. *Measurements and Control, 130*, 146-147
2. Asai K., & Takashima S., (eds)., (1994). *Manufacturing Automation Systems and CIM Factories*. New York: Chapman and Hall.
3. Biekert R., (1998). *CIM Technology Fundamentals and Applications*. Tinley Park, IL: The Goodheart-Willcox Company, Inc.
4. Butterworth, D., (1996). Developments in Shell-and-Tube Heat Exchangers, in New Developments in Heat Exchangers, N. Afgan, M. Carvalho, A. Bar-Cohen, D. Butterworth, & W. Roetzel., (eds)., Gordon & Breach, New York, pp.437–447.
5. Earle, R., & Earle, M., (2012). Unit Operations in Food Processing. [online] Available at: https://viden.io/knowledge/462/attachments/2153?name=viden-clip-K441-WINSEM2016-17_CHE1017_ETH_5180_RM002_Unit+operations+in+food+processing+-+Ea.pdf[Accessed 20 April 2018]
6. Fellows, P. J., (2000). Food Processing Technology. CRC Press. Woodhead Publishing Limited. Cambridge, England. [ebook] Available at: *http://www.webpal.org/SAFE/aaarecovery/2_food_storage/Food%20Processing%20Technology.pdf* [Accessed 20 April 2018]
7. Caldwell, G., & Darwin., Robotics and Automation in the Food Industry: Current and Future Technologies. Woodhead Publishing.1st edition, 2012
8. Hui, Y. H., (2006). “Handbook of fruits and fruit processing.” Blackwell Publishing
9. Leniger, H. A., & Beverloo, W. A., (1975). Food Process Engineering. D. Reidel, Dordrecht, pp. 236–268.
10. Gunasekaran, S., *Automation of Food Processing* Food Engineering– Vol. IV
11. Singh R., & Heldman, D., (2009). Introduction to Food Engineering, 4th (ed). [PDF] San Diego, CA: Academic Press. Available at.http://197.14.51.10:81/pmb/AGROALIMENTAIRE/Introduction%20to%20Food%20Engineering%20Fourth%20Edition.pdf [Accessed 20 April 2018]

CHAPTER 2

ROBOTICS IN THE FOOD INDUSTRY

"While some people may think being a chef only entails making enticing dishes and pushing the culinary boundaries, being a part of the food industry involves much more."

—Marcus Samuelsson

CONTENTS

This chapter concisely reviews the European food and beverage market and diagrams current patterns including client and administrative requests, commodity costs, statistic and ecological issues, which will shape its future advancement. Current manufacturing producing systems are examined together with the drivers for change with accentuation on the need to embrace cutting-edge designing innovation. The difficulties and opportunities displayed by the implementation of robotization are underlined and some food-factory-of-the-future concepts are laid out.

2.1. INTRODUCTION

The utilization of robotics in the advanced food industry has expanded over late years as a result of numerous favorable circumstances. Be that as it may, the industry has not taken to the innovation with an indistinguishable pace from the car and different enterprises because of specialized and some different reasons. The high speculation rate, variable nature of food items, delicate and perishable attributes of food things, clean and cleansing necessities, high creation volume rates are influencing impediment in use of robotics in industry yet the chance to even now exists to convey huge advantages regarding expanded food time span of usability, cost diminishments and adaptability. The test is to grow ease, adaptable, sterile and canny machines for the food business. This is a not intense as the innovations now exist to accomplish it. The advantages that could be picked up if frameworks are outlined, introduced and worked well could choose will's identity the real food makers of comfort foods later on.

The utilization of robotics has changed assembling in relatively every modern part, and by expanding proficiency and authorizing consistency has brought about the arrangement of the tremendous scope of solid items at reasonable costs that supports every cutting-edge economy. Up to this point, food manufacturing has been a special case to this pattern and numerous reasons can be referred to for this including as a result of their common inconstancy inconsistency and shade they are not manageable to computerized handling strategies. Be that as it may, progresses in innovation and statistic and market powers are currently quickly changing the European scene in the food manufacturing division and a layout of their variables and the opportunities created is quickly reviewed beneath.

Robot, from the Czechoslovakian word, "robota" which means constrained work. A robot can be characterized as a programmable, self-controlled gadget comprising of electronic, electrical, or mechanical units.

All the more, for the most part, it is a machine that capacities set up of a living operator. As indicated by British Robot Association, "A modern robot is a reprogrammable gadget planned both to control and additionally transport parts, instruments, or determined assembling executes through factor modified movements for the execution of particular assembling undertakings."

The International Standards Organization (ISO) characterizes a robot as, "A naturally controlled, re-programmable, multi-reason, manipulative machine with a few degrees of opportunity, which might be either settled set up or versatile for use in mechanical computerization applications." Robots are particularly attractive for certain work capacities on the grounds that, not at all like people, they never get drained; they can work in physical conditions that are awkward or even hazardous; they can work in airless conditions; they don't get exhausted by redundancy, and they can't be diverted from the job needing to be done. The robot is capable, dependable and can be utilized as a part of hot temperature territory where a human in the wake of working for so long can end up wiped out and depleted.

Explanations behind automating processes:

- Need to decrease coordinate work.
- Can't inspire individuals to carry out the activity.
- Need to expand quality.
- Difficult to carry out the activity physically.
- Need to build generation.
- Difficult to meet details reliably.
- Need to give adaptability in forms.
- Hazardous to workforce.
- Eliminates a defilement source.

Robotics is the branch of innovation that deals with the plan, development, activity, and use of robots, and also PC frameworks for their control, tangible criticism, and data preparing. These advancements manage mechanized machines that can replace people in risky situations or assembling forms, or look like people in appearance, conduct, or potentially cognizance.

2.1.1. The Benefits of Automation in the Food Industry

Food makers and merchants frequently work inside limited edges and top notch principles. With regards to raising benefit while furnishing shoppers with a far and away superior item, robotic automation is a characteristic fit. In spite of this, it presently can't seem to achieve its maximum capacity.

As of not long ago, leaving food cultivation, planning, and distribution to machines was about unbelievable. Presently, be that as it may, robotic systems are emerging that can quicken and streamline every one of these procedures. From soil to plate, finish mechanization is inside reach.

In the course of the most recent decade, robotized cultivate gear and harvest checking frameworks have made farmers employments less demanding. Robots help guarantee the land is utilized effectively–limiting irritations, soil depletion, and the most burdensome human work.

However, as food gets nearer to the plate, robots have customarily retreated as of not long ago.

Robotization frameworks are presently being produced, tried, and guided that will think about food science challenges additionally down the store network. Albeit boundless selection may even now be far off, promising signs have been found in these zones:

2.1.1.1. Quality Control

Quality is the overall worry for both food makers and purchasers. When handling and bundling are robotized, deformities can be seen and segregated all the more effortlessly. This guarantees flawed items don't enter the commercial center and fundamental assembling shortcomings are adjusted.

2.1.1.2. Enhanced Safety

Mechanization and machine vision can cooperate to give a more granular perspective of each progression in food preparing. Pioneers can adjust all the more rapidly to changes in wellbeing gauges on the grounds that an adjustment in one process won't present sudden outcomes in different regions.

2.1.1.3. End-to-End Traceability

Traceability is a developing worry in numerous industries. At the point when items are traceable, organizations can catch better quality, wellbeing, and benefit on the double. By streamlining traceability, nourishment makers can eliminate administrative consistence costs and ceaselessly enhance key procedures.

2.1.1.4. Brand Protection

Food makers depend upon their image notoriety to keep up a decent association with the end buyer. A solitary item review or infection episode can be sufficient to screen a business–and it can toss clients' lives into disarray. At last, mechanization implies mark insurance.

2.1.2. The Industry

The food-producing sector is one of the biggest manufacturing sectors in most European nations and is a fundamental key component in every single national economy. Generally speaking, food manufacturing represents approximately 13% of all assembling in the European Union, contributing about €900 billion to the economy and connecting with somewhere in the range of 4 million representatives in its activities. It is a differing and to some degree divided industry including around 300,000 organizations, by far most of which can be named little to-medium endeavors (SME). Assembling activities are for the most part manual in nature, especially in the SME division, while islands of perfection in best in class computerization are progressively normal, the specialized foundation of the area is, for the most part, powerless and use on designing innovative work is lower than in most other assembling segments.

This chapter concisely analysis the present European market (which is a regularly different market) and how it impacts producing procedures. Drivers for change is outlined, together with financial and specialized difficulties they present, and some conceivable future assembling ideas will be acquainted that may help with address a portion of the financial market requests visualized for the following decade.

2.1.3. The Market

Both the business and the retail industry are extremely different, however as appeared in Figure 2.1 piece of the overall industry in Europe has a tendency to be focused on a similarly little arrangement of real retailers and it is unavoidable that they impact the market shape, item range and estimating theories. There are however some larger patterns, including:

- a descending weight on costs coordinated by an upward weight on ware costs.

- an expanding limit in non-European nations that straightforwardly contend with European items, through utilization of worldwide coordination's and financial administrations.
- an interest for a more noteworthy assortment of items with shorter run times, which embeds instability into the market.
- a necessity for a speedier reaction in the production network.
- more new nourishment with an attendant shorter time to showcase.
- an expanding open consciousness of nourishment content and clean norms and in this way a prerequisite for producers to meet the most noteworthy Hazard Analysis and Critical Control Points (HACCP) Standards.

The market will likewise be affected by a steady change in European demography and the want for a 'green' market picture, which will think about assembling techniques, bundling and coordination's to accentuate the limiting of waste and vitality. In an inexorably focused market retailers will require a consistency of item quality at most reduced cost, high time span of usability, and an adaptable, quick reaction to client request, to keep up and create piece of market share (Figure 2.1).

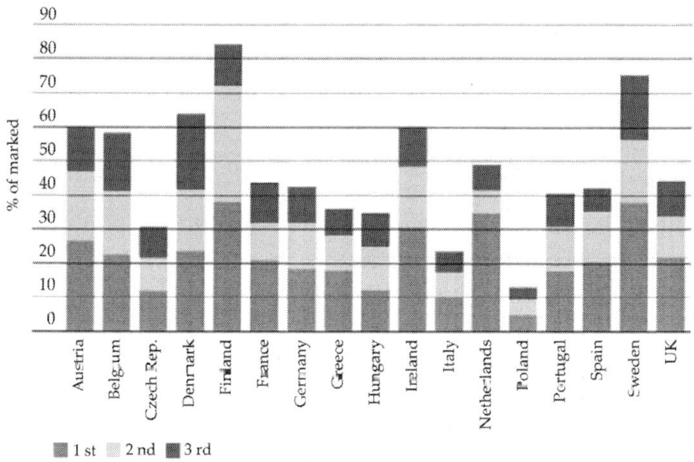

Figure 2.1: European retail market share: Current market share of the three largest retailers in various member states, (*Source: CAA calculations based on data provided by CIAA members*).

2.2. CURRENT MANUFACTURING PROCEDURES

The food manufacturing industry has its roots in the broadened kitchen idea where there was an emphasis on food outline and readiness, with the processing plant being an intensified adaptation of household methods of manual arrangement and supervision. As creation expanded to take care of demand the bigger organizations saw the advantage of utilizing robotization, especially toward the finish of-line bundling and palletizing.

The accessibility of profoundly powerful machines, for example, the Delta class of robot, has enabled robotization to move upstream and embrace quick pick-and-place activities with food items on the creation lines. Various noteworthy frameworks are right now introduced, by and large, focused on the high-volume, long-life, and single-product offerings. Small organizations, which constitute more than 90% of the European food producing base, have been ease back to copy the bigger partnerships and there are various reasons why this may be so:

- As of not long ago, there has been a satisfactory supply of accessibility of work that could be conveyed on an agreement premise and that could rapidly be sent to take care of fast changes in demand. Manipulative errands are for the most part basic dreary tasks, so preparing prerequisites and related expenses were correspondingly negligible.
- The commercial center is especially unpredictable in nature with short-term orders being the standard as opposed to long-haul contracts. This has debilitated capital interest in robotization, which is frequently wrongly thought to be unyielding in task and therefore inadmissible for managing short running of items, which are progressively an element of the market design.
- It was broadly suspected that robotization was quite unsatisfactory for the get-together and setting of delicate, variable-dimensional, delicate, tricky/sticky characteristic items, a view bolstered by the omnipresent pictures of robot activities on substantial scale auto-generation lines.
- The proficient building accessible inside existing SME administration structures is for the most part frail in this area. This not just effects on the take-up of the innovation at line level, yet in addition denies organizations of educated inside direction on capital speculation choices at Board level, and this builds the apparent danger of embracing propelled producing innovation.

The above contemplations have impacted the development of the business, especially in the dominating SME part, with the outcome that there is as yet a solid dependence on manual assembling techniques. Despite the fact that the impression given is now and again that of a nineteenth century as opposed to twenty-first-century endeavor, the dynamic and lively administration structure that portrays the segment has so far guaranteed that the requests of the commercial center have been basically fulfilled.

In any case, in the most recent decade or somewhere in the vicinity, various market and different patterns have developed that truly challenge the conventional manual serious assembling approach.

- For various statistic and different reasons, the accessibility of appropriate minimal effort work inside the European Union has decreased, which straightforwardly impacts on the adaptability of the work concentrated approach.
- Existing and approaching EC enactment on work law and wellbeing and security mandates are putting upward weight on work expenses and rendering elective arrangements more alluring.
- The expanding interest for guaranteed clean techniques, consistency of item quality and line security, are giving powerful contentions to isolating human administrators from the generation procedure.
- Increased product costs, the requirement for decreases in vitality and water utilization, potential zero access to landfill destinations, and general carbon impression issues, are among a scope of variables now inciting a radical audit of assembling forms inside the division.
- Modern prerequisites for item traceability and the conspicuous business favorable circumstances of utilizing a totally incorporated assembling, appropriation, robotized warehousing and advertising frameworks are basically incongruent with conventional manual assembling systems.

In this way, albeit a portion of the drivers for change are set by administrative and statistic factors, the key drivers are being set by the market itself. Consumer loyalty and anticipated duplication of sales must be ensured by results of top notch, which are totally sheltered to expand, are alluringly bundled, and exist in adequate cost limits. It is advantageous if the retailer can assert that his items are sourced from makers who meet the most astounding gauges in clean creation, and grasp a green plan, suggesting high proficiency, low ecological effect, moral sourcing of material, and utilize talented expert workforce. Retailers now require a completely coordinated production network with traceability through assembling, coordination, stockpiling, rack arrangement to purpose of offer and past. This gives a capable responsive instrument that can strip costs out of the procedure and must be compelling if implanted inside a propelled fabricating condition.

2.3. AUTOMATION IN THE FOOD SECTOR

The above drivers have prompted a slow take-up of robotization inside the segment, especially inside bigger makers. At present, the division represents the biggest market for mechanical technology inside the UK, which is maybe demonstrative of a general pattern inside the business. Due to the assorted variety and nature of the product offering, various assembling methodologies have been embraced.

In high-volume, long-life marked single-product offerings, an alleged hard computerization arrangement is fitting, being executed by robots or blends of less difficult electro-mechanical or different gadgets. Of the item is general fit as a fiddle and very much situated on hold, at that point straightforward electro-mechanical arrangements will get the job done. Poor line localization will more often than not include visual-serving mechanical systems, especially if the item is additionally sporadic fit as a fiddle. It has been exhibited that a mix of apply autonomy and electro-mechanical procedures can address the robotization of most sustenance results of either customary or sporadic shape.

At first in the segment, standard mechanical robots were utilized for end-of-line assignments including bundling and palletizing. These step by step climbed the line to attempt straightforward item position assignments and exhibited their capacities in a scope of uses one of which is appeared in Figure 2.2.

In any case, there was an interest for speedier and more agile machines reasonable for quick pick-and-place activities on singular food items. This test was met by the presentation of the Delta group of robots now accessible from a scope of makers. These robots are streamlined for quick task (regularly 100–120 picks for every moment) with lightweight payloads (commonly 1-2 kg) and have been especially effective in handling food items. An average application including putting items in appeared in Figure

2.3. Establishments can be fell to produce a workspace without human administrators as appeared in Figure 2.4, a design that is characteristic of future assembling patterns.

A key favorable position of apply autonomy over 'hard-wired' electro-mechanical frameworks is the capacity to reconstruct their activities for various assignments utilizing straightforward, effectively executed systems, subsequently addressing the requirement for adaptable creation ability inside the business. Another extremely positive pattern is the steady diminishment in the expenses of robots versus the unavoidable ascent in labor costs, in the course of the most recent decade (as appeared in Figure 2.5), which renders this approach progressively appealing.

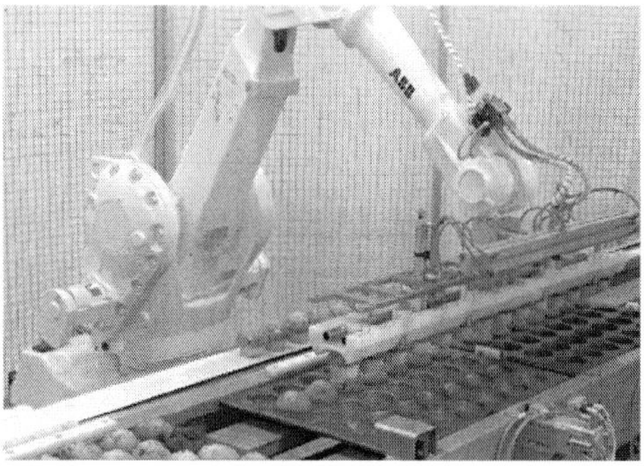

Figure 2.2: Batch processing on a bakery assembly line, (*Source: G. Caldwell Darwin (2013), Robotics and Automation in the Food Industry: Current and Future Technologies, Woodhead Publishing*).

Figure 2.3: Rapid handling on a high-volume pancake line, (*Source: G. Caldwell Darwin (2013), Robotics and Automation in the Food Industry. Current and Future Technologies, Woodhead Publishing*).

2.4. SPECIFICATIONS FOR A FOOD SECTOR ROBOT

Automated frameworks have been produced over decades for fruitful application in an immense scope of modern divisions and they support the progressed in overall monetary improvement. An assortment of automated frameworks, created for such divisions, are promptly accessible from global organizations and are frequently offered as assembling answers for the food business.

Be that as it may, as demonstrated, the food industry is an uncommon market and it is valuable to plot a few rules for the outline of gear for the segment. The utilization of mechanical autonomy in the food business was checked on by Wallin (1995, 1997) and Purnell (1998) and all the more as of late by Moreno-Massey et al. (2010) who delineated a progression of attractive plan includes that ought to be considered in the improvement of a food review robot.

Any outline should meet the rules for food dealing with gear given, for instance by and Lelieveld et al. (2003, 2007) and be in a perfect world of stainless-steel development with entrance insurance rating to IP67, and all parts obvious and available for review and manual cleaning. The benchmarks of sterile plan required will rely upon the application, being higher, for instance, in preparing poultry and fish items than in handling dry-food items, for example, rolls or bread. A universally useful robot, which could possibly be utilized to process any kind of item, should meet the most astounding models of sterile plan.

For use on standard-width, high-volume creation lines there is a necessity for moderately quick exchange rates of in the vicinity of 80 and 120 picks for each moment with a payload of 1-2 kg in a working space of around 1 x 2 m. Wellbeing and power factors propose a lightweight outline deployable, if fundamental, with insignificant guarding in nearness to human administrators in a blended mode generation situation. In a perfect world, it ought to possess an insignificant volume, be moveable without noteworthy recalibration, and situation ought not to encroach on the uprightness of the industrial facility floor.

The specialized abilities required for proficient activity should coordinate the application aptitude base of the business and since fundamental IT aptitudes are by and large restricted in the division, programming and reconstructing assignments ought to be made basic and utilize natural, human-based techniques that limit specialized pressure. The mechanical autonomy industry has gained huge ground intending to this issue by creating graphical and educate by demonstrating strategies that limits the IT abilities required to work. Future advancements are conceivable utilizing symbol based programming touch screens, motion and discourse acknowledgment, or completely instrumented gloves that straightforwardly control the robot's activity.

Cost will dependably be a factor impacting the take up of mechanization, especially in the SME area. As appeared above, the genuine cost of robots has fallen throughout the years, and the pattern will no uncertainty proceed to the benefits of the assembling area. Another positive factor that can be abused is the decreased exactness necessities in dealing with food items, instead of the accuracy requested, for instance, in the car business. This plan unwinding can be converted into a prerequisite for less hardened and in this manner lighter arms, with little drive systems, requiring less idleness and perhaps bringing about lower costs (Figure 2.4).

Figure 2.4: A progression of Delta robots performing 'pick-and-place' activities in the substantial scale creation of fragile pretzel items, (*Source: G. Caldwell Darwin (2013), Robotics and Automation in the Food Industry: Current and Future Technologies, Woodhead Publishing*)

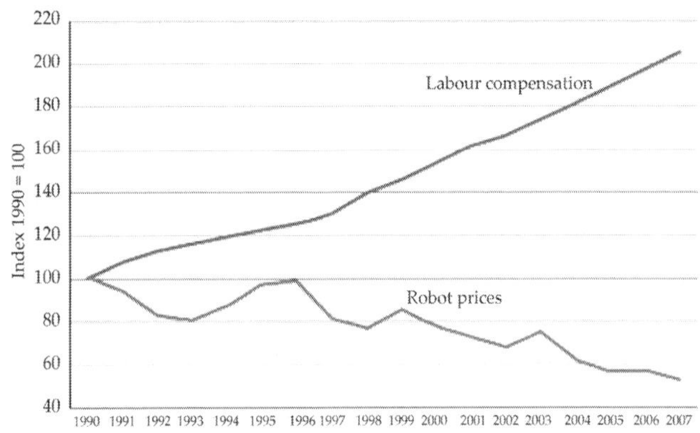

Figure 2.5: Patterns in capital expenses of robots versus European work costs.

Source: International Federation of Robotics, Annual Review World Robotics Magazine, 2008

2.4.1. Gripper Technology

Food items by their temperament have a tendency to be unpredictable fit as a fiddle, sticky/elusive, delicate or delicate. It was, by and large, trusted that sheltered and productive working must be given by the intercession of the human hand, and such an observation has stayed unchallenged until late years. Obviously, any computerized methodology will require the arrangement of an appropriate and effector or gripper and given, the changeability of food items, the outline will by and large be extraordinary to the objective item, however, should meet non-specific sterile prerequisites. Gripper configuration is accordingly more application focused in this area than somewhere else in assembling. Given this application-situated approach, various robotized arrangements have been created as of late that have conveyed grippers in forms that up to this point have been viewed as the elite area of human administrators. The accomplishment here, especially with troublesome food items, seems to show that gripper configuration should represent no significant issue in the automated applications to food items. A couple of cases of these arrangements are laid out quickly beneath.

An early issue postured in the mechanization of a sandwich creation line was the precise situation of thin cucumber and tomato cuts. A sterile non-contact gripper arrangement in light of the Bernoulli guideline was produced by Davis et al. (2008) and appeared in Figure 2.6 that met the full operational speed expected of the creation line. The idea is currently broadly sent in the area, for instance in the pressing of fragile poppadum items and the expansion to different applications, for example, new vegetables has been investigated by Petterson et al. (2011).

The precise and quick position of thin, delicate pasta cuts in the association of lasagne partitions has ordinarily required the utilization of talented human smoothness. A basic component created by Moreno-Massey and Caldwell (2007) and appeared in Figure 2.7, effectively copies the human administrators both in expertise and speed. The guideline is material to the gathering of comparable food items.

Certain food items require the collapsing of thin, delicate shapes, an errand for which human fingers are very much adjusted. Gripper plans created by analysts, for example, Yao and Dai (2008), exhibit how this human smoothness can be copied in precision and speed. A case of such a gripper creating a perplexing origami shape in light paper material is appeared in Figure 2.8; the augmentation of this strategy to find bread kitchen and sweet shop items is evident and investigated all the more completely beneath.

It would thus be able to be outlined that gripper configuration is currently an all-around created innovation, can address the difficulties exhibited by the food business, and need not set breaking points to the scope of mechanization arrangement conceivable in food manufacture.

Figure 2.6: Picking and placing of thin cucumber and tomato slice on a sandwich production line, using a Bernoulli non-contact gripper, (*Source: G. Caldwell Darwin (2013), Robotics and Automation in the Food Industry: Current and Future Technologies, Woodhead Publishing*).

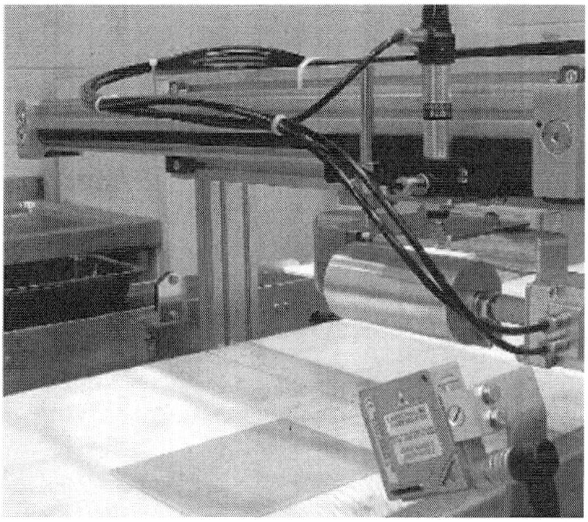

Figure 2.7: A cylindrical gripper for the acquisition and precision placing of thin pasta slices on a Lasagne Ready Meal production line, (*Source: G. Caldwell Darwin (2013), Robotics and Automation in the Food Industry: Current and Future Technologies, Woodhead Publishing*).

Figure 2.8: The generation of a complex 'origami' shape in thin cardboard using a dexterous gripper design, (*Source: G. Caldwell Darwin (2013), Robotics and Automation in the Food Industry: Current and Future Technologies, Woodhead Publishing*).

2.4.2. Sensor Technology

Manual task has the benefit of giving a grouping of human activities to amass the items as well as to acquire visual input for quality confirmation and, if essential, for remedial methods. Expelling human administrators from the generation line puts an accentuation on programmed examination frameworks, which must screen the nature of the all through at each phase all the while. Cleanliness, wellbeing, unwavering quality are key ideas here, requiring the arrangement of proper sensors, which must be of a sterile outline, non-contact, low-control, non-unsafe and verification against visit, thorough cleaning administrations. Current methods include the utilization of pervasive vision frameworks with metal indicator and X-beam examination frameworks for contaminant discovery and expulsion. Notwithstanding, in completely mechanized frameworks there is a need to expand the scope of accessible sensor innovation to screen always, for instance, the nature of information crude materials and inner nature of every nourishment item and the outer surface attributes. Some encouraging advancements have been accounted for by Wu (2011), in view of the utilization of low-control microwave innovation, which have brought about the outline of sensors fit for the fast imaging of the inward picture of an egg is appeared in Figure 2.9 and contrasted and the real egg cross area got by dismemberment. The imaging procedure is non-contact, control dispersal is asserted to be a small amount of that of a cell phone, and speeds of up to 100 pictures for every second are conceivable. Plainly this kind of sensor advancement that must continue in parallel with the augmentation of completely robotized frameworks to guarantee nature of yield and client trust in the creation methods.

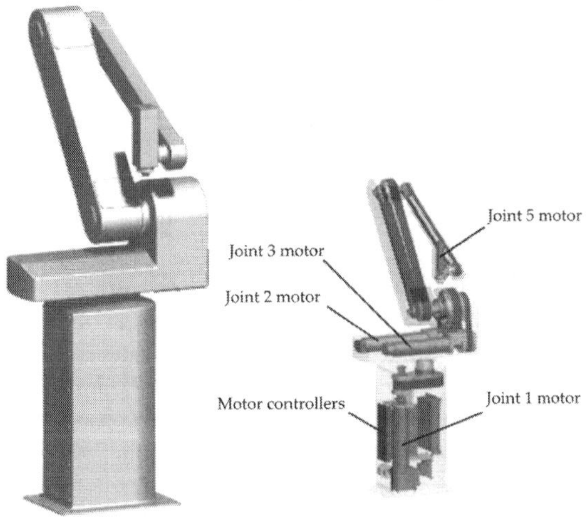

Figure 2.9: The 'Grail' food-grade prototype robot, (*Source: G. Caldwell Darwin (2013), Robotics and Automation in the Food Industry: Current and Future Technologies, Woodhead Publishing*).

Another perfect future prerequisite for robotized generation is to follow every item from moment of pressing through transport, stockpiling, to purpose of offer and past. Current radio-recurrence distinguishing proof (RFID) innovation is unquestionably actually equipped for playing out this capacity, yet the current cost model and hardware configuration don't yet seem proper for far-reaching sending in the segment. The effect of fruitful arrangement of this innovation on quality control is clear, yet it will likewise speak to an intense showcasing apparatus by coordinating traceability into a total creation coordination control plot that could improve reaction to quick changes in the market request. As of now, a Department of Food and Rural Affairs (Defra) supported, attainability consider into the use of this innovation (Wu, 2010) is being embraced at the University of Manchester, which if fruitful could quicken the take-up of this vital sensor innovation inside the business.

2.5. FUTURE TRENDS

Market powers will keep on encouraging the take-up of mechanization in food generation with a scope of assembling procedures. Standard mechanical robots will be unavoidable in the bundling and palletizing administrations, while Delta-type robots will keep on dominating in the high-volume, get together/pick-and-place part, especially where item arrangement is poor requiring remedial visual sequencing and where adaptable reinventing might be a required choice.

For high-volume, single product offerings with great limitation, hard-wired electro-mechanical frameworks may give more sparing arrangements. The test is to acquire moderate computerization arrangements in low-to medium-volume blended product offerings with for the most part poor item confinement. The application space would be dominatingly in the SME part, where there will be an interest for a blend of manual and robotized producing arrangements. Preferably this prerequisite could be met by a lightweight, low-idleness, seat/stage mounted sensor driven robot that meets the sterile necessities of the food business.

A conceivable contender for this part is the 'Grail' robot, being right now created at the University of Sheffield, with the help of 'Defra' Link Research ideas. This machine, which is right now experiencing modern trials, is controlled by brushless DC engines with straightforward belt-driven transmission instruments, designed to limit arm inactivity. All joints are back-drivable and the packaging is built from lightweight stainless-steel boards, with unique consideration given to the outline of joint seals to guarantee that the robot meets the required sterile working norms and security against dilute entrance amid wash. An upgraded direction arranging calculation has been produced to augment pick-and-place execution speeds and give uneven quickening profiles to limit required pinnacle joint torques. The general blueprints of the robot are appeared in Figure 2.10, and the specialized outline details are recorded underneath:

2.5.1. Design Specifications

- Main application pick-and-place.
- Material 304 or 316 stainless steel.
- Degrees of flexibility 4.
- Payload 1 kg.
- Reach 720 mm.
- Repeatability ± 2 mm.
- Weight 20 kg.
- Mounting floor/roof.
- Ingress protection rating IP 67.
- Operating life 40 000 h.
- Pick-and-place speed (25 x 300 x 25 mm cycle).
- 90–100 cycles/min (0.5 kg).
- 75–85 cycles/min (1 kg).

The mix of a lightweight structure, low-inactivity drive geometry, and obliged elbow joint speeds and torques add to the general wellbeing of a gadget that could work in nearness to human administrators with at least boundary security. Future advancements will investigate the fuse of fluidly consistent joints to improve the security of task.

A creative tech-by-demonstrating procedure is proposed for introductory programming and quick program changes will be affected by a touch screen graphical choice board that gives access to a menu of conceivable activities and requires no IT abilities to execute.

This is an unmistakably a practicality examine that will investigate the likelihood of growing, ease, moderately protected, exceedingly utilitarian automated gadgets upgraded for low to medium keeps running of variable geometry food items, inside a workplace where the IT expertise base is restricted. On the off chance that the idea is fruitful the advancement of such gadgets will be valuable expansion to the developing exhibit of profoundly viable robotization hardware now accessible to the sustenance business, to meet the extremely assorted nature of its assembling profile.

Most food manufacturing processes are embraced in standard processing plant units with high roofs to suit an extensive variety of assembling hardware. In any case, most by far of food assembly procedures are typically limited to a working volume that reaches from 2 to 3 m in tallness, bringing about a generally vast volume of unused manufacturing plant space which must be cooled, sifted and cleanly treated. The idea of containing the assembling procedure inside a generally little, fixed condition was proposed by Gray (2004) and the thought investigated in more detail with the help of Defra explore subsidizing. The supposed 'food plant in a pipe' idea would comprise of a progression of fixed, standard dimensional, connected creation modules of (for accommodation of make) round cross area, that are lightweight in development, effortlessly moved, and secured position with standard mechanical fixing and correspondence interfaces. A schematic of the approach is represented in Figure 2.10. Variable setups are conceivable; the framework is saleable and flexible to meet shifting creation modes. The idea is thought to be most reasonable for medium-to-long haul creation keeps running with the approach limiting inside natural vitality and water necessities and improving clean task, through the total expulsion of administrators from contact with the item–an element that permits the utilization of dormant conditions, perhaps with capable sterilizing methodology that could upgrade the timeframe of realistic usability of the item. Obviously, there are building difficulties to be met in the down to earth usage of such a plan, for example, the advancement of programmed quality-control frameworks, and an office for intercession on account of framework disappointment. Nonetheless, such issues have been as of now tended to in other assembling segments where administrators must be physically isolated from the procedure for wellbeing or cleanliness reasons.

Such an idea seems to address various assembling and ecological issues that are right now affecting on the division and which will impact its future advancement. Full computerization will create yields of reliable quality, with less waste and least danger of sullying through human contact. Limiting assembling to a fixed domain will bring about less profluent transfer, more noteworthy sterile potential, and a general littler carbon impression. The required PC control framework will be perfect with cutting-edge traceability innovation, coordination and robotized warehousing strategies, enabling administration to receive a quick and adaptable reaction to showcase request, lessen squander and expand degree of profitability.

2.6. CHALLENGES AND OPPORTUNITIES

The unmatched execution of robots to achieve different undertakings in food industry accompanies the difficulties which specialists are as yet endeavoring to determine Mueller et al., 2014. An exceptionally late pattern is to apply the idea of Cyber-Physical System (CPS) in food industry. Crossing over the physical world with the virtual world, CPS is a current multi-disciplinary research area in view of the idea of Internet of Thing (IoT) that discovers potential to streamline end-to-end inventory network in food part. CPS can assume its part to accomplish the most abnormal amount of conviction in food security Khan et al., 2014b. European Commission as of late distinguished seven key spaces which can possibly profit by foundation and apparatuses related with digital physical building Henshaw and Barneveld, 2016. Food industry together with rural area is recorded as one of the needs where CPS is expected to have critical effect in future. The fleeting points of reference for CPS contribution include: (i) Improved food security by sensors arrangement to check for infections and to get to item's freshness (ii) Hygienic help utilizing independent machines (iii) Precision cultivating by utilizing automatons, sensors and cutting-edge cultivating machines. In the long haul, the entire creation and inventory network will witness correspondence of shrewd food names to give inside and our understanding of where precisely the nourishment is originating from. Likewise, future CPS in rising areas like food industry will be beneficiated by cloud mechanical technology as featured in Chaâri et al., 2016.

A run of the mill CPS-based framework for food manufacturing comprises of three essential components; creation machine process, field gadget process and assembling control process. From equipment point of view, such a food fabricating framework may include robots, Programmable Logic Controllers (PLC), actuators, sensors, systems and different parts to acknowledge movement control and machine vision frameworks. Programming calculations may depend on Artificial Intelligence (AI), neural systems, fluffy rationale and other machine learning standards.

CPS-based food traceability frameworks can limit low quality or hazardous items in store network. A current report announced in Chen, 2017 proposed a food traceability framework acknowledged through mix of CPS and undertaking structures. Propelled by mist figuring, the oddity of the displayed arrangement lies in savvy esteem stream-based CPS which intends to advance occasions for following and following procedures with the assistance of a community component. Figure 2.10 delineates applied system of the proposed traceable instrument.

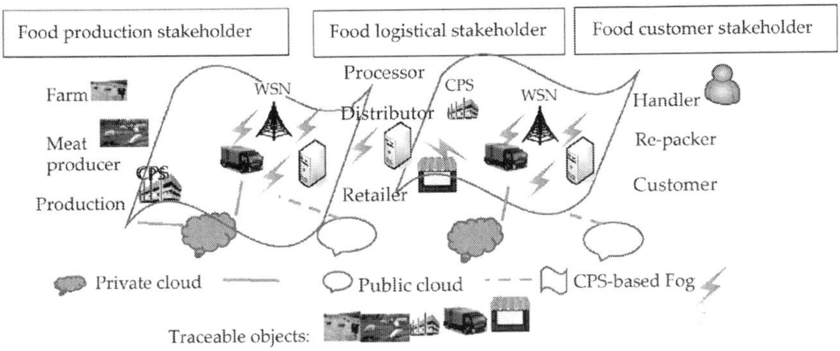

Figure 2.10: CPS-based food traceability system, (*Source: Chen, 2017*).

2.7. TOP FIVE WAYS ROBOTICS IN CHANGING THE FOOD INDUSTRY

Robotics is influencing each connection in the food production network. Here are five ways it's changing the business.

The food business is evolving. As the worldwide populace comes to more than 7.5 billion, food demand keeps on rising. Food providers are feeling the squeeze to work all the more effective. Not just that, purchasers need better quality, manageable food accessible whenever it might suit them.

Robotics and automation are a key piece of the arrangement. Contrasted with different sectors, the food-producing part has been generally ease back to receive robotics. Yet, finished the most recent few years robotics has begun to advance into relatively every connection in the food inventory network, from the field to the kitchen.

2.7.1. Robotics in Agriculture

2.7.1.1. Yield Based Agribusiness is the Beginning of the Food Journey

The exactness agribusiness industry is relied upon to be worth $7.87 billion by the year 2022. Robotics is an enormous piece of this. Horticultural automatons alone are anticipated to be worth $3.9 billion by the same year. Robotic applications incorporate planting, recognizing and arranging seedlings. There are additionally self-sufficient tractors, weeding robots and reaping robots. Automatons and self-governing ground vehicles are being utilized to screen and investigate crops. One late research venture from Harvard means to take care of a pivotal issue for worldwide horticulture—the decrease of honeybee populaces. The specialists propose to utilize a swarm of modest automatons to fertilize crops.

For non-plant horticulture, robotics is additionally being acquainted with the dairy, poultry, and hamburger cultivating enterprises. Applications incorporate self-ruling encouraging and draining, egg gathering and arranging, and self-sufficient cleaning.

2.7.2. Robotics in Food Manufacture

Independent food manufacturer might be the way to handling the rising food demand. The estimation of the worldwide food mechanization industry is relied upon to twofold in the following five years, achieving $2.5 billion by 2022. The Asia-Pacific market is a gigantic driver because of the prevalence of prepared to-eat foods in that piece of the world.

Food manufacture can be part into two distinct stages:

- Essential preparing—Raw food items are cleaned, arranged, transported and mixed. Mechanical applications incorporate butchery, and foods grown from the ground arranging.
- Auxiliary preparing—Ingredients are consolidated to shape new sustenance items by cooking, heating, chilling and so on. Mechanical applications incorporate item arranging, deformity expulsion and blending.

Robotics applications have a tendency to be more qualified to auxiliary preparing, as the sustenance is more institutionalized by at that point. Be that as it may, we are beginning to see more essential handling robots.

2.7.3. Robotics in Food Packaging

Food bundling or packaging robots have been consolidated into parts of the food store network for quite a while. Nonetheless, the most recent advancement is that the whole bundling procedure can be mechanized.

Food packaging can be part into three phases:

- Essential packaging—individual foods are bundled. For instance, a pick-and-place robot places desserts into plastic tubs.
- Auxiliary packaging—Individual bundles are assembled together. For instance, another pick-and-place robot stacks those plastic tubs into a bigger box.
- Tertiary packaging—Secondary bundles are assembled for delivery. For instance, a palletizing robot puts numerous cases onto a bed.

It appears to be likely that robotic packaging will keep on being one of the fundamental applications in the food business.

2.7.4. Robotics in Food Delivery

Robotic food delivery has been getting a ton of press as of late. Recently, Domino's Pizza reported that it would acquaint self-governing ground vehicles with convey pizzas, following their first fruitful (and exceptionally loud) conveyance by ramble toward the finish of a year ago.

While self-sufficient food delivery may appear somewhat like "only the most recent prevailing fashion" it really addresses a rising business sector drift. In the course of the most recent few years, there has been a gigantic spike sought after for eatery quality, home-conveyed, prepared to-eat food. Regardless of whether self-sufficient delivery will end up boundless is easily proven wrong, yet our preference for takeaway is certainly changing the food business.

2.7.5. Robotics in Cookery

The last phase of the food inventory network is cooking. We talked about cookery robots in 2015 in the article Modern Cooking: Are Robot Chefs Really Here? In it, we presented the Moley mechanical kitchen and other automated gourmet specialists. Up until now, Moley's site still classes the undertaking as "being developed" however they say the robot will be prepared for advertise before the current years over.

Be that as it may, cookery robotization does not really mean cooking dinners on request. The APRIL robot, for instance, tries to convey eatery quality to mass created food. Customary large-scale manufacturing techniques diminish nature of sustenance due to some degree to the expansive clump sizes. By utilizing a KUKA robot to cook in little clusters, the creators assert that the nature of the food is better.

Contrasted with alternate phases of the food production network, cookery and conveyance robots are still in their earliest stages. Nonetheless, it is obvious that apply autonomy is changing the food business.

2.8. CONCLUSION

Food producing is, by its inclination, very assorted and its techniques will no uncertainty remain along these lines, yet the predisposition against mechanization has now to a great extent vanished. Manual activity will remain, especially to meet neighborhood or bespoke markets, however, will be blended with island of suitable computerization. Be that as it may, in vast scale generation, robotization w-ill command, with the rise of completely computerized plant, intended for adaptable creation in a vitality effective aseptic condition and connected to an absolutely traceable coordination affix to the point of offer and maybe past. The general shape and nature of the business will be controlled by showcase powers, which thusly will be impacted by statistic, environmental and monetary weights. The computerization business has a critical part to play in giving the innovation that will guarantee that the area keeps on taking care of the demand for promptly accessible, high caliber and reasonable food items.

Maybe the best test the business faces in the take-up of robotization innovation is its constrained access to proficient designing and IT abilities that are required to support the appropriation of cutting edge producing strategies. In this manner, instructive foundations, preparing focuses and proficient building organizations have a similarly vital part in making a veritable twenty-first-century European food producing industry.

REFERENCES

1. Chua, P. Y., Illner, T., & Caldwell, D. G., (2003). Robotic Manipulation of Food Products A Review. Industrial Robotics: *An International Journal, 30*(4), 345–354.
2. Davis, S., Gray, J. O., & Caldwell, D. G., (2008). An end effector based on the Bernoulli principle for the handling of fruit and vegetables. International Journal of Robotics and C.I.M.
3. Caldwell G. Darwin, (2013). Robotics and Automation in the Food Industry: Current and Future Technologies. Woodhead Publishing1st edition
4. Nayik, G. A., Muzaffar, K., & Gull, A., (2015). Robotics and Food Technology: A Mini Review. *J Nutr Food Science, 5,* pp. 384.

CHAPTER 3

SENSORS FOR AUTOMATED FOOD PROCESS CONTROL: AN INTRODUCTION

"The food industry profits from providing poor quality foods with poor nutritional value that people eat a lot of."

—Mark Hyman

CONTENTS

The estimation of process factors gives not just the way to observing and controlling a procedure, and subsequently to provide steady quality unaffected by the administrator; it additionally can be vital to lessening capital tied up in stock and to utilizing vitality all the more productively. Parallel to this, the administration of advantages both at the process and venture level is picking up in significance. This chapter surveys the prerequisites of the food business on field instrumentation, the instruments accessible to quantify process factors, and the incorporation of these instruments into robotization frameworks. A couple of pragmatic cases and an attitude toward future improvements finish the part.

3.1. INTRODUCTION

Food, similar to water, is a fundamental prerequisite of life. For some on the planet who live where sustenance is rare, need overrules quality in the day-by-day battle forever. In the created world, be that as it may, quality is an immense factor in figuring out what we purchase. Quality, obviously, lies in the eyes and taste buds of the buyer: for the greens among us in 'naturally developed' leafy foods, for the dietician in dietary esteem, and for the voracious in the span of the segment. For the maker, other criteria must be satisfied to accomplish essential quality. What nobody needs is snails in their lettuce and flies in their soup, or to the extent handled nourishment is concerned, hair, glass, metal, mortars, chemicals, and microorganisms in their tins or bundles. Also, obviously, much the same as a Warhol painting, the present tin of soup should pose a flavor like yesterday's tin of soup, and tomorrow resembles today's. It isn't just the assortment of processed food available accessible today that inspires yet, in addition, the predictable quality with which it is delivered.

A second factor in food choice is, obviously, taken a toll–not only for the buyer who likes to purchase his food as sensibly as could be expected under the circumstances, yet in addition for the grocery store where he purchases. Looked with savage rivalry from equal organizations, value cutting directs the market. This thusly powers the food producer to deliver as inexpensively and productively as could reasonably be expected, however without lost quality. Cost diminishment is generally connected with an expansion in mechanization and decrease of labor inside a production facility. Steady quality has much to do with controlling processes. 'First measure, at that point control' is a familiar proverb, yet depending where it is connected along the inventory network, estimation brings extra advantages.

Food production facilities share a common infrastructure: inbound raw materials are stored ready for the production process; outbound products are stored before delivery. Here measurement of stock level and concepts such as vendor managed inventory, can lead to significant savings through reduction of stock to

The minimum required for reliable production. The production process itself is often controlled by recipes to make products that frequently change. Then, the finished product must be bottled or packed, a procedure that requires an entirely different kind of control from that of the making of the product. Finally, there are the so-called utilities: electricity, air, heat, cooling, and water. Efficient use of energy has become a byword in manufacturing, and key to optimal use is measuring what is consumed.

Food production facilities share a typical framework: inbound crude materials are put away prepared for the creation procedure; outbound items are put away before conveyance. Here estimation of stock level, and ideas, for example, seller oversaw stock, can prompt critical reserve funds through lessening of stock to the base required for solid generation. The creation procedure itself is regularly controlled by formulas to roll out items that as often as possible improvement. At that point, the completed item should be packaged or stuffed, a technique that requires an altogether extraordinary sort of control from that of the making of the item. At last, there are the purported utilities: power, air, warmth, cooling, and water. Proficient utilization of vitality has turned into a precept in assembling, and key to ideal utilize is estimating what is expended.

3.2. SPECIAL CONSIDERATIONS FOR FOOD INSTRUMENTATION

The procedures engaged with delivering food find numerous parallels in different businesses. Where they contrast is that food is delivered for eating and drinking by people or creatures: thus, the item should be ok for utilization. For process instrumentation, this prompts various unique contemplations as to its outline and utilize.

3.2.1. Regulatory Agencies

The food industry varies from numerous others in the degree to which the whole inventory network, from creation to appropriation, is checked by government organizations. The business itself has additionally set up its own guard dogs to evaluate progresses underway building and make proposals with respect to great assembling practice. While picking process instrumentation, hence, it pays to check whether a comparing endorsement or authorization has been allowed. Table 3.1 records a few bodies that are of specific significance to safe food manufacture. The prime point is to guarantee the wellbeing of an item, in this way shielding shoppers from presentation to, for instance, substance contaminants and poisons, microscopic organisms of outside bodies. Once recognized in that capacity, a faulty item must be expelled from the market instantly, requesting large amounts of traceability at all stages, of generation. To the extent instrumentation is concerned, the impact of direction is twofold. From one viewpoint, the directions manage how and of what materials an estimating gadget might be developed, on the other, their estimations help satisfy the directions by giving the data important to control and screen the procedure.

3.2.2. Wetted Parts

A very much planned instrument for use in the food business must think about three elements: how does the gadget carry on in typical activity, what happens if a segment part turns out to be mechanically faulty and how might it be cleaned and sanitized. The alleged wetted parts, that is, those parts of the instrument in coordinate contact with the material being estimated, are the most basic in the two regards. Indeed, even non-contact gadgets must be considered to have wetted parts when they barge in into a vessel. In spite of the fact that they are not specifically in contact with the medium, they may have hole in which food can assemble and decay, or be presented to high temperatures or vapors that discharge undesirable substances.

Taking the three cardinal sins, substance defilement, microscopic organisms and remote bodies as criteria, each wetted part should be verified whether it represents a peril. Compound sullying might be caused by relocation, wear and tear of a moving part, scraped spot by streaming material, spillage into the procedure or just by picking the wrong material for the foodstuff being prepared. It is essential to recall that, regardless of whether we do eat them, foods are just blends of natural and inorganic chemicals, which might be corrosive or antacid and which may respond with the materials with which they are in contact.

To maintain a strategic distance from compound pollution through the wrong decision of material, it is critical that instrument makers utilize substances thought about safe for sustenance creation. To this end, the Food and Drug Administration of the United States of America (FDA) keeps up a rundown of Food Contact Substances (FCS) (FDA, 2010). A comparable archive for metals is Guidelines on Metals and Alloys utilized as Food Contact Materials (Council of Europe, 2002), and for stainless steels the Sheffield Corrosion Handbook (Avesta, 1994). Most instruments utilized as a part of food production are made of steel grades 1.4301/304 (18% Cr, 10% Ni) or 316 (17% Cr, 12% Ni +Mo), albeit different compounds or titanium might be utilized. Table 3.1 records the properties of different stainless steels utilized as a part of sustenance creation.

Table 3.1: Regulatory and Standardization Authorities

3-A sanitary standards	American standards service with a long history of producing hygienic standards for the dairy industry. Also tests and authorises process instrumentation regarding its fitness for use in food production. 3-A standards are binding in the USA, but nowadays have less impact worldwide.
European Hygienic Equipment Design Group	Formed in 1989 with the primary aim of the safe manufacture of food products, is supported by research institutes, equipment manufacturers and end users. Issues guidelines on hygiene, tests and certifies equipment and publicises state-of-the-an technologies. An EHEDG approval means that the equipment has been successfully tested with respect to its suitability for food applications.
International Dairy Federation US Food and Drug Administration	Responsible for standards and recommendations in the dairy industry which are often published as ISO standards. US government agency charged with ensuring the safety of food production. Issues licences for products, whereby the processes, constituents, materials and constructional details are subject to examination. The use of an FDA-approved material is a guarantee that the component concerned will not be attacked by the food product.
EU and national governmental agencies	Play a similar role to the FDA in the European Union or the country to which they belong. Lay down the permissible levels of trace organic and inorganic substances in food products. Since *this* is an area of permanent research, it is possible that regulations vary from country to country
Retail organisations: IFS GFSI, BRC, SQF, ISO 22 000	Retail associations, certification bodies and quality standards set up to ensure the quality of food in supermarkets and other retail outlets by auditing their suppliers. Although not the primary focus, good manufacturing practice and correctly installed technology are factors in certification.

Source: G. Caldwell Darwin (2013), Robotics and Automation in the Food Industry: Current and Future Technologies, Woodhead Publishing

Table 3.2: Various Stainless Steels Suitable for the Food Industry

Material	AISI*	Properties	Uses
1.4301	304	Good resistance against organic acids, salt and alkalis at moderate temperatures	Pipework, vessels, instrument housings
I. 4404	316 L	Increased resistance against non-oxidising acids such as acetic acid, tartaric acid, phosphoric acid Increased resistance against pitting and imercrystalline corrosion	Pipework, vessels, instrument wetted parts
1.4435	316 L	Better corrosion resistance than Type 1.4404	Pipework, vessels, instrument wetted parts
1.4571	316 Ti	Increased corrosion resistance against particular acids and salt water Resistance against pitting corrosion	Pipework, vessels, instrument wetted parts
1.4362	2205	High strength duplex steels	Retorts, storage vessels
1.4462	2304	with good resistance to stress corrosion cracking in salt solutions at elevated temperatures Used with, e.g. hot brine (>50°C) with solids, stagnant and slow moving salty foods	where light construction and strength is important, instrument wetted pans (for the oil industry)

The AISI steels are equivalents but do not have identical compositions.

Source: G. Caldwell Darwin (2013), Robotics and Automation in the Food Industry: Current and Future Technologies, Woodhead Publishing.

Taking the three cardinal sins, substance tainting, microscopic organisms and remote bodies as criteria, each wetted part should be verified whether it represents a peril. Synthetic defilement might be caused by relocation, wear and tear of a moving part, scraped spot by streaming material, spillage into the procedure or essentially by picking the wrong material for the foodstuff being handled. It is essential to recall that, regardless of whether we do eat them, foods are simply blends of natural and inorganic chemicals, which might be corrosive or basic and which may respond with the materials with which they are in contact.

To stay away from synthetic tainting through the wrong decision of material, it is essential that instrument makers utilize substances thought about safe for food creation. To this end, the Food and Drug Administration of the United States of America (FDA) keeps up a rundown of Food Contact Substances (FCS) (FDA, 2010). A comparable record for metals is Guidelines on Metals and Alloys utilized as Food Contact Materials (Council of Europe, 2002), and for stainless steels the Sheffield Corrosion Handbook (Avesta, 1994). Most instruments utilized as a part of sustenance generation are made of steel grades 1.4301/304 (18% Cr, 10% Ni) or 316 (17% Cr, 12% Ni +Mo), albeit different combinations or titanium might be utilized. Table 3.2 records the properties of different stainless steels utilized as a part of food production.

Various variables may help diminish the danger of bacterial defilement. Customary cleaning and whole free outline decreases the expansive danger of contamination, yet the best possible material, plan and completing of the wetted parts is similarly as imperative. Deadlocks ought to be kept away from since they are harder, if certainly feasible, to clean. Notwithstanding item wastage, they mean longer stage shifts with all the more flushing water, longer warming circumstances for cleaning, and an enormous warmth interest for steam sanitization. Wetted parts must display a high level of erosion protection, both against foodstuffs with which they are in contact and in addition the specialists used to clean and disinfect them. Furthermore, high temperatures, solid vibration, and mechanical pressure may improve the danger of electrochemical and intergranular consumption of stainless steel. The one outcomes in surface setting, giving a perfect rearing spot to microbes, the other in the exhaustion of the nickel and chromium at the grain limits, which implies a part will rust.

Configuration factors that decrease the danger of bacterial contamination are streamlined shapes, clean welding, no alcoves and corners, and no obstacles that may make the item accumulate and decay. It is additionally standard practice that wetted stainless steel parts are electrically or mechanically cleaned. A surface harshness Ra ≤ 0.8μm/150 coarseness is viewed as protected in any case, for certain procedure associations, even lower esteems are typical. Cleaning has the twofold impact of lessening vulnerability to surface setting and keeping items from staying.

The introduction of foreign bodies happens more coincidentally than plan, yet great hazard investigation can relieve the impacts, should the mischance happen. What could be the reasons for disappointment, and what happens when a part fizzles? Are streaming gases, fluids or solids causing scraped area or creating high mechanical powers that, joined with high temperature or vibration, are expanding electrochemical assault or mechanical weariness? Is the back weight of a stream meter sufficiently high to keep fluids from bubbling and causing cavitation? Has a thermowell been intended to withstand the weights to which it is subject? Does its situation inside a pipe subject it to more than the outline powers? Are instruments ready to withstand the powers and temperatures created amid outer cleaning or cleaning set up? What happens if the stomach of a weight transmitter is cracked? Is the fill liquid FDA affirmed–does its discharge imply that the whole group must be discarded? These and numerous other criteria ought to be surveyed while picking an estimating instrument.

It is the consequence of such contemplations that weight transmitters with metal stomachs, as opposed to fired estimating cells, are utilized as a part of direct contact with sustenance. Mercury-in-glass thermometers are really restricted by FDA Rule 21 CFR Part 113 for low-corrosive canned-nourishment answers, yet here worries by the makers about mercury spills have lead the FDA to propose a run change enabling proportional electronic gadgets to be utilized. In sustenance filling applications, there is an unmistakable pattern towards supplanting more seasoned mechanical cylinder and glass fillers by electromagnetic and Coriolis flowmeters. Here wear and tear, upkeep, and in addition estimation precision, are the driving variables for restoration.

3.2.3. Process Connections

A procedure association is the methods by which an instrument is introduced in a pipe or vessel that contains the procedure material to be estimated. Preferably, it should offer the item no plausibility to end up ensnared. One approach to do this is to weld the instrument association set up and after that granulate and clean the weld. This is, obviously, not down to earth for little pipe widths and substitution is troublesome, should the instrument come up short. By and by welding is frequently experienced for thermowells, where the sensor embed is effortlessly supplanted, and also for a few sorts of flowmeter.

The larger part of instruments in food preparing, in any case, are introduced by methods for supposed sterile couplings. These give whole free association and are intended to be effortlessly gotten off, enabling instruments and pipes to be immediately expelled for cleaning. The couplings themselves are typically held together with clasps or outside screw associations. Throughout the years various diverse outlines have gone onto the market, a determination of which are abridged in Table 3.3. Producers likewise offer extraordinary connectors for mounting instruments on tanks and ports for the addition of, for instance, pH sensors into tanks and pipes. Here it is imperative that these have a sterile endorsement issued by the European Hygienic Equipment Design Group (EHEDG) or comparative body.

3.2.4. Instrument Housings

Contingent on the plan of the instrument, a lodging may contain just the interfacing terminals, for instance, with respect to a temperature test, or the whole transmitter gadgets. In the two cases, it must give satisfactory security against the entrance of dampness or clean and when the instrument is utilized as a part of an unstable hazard region, the departure of a start or fire. The previous is represented by the level of security gave by the lodging, the last by a reasonable kind of assurance gave by the lodging or the instrument yield circuit.

To the extent the entrance of dampness and tidy is worried, there are two noteworthy characterization frameworks, IP appraisals, and NEMA evaluations. Tables 3.4 and 3.5 clarify the significance of each evaluating write. It is normal that producers quote both in their specialized details. In the IP standard, the level of security is demonstrated by a two-section code. The principal number is worried about the security from the entrance of strong issue, the second with water: the higher the number, the better the insurance against clean, water planes or submersion. Keeping in mind the end goal to withstand the successive cleaning in a nourishment generation office, lodgings with evaluations of IP 65 or better are required. The NEMA standard involves 14 write codes that arrangement with useful necessities on lodgings reasonable for indoor and open-air utilize. It likewise creates an impression about the insurance from outer impacts and conditions, for example, mechanical effect, consumption, mugginess, form, bugs, tidy, and so forth. A NEMA 4X fenced in area is most appropriate to the necessities of the food business.

Explosion protection isn't a noteworthy issue in the generation of food. On the off chance that combustible fluids or effectively ignitable gases are available, for example in the refining of spirits, at that point instrumentation must be affirmed for use in explosion-hazardous regions. Powders can likewise be an issue since dust storms are effortlessly flammable under specific conditions. For processing, stockpiling, movement and packing tasks, along these lines, the Dust-Ex gear ought to be utilized.

Albeit appropriate kinds of protection have been long agreed for quite some time, as of not long ago Europe and North America had their own terminology for arranging an unsafe region. The circumstance has been facilitated by the most recent IEC standard, which has likewise been received by North America. Table 3.3 records and clarifies the sorts of protection of interests to food processing. Perusers needing a more full clarification of blast security are alluded the booklet 'Basics of Explosion Protection' (Stahl, 2007).

Table 3.3: Sanitary Couplings for the Food Business

Type	Use	Description
Dairy coupling to DIN 11851	Pipes and tanks	Low cost coupling with threaded boss and slotted sleeve. It does not allow flush mourning and is no longer considered to be hygienic; however, a special gasket set can be purchased which brings the coupling up to standard.
Aseptic screwed, flange (DRD) and clamp unions to DIN 11864-1, DIN 11864-2 and DIN 11864-3	Pipes and tanks	Designed to EHEDG recommendations, as a replacement for the dairy coupling, these unions offer better hygiene thanks to a flush sealing construction. The mechanical coupling is via threaded sleeve, bolts or clamp, the seal being flush with the pipe wall. Can be used for in-line and insertion instruments.
Varivent® coupling	Pipes	In-line housing that allows the flush mourning of a sensor, which is attached to the housing by means of a screw clamp. Three housing types cover a wide range of pipe diameters. A tank adapter is also available.
APV coupling	Pipes	In-line housing of similar construction to the Varivent coupling. The sensor, however, is bolted in position.

SMS coupling	Pipes and tanks	Low cost, Scandinavian standardised screw coupling to SMS 195346. Its weakness lies in the hygienic adaptation to the process, which does not allow flush mounting, and for this reason it is no longer considered to comply with modem hygienic standards.
IDF coupling	Pipes and tanks	International Dairy Federation (IDF) screw coupling standardised in ISO 2853
Tri-clamp® coupling	Pipes and tanks	Sanitary coupling with bevel seating produced **by** the Tri-Clover Company in America. Instruments are quickly mourned and fixed with snap-on clamps. The couplings find widespread use in America.

Source: G. Caldwell Darwin (2013), Robotics and Automation in the Food Industry: Current and Future Technologies, Woodhead Publishing

Table 3.4: Ingress Protection Classifications as per IEC 60 259

Code	Ingress protection against solids	Code	Ingress protection against water
0	Not protected	0	Not protected
I	>50 mm diameter, e.g. hand	I	Vertical dripping
2	a.12.5 mm diameter, e.g. forger	2	Dripping (15° inclination)
3	>2.5 mm diameter, e.g. tool	3	Water spray
4	>**1** mm diameter, e.g. wire	4	Splash water
5	Protected from dust	5	Jet of water
6	Dust-proof	6	Strong jet of water
		7	Temporary submersion
		8	Total submersion
		9	High pressure water, 80°C, 120 bar, 30 cm distance from nozzle

Source: G. Caldwell Darwin (2013), Robotics and Automation in the Food Industry: Current and Future Technologies, Woodhead Publishing

Table 3.5: Kind of Protection Enclosures as per NEMA Standard 250

Type	Indoor	Outdoor	Degree of protection
1	Yes		Protection against contact with equipment within the housing
2	Yes		Protection against a specified quantity of water droplets and din
3		Yes	Protection against blown dust, rain, sleet, rain and snow, and external ice formation
3R		Yes	Protection against rain, sleet, rain and snow, and external ice formation
3S		Yes	Protection against blown dust, rain, and operation of an external mechanism is guaranteed if this is iced up
4	Yes	Yes	Protection against blown dust, jets splashes and jets of water
4X	Yes	Yes	Protection against corrosion, blown dust, jets splashes and jets of water
5	Yes		Protection against dust falling, dirt and lubricating non-corrosive fluids
6	Yes	Yes	Protection against water penetration during occasional temporary submerging in limited depth
6P	Yes	Yes	Protection against water penetration during prolonged submerging in limited depth
11	Yes		Protection against the corrosive effects of fluids and gases of operating material inside the enclosure submerged in oil
12	Yes		Protection against dust, din and lubricating non-corrosive fluids
12K	Yes		Protection against dust, din and lubricating non-corrosive fluids for enclosures subjected to vibration
13	Yes		Protection against dust, water spray, oil and non-corrosive coolants

Source: G. Caldwell Darwin (2013), Robotics and Automation in the Food Industry: Current and Future Technologies, Woodhead Publishing

Table 3.6: Types of Protection for Gas and Dust (D) Atmospheres

Type of protection	Standard	Significance
Flameproof enclosure 'd' Protected by enclosure 'tD'	IEC 60079-1 IEC 61241-1	The enclosure (housing) is designed such that any spark or explosion is retained within it. Frequently met in America where the entire cabling is routed in conduits. Also used in Europe for Dust-Ex protection and system components.
Increased safety 'e'	IEC 60079-7	Electronics are designed such that high temperatures and sparking cannot occur during normal operation. This type of protection is met e.g. in 4-wire devices and power supplies.

Intrinsic safety 'i' Intrinsic safety 'iD'	IEC 60079-11 IEC 61241-11	Protection is achieved by limiting the current flowing in the device signal circuits. Restrictions apply to installation and cables used Favoured method for European installations.
Type of protection 'n'	IEC 60079-15	Device is designed so that, under normal and defined abnormal operating conditions, *it* cannot ignite a surrounding explosive atmosphere. Used for all Zone 2 electrical equipment.
Encapsulated 'm' Encapsulated 'mD'	IEC 60079-18 IEC 61241-18	The entire electronics of the device are potted, so that no spark can enter the surrounding

Source: G. Caldwell Darwin (2013), Robotics and Automation in the Food Industry: Current and Future Technologies, Woodhead Publishing

3.3. MEASUREMENT METHODS

Keeping in mind the end goal to guarantee steady quality and financial creation of food, the hidden procedures must be controlled. This implies thusly that the factors influencing the procedure, for instance, temperature, or the parameters showing a specific quality level, for instance, particular gravity, must be estimated. Regularly in process control, it is adequate to gauge the factors weight, temperature, stream, and level. Factors, for example, pH, turbidity, consistency, and thickness give a knowledge into the nature of an item. There are other vital amounts, in any case, that are not shrouded in this part. Follow dampness and mugginess sensors assume a vital part away and molding of items, for example, put away organic product, meat or other crude materials. The peruse keen on dampness estimation techniques is alluded to other writing regarding the matter (Bentley, 1998) (Figure 3.1).

Figure 3.1: Pressure transmitter on a keg-filling line, (*Source: Courtesy of Endress+Hauser*).

Perusers needing a more extensive diagram of the instrumentation accessible to gauge weight, temperature, stream, and level are alluded to 'Instrumentation and sensors for the food industry' (Berrie, 2001). This section confines itself fundamentally to electrical instruments that are effectively incorporated into control and plant resource administration frameworks.

3.3.1. Pressure

In food processing, the primary applications for weight estimations are in channeling, crosswise over channels, and in addition in shut tanks. Figure 3.1 demonstrates a commonplace application. The hardest conditions are to be found in channeling, where instruments must be intended to withstand over-burden weights a long ways past their typical working extent. Pumping activity and opening and shutting of valves makes surges in weight up the most extreme working weight of the pump. High weight crests are additionally produced if a valve closes unexpectedly–a specific issue in the low weight locale. Weight instruments should likewise be safe to vibration and have the capacity to withstand the temperatures and stresses incited by inner and outside cleaning.

Weight instruments may quantify outright, check and differential weight:

- Absolute weight gadgets measure the genuine weight following up on the sensor, showing around 1000 millibar for barometrical weight

- Gauge-weight gadgets measure the weight with respect to barometrical weight, showing around 0 millibar for environmental weight–bring down weights are yield as negative esteems
- Differential weight gadgets measure the weight distinction between two tapping focuses, for instance, in a pipeline or a tank. They are generally used to gauge level and, regarding an essential component, stream

There are three fundamental kinds of weight estimating instrument: manometers, mechanical weight measures and electrical weight transmitters. Of these, electrical transmitters are most suited to computerization, in spite of the fact that alternate sorts are still experienced in numerous plants.

3.3.1.1. Electrical Pressure Transmitters

Electrical weight transmitters yield an institutionalized signal that speaks the amount being estimated, for instance, a 4– 20 mA current signal, or a deliberate esteem transmitted over a computerized correspondence interface. Most electrical weight transmitters utilize an adaptable, earthenware or metallic stomach as the weight detecting component, which frames the front detaching component of a detecting chamber. By utilizing the proper procedure association, it is conceivable to create flush-mounted weight gadgets that have no holes and that are anything but difficult to clean.

When all is said in done there are three estimation standards: resistive, capacitive and inductive. The plan of the cells relies upon the maker and estimation standard utilized. Weight can likewise be estimated by the reverberation wire or vibrating-bar strategy, whereby the move in reverberation recurrence or change in vibration recurrence is corresponding to the weight following up on the detecting component.

Table 3.7 abridges run of the mill working conditions for weight estimating instruments. No correctness's are incorporated since they rely upon the individual transmitter. Most information sheets quote the deliberate blunder, the hysteresis, and reproducibility estimated at a reference temperature, typically 20°C albeit now and then other 'in general' correctness's are given. The temperature impact, which demonstrates by how much the exactness changes for a given temperature rise, is a critical factor in weight estimation. Contingent upon the fill liquid, which is required when the sensor is working with a stomach seal (an expansion which permits flush mounting) or a remote seal and vessels, this can be very huge. The over-burden weight shows the capacity of a gadget to withstand weight crests without mechanical harm to the sensor (a recalibration might be important).

3.3.2. Temperature

Observing and control of temperature is a critical factor in guaranteeing the nature of the finished result. For instance, a wrong temperature amid the fabricate of dairy items genuinely influences their time span of usability, while the wrong temperature in a counter to produce canned foods may bring about bacterial contamination. In sanitization, solidify drying and comparative procedures, temperature control is fundamental if the quality and wellbeing of the item are to be ensured. Any unaccountable ascent in temperature in a capacity storehouse might be the main sign that the crude material in store is breaking down. Temperature is additionally estimated in utility applications, for instance in refrigeration, steam warming or cleaning set up (CIP) plant, see Figure 3.2. The temperatures to be recorded range from about −50°C (−58°F) in chilly stockpiling to +150°C (+302°F) in cleansing set up (SIP) applications. Just in steam age will higher temperatures up to +250°C (482°F) be found? Precision, contingent upon application regularly ±1°C or better, is imperative, and obviously, cleanliness is fundamental where there is contact with food.

3.3.2.1. Estimation Standards

Temperature can be controlled by estimating one of three amounts on which it has an impact: force, power, and radiation. Radiation is the wonder utilized by pyrometers, and since they are essentially utilized as a part of high-temperature applications, for instance, in heaters, they won't be viewed as further. Power write temperature gadgets incorporate bimetallic strips, which might be utilized as ceaselessly perusing checks or exchanging gadgets, and in addition filled warm frameworks, that is, thermometers. Mercury-in-glass thermometers are as yet restricted for specific applications by the FDA, see Section 3.2.1, however in these cases, a second temperature sensor can be introduced for control. Generally their utilization today is constrained to optical go down of electrical frameworks (Figure 3.2).

Figure 3.2: Temperature measurement in a CIP pipe, (*Source: Courtsey of Endress+Hauser*).

Table 3.7: Typical Operating Conditions for Pressure-Measuring Devices. Values are Guidelines only: for Specific Values see the Manufacturer's Data Sheet

Type	Typical temperature range	Min. pressure range	Max. pressure range	Differential pressure/static pressure	Remarks
Manometer	-10°C to +60°C	0-1 mbar	0-20 bar		Inaccurate if not operated within the rated temperature range
Pressure gauge (diaphragm, capsule, Bourden gauges)	-25°C to +60°C, diaphragm gauges up to 100°C	0-10 mbar	0-25 bar		Significant temperature effect. Diaphragm gauges can withstand up to 5x overload (maximum 40 bar)
Electrical pressure transmitter	-30°C to +200°C	0-100 mbar	0-40 bar (ceramic) 0-600 bar (metal)	Up to 3 bar for static pressures up to 420 bar (ceramic) Up to 500 bar for static pressures up to 500 bar (metal)	Resistant to vibration and overload. Measuring range of 4-20 mA/HART devices can be turned down by 40:1 or more. Differential pressure transmitters with remote seals and capillaries may have a significant temperature effect

Source: G. Caldwell Darwin (2013), Robotics and Automation in the Food Industry: Current and Future Technologies, Woodhead Publishing

3.3.2.2. Electrical Temperature Gadgets

Electrical temperature gadgets utilize the temperature reliance of electrical properties of specific materials to give an estimation. There are four distinct types: semiconductors, resistance temperature detectors (RTD), thermocouples and silicon resistors, albeit just RTDs and thermocouples are of genuine enthusiasm for the generation of food.

RTDs, otherwise called protection thermometers, include a thin-film or wire resistor with a standard protection of 100 Ω, 500 Ω or 1000 Ω. The resistor material might be platinum or nickel, the institutionalized assignments being Pt100, Pt500, Pt1000 or Ni100. The sensors are exceptionally steady, have a wide working reach from illustration −200°C to +850°C (−358°C to +1562°C) contingent on sort, and show a very much characterized connection amongst protection and temperature. The conditions are indicated in various universal benchmarks (IEC, 2008). RTDs adjust to one of two institutionalized exactness classes: Class A = 0.15°C + 0.002 (t °C); Class B = 0.30°C + 0.005 (t °C); where (t) is the unsigned numerical estimation of the temperature in °C. Higher correctness's are likewise accessible.

RTDs regularly have a generally moderate time reaction and deliver just a little change in protection per unit increment in temperature. To confine or stay away from blunders where detecting component and assessing hardware are isolated by some separation, 3-wire or even 4-wire associations are utilized. RTDs are likewise delicate to vibration and stun, with the goal that care must be taken in their plan.

A *thermocouple (TC)* involves two wires of various metal consolidated at their closures. On the off chance that the two intersections are at various temperatures, a potential contrast is made that makes a present stream around the circle. In the event that one of the intersections is kept at a consistent temperature, the greatness of the present that streams is a measure of the temperature at the other.

TCs have a wide working temperature run: contingent on sort and conductor distance across, they can make the grade regarding +1800°C (+3472°F) and are institutionalized (IEC, 1995). They, as a rule, have a speedier reaction than RTDs, however, are less exact (+/- 0.5°C to +/- 1.5°C for IEC 60584 Class 1, contingent upon type). In Europe, they are typically just utilized as a part of inclination to RTDs for temperatures over 600°C. In America, notwithstanding, they are in effect progressively utilized for bring down temperature ranges where exactness isn't a main consideration in estimation. TCs are basic, tough, and reasonable and require no outside power supply. Then again, the flag is non-direct, they show low affectability, and have moderately low steadiness. A reference intersection is required and they should be adjusted. Additionally, metallurgical changes and maturing in some cases causes lost execution. A more major issue is that the low voltage yield is helpless to electromagnetic obstruction (Figure 3.3).

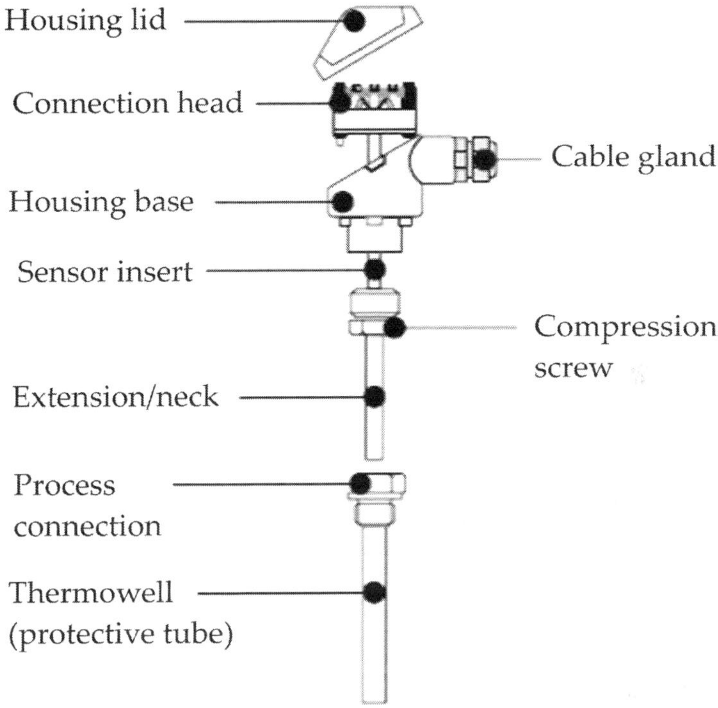

Figure 3.3: Components of a temperature-measuring assembly, showing a sensor insert with terminal connections, (*Source: G. Caldwell Darwin (2013), Robotics and Automation in the Food Industry: Current and Future Technologies, Woodhead Publishing*).

In spite of the fact that a total temperature transmitter with thermowell can be acquired from most makers, numerous clients want to purchase the different segments from various providers. Figure 3.3 shows schematically a run of the mill temperature get together:

- The sensor embed is the core of the get-together. It involves a metal sheath containing the detecting component and association wires. The association terminals ordinarily provide food for 2-wire, 3-wire and 4-wire association
- The lodging secures the sensor embed, the terminals and, where suitable, the head transmitter from the entrance of water and clean. Numerous lodging plans are institutionalized

- The thermowell might be isolated or a fundamental piece of the temperature get together and furnishes the contact with the medium. It is intended for a specific medium and scope of stream rates and should be utilized as depicted in the proper standard or information sheet
- For high-temperature applications, an augmentation (or neck) is embedded between the lodging and thermowell

To lessen mechanical load when thermowells are mounted in funnels, a position or introduction ought to be picked that decreases the surface region confronting the bearing of stream. Moreover, a thermowell must be secured with medium consistently when the flag is required. Since the thermowell produces turbulence in the downstream stream, care ought to be taken that it is not situated too intently to some other instrumentation in the pipe. In tanks, there is generally space for vertical or level mounting. Care ought to be taken not to position it excessively near moving parts, for example, instigators, pumps or valves, since the powers produced by the development of liquid may surpass the mechanical quality of the thermowell or prompt expanded scraped spot or erosion.

Table 3.8: Areas of Application of Electrical Flowmeters

Measurement principle	Application
Electromagnetic	Electrically conductive liquids (>5 gS/cm) with or without solids, e.g. water, wastewater, sludge, slurries, pastes, acids, alkalis, juices, fruit pulp, etc.
Coriolis	Virtually all fluids: cleaning agents and solvents, fuels, vegetable oils, animal fats, alcohol, fruit solutions, vinegar, ketchup, mayonnaise, gases, etc. Mass flow, density and temperature (the primary measured variables) can be used to derive other variables such as volume flow, solid contents, concentrations, and complex density functions. Viscosity can also be measured.
Vortex	Volume flow of liquids, gases and steam.
Ultrasonic	Volume flow of any liquid, regardless of electrical conductivity.
Differential pressure	Flow measurement of gases, steam and liquids.
Thermal mass	Mass flow of gases.

Source: G. Caldwell Darwin (2013), Robotics and Automation in the Food Industry: Current and Future Technologies, Woodhead Publishing

3.3.3. Flow

Flow is an imperative control variable that additionally conveys amount data to the plant administrator. There are numerous strategies for estimating flow, and choice can be troublesome. It must consider not just the physical properties of the medium to be estimated yet, in addition, the required precision, line measure, condition, buy expenses and establishment. Except for brace on ultrasonic instruments, all flowmeters are introduced in-line, so cleanliness is likewise a vital factor. The detecting component must have the capacity to withstand the temperatures met in CIP or SIP methodology, which most sterile renditions do, or if the instrument can't be cleaned in situ, it must be effectively destroyed to encourage cleaning.

Liquids with high thickness or containing entrained solids call for non-obstructive estimation. On the off chance that liquids set at moderately high temperature, similar to the case for hydrogenized oils, or are sticky, as are chocolate and sugar syrup, the flowmeter must be warmed. Build-up may be an issue when the item is chilled: for this situation, a remote lodging, which can be mounted at some separation from the estimating point, must be chosen. Mechanical flowmeters are from time to time utilized as a part of sterile procedures these days, however, two mechanical flowmeters, positive relocation, and turbine are still experienced in more seasoned plant. A portrayal can be found in 'Instrumentation and sensors for the sustenance business' (Berrie, 2001).

Table 3.8 shows the zones of utilization of each estimating standard canvassed in this part. For the reasons for stream estimation a conductive fluid has a conductivity ≥ 5 μS/cm for remotely controlled flowmeters, ≥ 50 μS/cm for circle fuelled gadgets. Cases of conductive fluids are well water, organic product juices, yogurt, drain, brew notwithstanding acids, salts and water-based emulsions. Nonconductive fluids are fuel oils, natural solvents, fluid sugar and demineralized water (Figure 3.4).

Figure 3.4: Electromagnetic Flowmeters Installed in a Brewery, (*Source: Courtesy of Endress + Hauser*).

3.3.3.1. Electrical Flowmeters

Electromagnetic flowmeters utilize the voltage initiated by a conductive fluid coursing through an attractive field to gauge volume stream (Faraday's law of electromagnetic acceptance). The prompted voltage is corresponding to the stream rate and free of changes in liquid thickness, consistency, and weight.

As the flowmeter is basically a straight pipe, not just brew, water and drain can be estimated, yet in addition items, for example, yogurt, molasses (when conductive) or rice pudding, which could free their constitution if constrained through a limitation. The flowmeter is appropriate for an assortment of uses, for instance, stream control, guardianship exchange, and fast filling. Figure 3.4 shows two electromagnetic flowmeters introduced in a bottling works.

Electromagnetic flowmeters have a general establishment necessity of just 3– 5 straight pipe breadths upstream of the estimating point. They work best in a turbulent stream administration. In the event that used to screen high consistency items in a laminar (plug stream) or transitional stream administration, the supreme precision of the meter may endure.

Coriolis mass flowmeters utilize the recurrence move caused by a flow of fluid through two metering tubes swaying at their characteristic recurrence to quantify mass stream. Since the normal recurrence is specifically identified with the mass of liquid in the tubes, it is likewise conceivable to quantify thickness. The tube temperature is likewise observed to make up for thickness and single-tube flowmeters can be utilized to gauge consistency. This adaptability joined with high exactness makes the Coriolis mass flowmeter an exceptionally important estimation gadget. The straight tube configuration is perfect for the nourishment business if titanium is an adequate material.

Coriolis flowmeters discover use in constant mixing, filling and process checking applications, especially when the blending is controlled by thickness or the last item is to be sold by weight. Their high precision makes them appropriate for dosing flavors and comparative nourishment added substances. They measure gooey media and compound items, for example, soup, chocolate, nectar, and mayonnaise. There is no necessity for here and there stream lengths and the gadgets work freely of changing liquid properties. Figure 3.5 demonstrates a bank of Coriolis flowmeters uniquely intended for dosing applications on a filling machine.

Vortex flowmeters measure flow speed by tallying the quantity of vortices shed every second from a feign body (Karman vortex road). The vortices are recognized by detecting the vibration of the feign body caused when they are shed. An accurately measured meter gives a straight yield, low weight misfortune and high precision over a wide stream extend for low thickness fluids, steam, and gas. At the point when the base indicated stream rate is come to, in any case, the gadget will remove, that is, the yield drops to zero.

Present day vortex meters are practical in giving quick flow rate and totalized flow. The gadget has no moving parts, and embeds specifically into the pipeline, making establishment and cleaning set up clear, in spite of the fact that it isn't appropriate for sterile applications. Expanding numbers are to be found in utility applications, be that as it may, especially for steam-stream estimation.

Ultrasonic flowmeters measure stream by either the Doppler technique or the season of-flight (TOF) strategy. In the Doppler strategy, the adjustment in recurrence that happens when an ultrasonic heartbeat is reflected by an air pocket or molecule gives a measure the speed of a streaming medium. In the TOF strategy, the distinction in the season of spread between a heartbeats going with and against the stream gives a similar data. Doppler flowmeters are rarely found in the food business.

Figure 3.5: Coriolis flowmeters installed on a filling machine, (*Source: Courtesy of Endress+Hauser*).

TOF ultrasonic flowmeters might be in-line, brace on, or the sensors might be embedded into the funneling. Precision is low in standard applications, however, can be expanded by utilizing extra sensors to frame in various ways. In this design, unsymmetrical and laminar flows can be estimated. They are appropriate for liquids with next to no strong substance, exceptionally gooey liquids and very low flow rates. Their adaptability raises them valuable for troubleshooting and charging.

3.3.3.2. Differential Weight Flowmeters

Measure the weight drop over a confinement in a pipe, which is relative to the square base of the liquid speed (Bernoulli's law). The confinement is the supposed essential component. A wide assortment of essential components exist, yet by a long shot, the most well-known is the whole plate. The scope of a differential weight flowmeter relies on the measure of the hole. Turndowns of greatest 4:1 are conceivable. Wear of the whole edge will lessen the meter's total exactness after some time, so normal adjustment is essential.

The food business utilizes differential weight estimation only for flow estimation in benefit frameworks, for instance, steam or packed aircraft. The abrogating factors here are the high working weights and temperatures that can be experienced. The method is prevalent due to its vigor and straightforwardness, joined with an abundance of free information encompassing task and establishment.

3.3.3.3. Thermal Mass Flowmeters

Thermal Mass flowmeters are also popularly known as thermal dispersion which comprises a family of instruments that are used to measure the total mass flow rate of a fluid, primarily gases, flowing through closed conduits. They are ordinarily utilized for dry-gas estimation, having a wide turndown, up to 100:1, and irrelevant weight misfortune. Mass flow rate is estimated specifically without requirement for extra temperature or weight pay. In the food business, they can be utilized as a part of utility and process gas estimation, including CO_2 dispersion, nitrogen gas conveyance and cleanse streams, gaseous petrol metering, and packed air. The warm properties of the gas must be considered while indicating the meter. As they can distinguish to a great degree low flows, they are additionally reasonable for spill recognition.

3.3.3.4. Mass Solid Flowmeters

There are many applications in the food industry which involve monitoring the transportation of bulk solids to various parts of the plant, for instance, coffee beans, salt, potatoes, tea, cereals, fruit, and vegetables. The requirement ranges from simple switches to determine that a product is no longer falling from a conveyor belt, to a continuous measurement, for example, the flow of crisps to a flavor drum, to ensure the correct ratio of the flavoring.

There are numerous applications in the food business which include checking the transportation of mass solids to different parts of the plant, for example, espresso beans, salt, potatoes, tea, oats, foods grown from the ground. The necessity ranges from basic changes to establish that an item is never again tumbling from a transport line, to a ceaseless estimation, for instance, the stream of crisps to a flavor drum, to guarantee the right proportion of the seasoning.

Microwave hindrances, for instance, flag either the nearness of nonattendance of item between their transmitter and beneficiary units. A Gunn diode can be utilized to decide if an item is moving or not. For persistent estimation, affect plate weighs can be utilized. Here free-streaming item falls onto a detecting plate that measures its mass stream. The fall stature must be no less than 800 mm, and the estimating range is from 30 to 120 kg/h. The option is to utilize a heap cell mounted added to a transport repertoire. After alignment, such frameworks are equipped for correctness's of 0.25–2%, contingent on outline.

3.3.4. Level

The essential use of level estimation in the food business is to screen substance of capacity or process tanks and storehouses containing crude, middle of the road and completed items. In process control, persistent estimations are utilized to control the filling and discharging of tanks, answers, reactors, and so on. Anticipation of overspill and dry running of pumps are run of the mill applications where level switches are utilized. Table 3.9 records the level-estimating instruments regularly utilized as a part of nourishment handling. Except for ultrasonic transmitters, all are accessible in clean forms.

3.3.4.1. Capacitance Level Transmitters

Utilize the capacitance between the detecting component, a rope or bar, and the tank dividers, as a measure of item level. Contingent on plan, they might be utilized for both consistent level estimation and level-constrain identification of both conductive fluids and electrically non-conductive materials. For non-directing fluids, estimation is reliant upon the dielectric steady, and the transmitter must be recalibrated on a difference in item. Contingent on the transmitter, it might be conceivable to self-adjust by contrasting the perusing and a point estimation. Mass solids with grain size to 30 mm can likewise be estimated. Current applications are confined to constrain discovery. The strategy is reasonable for weights up to 100 bar and in the temperature go–80°C to +400°C (−112°F to +752°F). It is likewise harsh to weight surges caused by, for instance, stirrers. Since capacitance transmitters offer a quick reaction they are especially appropriate for little vessels, for instance, for dosing and filling machines.

3.3.4.2. Conductance Level

Switches are utilized for confine recognition in conductive fluids. The detecting component involves a stainless steel pole that is mounted in the best or the side mass of the vessel. The exchanging position is controlled by the length or the mounting purpose of the sensor separately. There is a sudden drop in protection (or ascend in current) when fluid comes into contact with the detecting component flagging that the farthest point level has been come to. The signal is be utilized to switch a transfer, NPN-or PNP-transistor, or to give a changing sign to a remote transmitter. The tests are basically the same as capacitance tests. The working conditions are additionally the same, whereby an adjustment in the conductivity of the fluid has no impact on the exchanging.

3.3.4.3. Hydrostatic Level Transmitters

Measure the weight applied by a head of fluid, which is relative to its stature. Estimations are made with exceptionally planned differential-weight or measure weight transmitters. The estimation is subject to the thickness, yet this can either be entered as an adjustment parameter or disposed of by aligning at the coveted void and full levels. On account of an open vessel, where barometrical weight is following up on the head of fluid, just a single estimating point at the base of the tank is required, see Figure 3.6.

In the event that the vessel is shut, an extra estimation of head weight is required. The distinction between the two estimations gives the level. For differential weight transmitters, the level is shown straightforwardly, however the fill liquid in the hair like to the head-weight estimating point may create a troublesome temperature impact. On the off chance that two measure weight transmitters are utilized, the head weight is accessible for control and the level is typically computed by the controller. On account of Fieldbus gadgets, the esteem can be downloaded for show at the lower gadget. Hydrostatic weight transmitters are the most utilized level estimation innovation in the nourishment business. They work persistently at temperatures up to +100°C (+212°F), are inhumane to develop of item and will withstand CIP temperatures of up to +130°C (+266°F) over brief periods without changeless harm or the requirement for rearrangement. Froth has no impact on the estimation. For measure weight transmitters, a significant overpressure go (20×) avoids harm because of weight surges caused for instance, by stirrers. The fill liquid must be appropriate for sustenance applications, and the lodging requires a rating of IP 65 or above.

3.3.4.4. Ultrasonic Transmitters

Measure level by emanating a sonic heartbeat and estimating the time it takes to get the reflection from the surface of the item, the purported TOF (time of flight). In all cases, the sensor is mounted off-kilter at the highest point of the tank, for mass solids, calculated towards the item surface. The transmitter is set up by entering the coveted vacant and full separations.

Where the air over the item isn't air, a sonic speed parameter should likewise be entered. It is additionally great practice to examine the vacant tank, with the goal that any deceptive echoes from tank fittings can be shifted from the approaching sign. Ultrasonic transmitters can be utilized to gauge fluids if they are not vanishing or degassing, and mass solids. When all is said in done they can be utilized for weights from 0.7 to 4 bar (sensor subordinate); the stomach can withstand temperatures up to +150°C (+302°F). Estimating ranges up to 70 m are conceivable. For mass solids, dusty conditions may briefly decrease this incentive by up to 25%.

For mass solids, the surface of the material must mirror a critical extent of the produced beat. An instrument will work autonomously of the point of the filling cone or if the granularity of the surface is more prominent than 1/4 of the wavelength utilized (>3–8 mm, contingent upon sensor). In the event that the granularity is under 1/6 of the wavelength, at that point, the surface becomes an integral factor. In the event that the surface has no articulated surface, it will go about as a mirror and the sensor gets a reverberate which has gone back by means of the storehouse dividers. Under these conditions, level can't be estimated precisely.

3.3.4.5. TOF Radar Level Transmitters

Measure in the very same route as ultrasonic transmitters, yet with a radar heartbeat. On account of recurrence adjusted consistent wave (FMCW) radar transmitters, which discover use in tank-checking applications, the stage move is estimated, and this is likewise relative to remove voyaged.

Figure 3.6: Hydrostatic Level Measurement on a Beer-Fermentation Tank, (*Source: Courtesy of Endress+Hauser*).

There are two kinds of TOF transmitter, a purported 'free space' adaptation, in which the beat is propelled from a stub, illustrative, bar or horn radio wire, and a 'guided' form, in which the beat goes down a steel rope or pole, see Figure 3.7. The utilization of 'free space' radar transmitters is typically limited to fluids that have a dielectric steady ≥1.8, which is the situation for most substances experienced in the food business. Guided radar transmitters are reasonable for the two solids and fluids. Solids must be fine-grained (≤ 20 mm) and like fluids, have a dielectric steady ≥1.8. Fine solids show no issues. The estimation is additionally autonomous of temperature and the weight, with medium temperatures up to 150°C introducing no issues. Estimating scopes of up to 70 m are feasible in the two solids and fluids.

The first *vibration switch* has a detecting component as a tuning fork (different renditions utilize poles) that is compelled to vibrate at a set recurrence. The recurrence changes and an exchanging signal is yield when fluid or strong comes into contact with the fork tines. This flag may switch a hand-off or transistor yield worked in the sensor itself or be passed on to a different exchanging unit.

Higher temperatures can be withstood for brief periods without harm to the sensor, for instance, when cleaning or disinfecting set up. The technique is reasonable for fluids having a thickness of 10,000 cSt or less and for solids with grain size up to 10 mm. It is unaffected by changes in physical properties of the item, including thickness and in addition in process conditions, for example, turbulence, froth, develop, gas air pockets or strong suspensions. The switch requires no adjustment and support is irrelevant.

Figure 3.7: Guided radar measurement in a flour silo (right). The sensor on the left is an older capacitance probe, (*Source: Courtesy of Endress+Hauser*).

3.3.5. Density

In-line density measurements are possible with Coriolis flowmeters and tuned hydrostatic pressure transmitters. New on the market is a continuously measuring vibration transmitter, which is essentially an adaptation of the vibration switch, with a density computer. This is capable of measuring densities between 0.3 and 2.0 g/cm^3 in liquids with maximum viscosity of 350 mPa.s and a flow velocity of up to 2 m/s. Process temperature may be in the range 0–+80°C (32–166°F), where +140°C (+194°F) is allowed for CIP, and process pressure from 0 to 25 bar absolute.

3.3.6. Analysis

Frequently the analysis or examination is done in the research center, yet the interest for in-line estimation has driven makers to build up an expanding number of detecting components. There has additionally been some advance on institutionalized electrical interfaces, enabling the client to choose tests autonomous of producer. Established in-line estimations are pH/ORP/Redox, conductivity and disintegrated oxygen. The related tests are for the most part associated with a solitary assessing unit, which can be designed to make the required estimation and is effectively coordinated into a control framework. Extra amounts that can be estimated are turbidity, shading (focus) and retention.

PH estimation has the largest scope of uses in the food business, for example in the generation of tomato sauce, mustard, curds, yogurt, and milk. It is utilized for quality control, for instance, the freshness of milk before it is prepared or mustard before it is packaged. PH may likewise be utilized to control and streamline chemical movement in certain food procedures or just to watch that no cleaning specialist has spilled into a refreshment. Together with conductivity estimation, it is additionally utilized as a part of the control of CIP cleaning frameworks, see Figure 3.8. Conductivity estimations are additionally made in the creation of different foods, sensors being introduced in funnels to distinguish stage shifts. Turbidity with the related shading and ingestion estimation discovers application at numerous phases in the fermenting of lager and making of wine and is additionally used to identify stage moves in milk. Broken up oxygen estimation is utilized to screen channel layer release, wort air circulation and yeast proliferation in blending. Another application is the checking of the degassing of water natural product squeezes and focuses.

3.4. DEVICE INTEGRATION

Contingent upon the degree of mechanization, gadgets must be incorporated into controllers and at times plant resource administration frameworks. The previous enables the procedure factors to be gained by the controller, the last enables the gadgets to be arranged, observed and analyzed by a plant resource administration apparatus. The apparatus might be an integral part of the control framework or be founded on an open innovation, for example, EDDL (Electronic Device Description Language) or FDT (Field Device Tool) and worked in parallel to the controller. For this situation, a parallel way to the framework is frequently given by a door or multiplexer (Figure 3.8).

Figure 3.8: pH measurement in a brewery, sensor mounted in a Varivent housing, (*Source: Courtesy of Endress+Hauser*).

The client must settle on joining over a 4–20 mA current interface or a Fieldbus framework. There are a wide range of fieldbuses however providers of estimating instruments bolster just a couple. These are ones utilized as a part of process computerization as opposed to industrial facility mechanization, albeit some traverse the two applications. This area investigates the highlights and advantages of each; the principal characteristics of each transport are to be found in Table 3.10. More data on systems all in all can be found in the Instrument Engineers Handbook (Liptak, 2012).

3.4.1. 4–20 mA/HART

The benefit of the 4–20 mA current interface is that it is easy to introduce and commission, if to some degree costly on wiring and bureau space. Instruments are associated with the controller through I/O cards, which notwithstanding present, additionally provide food for voltage, beat/recurrence, and parallel signs, and see Figure 3.9. The controller sees just the procedure esteem and a blame flag. The abundance of data contained in an advanced estimating instrument, for instance, extra estimated esteems or demonstrative data, goes to squander.

These days, the larger part of 4–20 mA gadgets additionally bolster the HART convention, which is superimposed on the present circle. HART gadgets can be designed and blames analyzed on a point-to-point premise by utilizing a handheld terminal or a PC running a FDT-outline program. On the off chance that HART computerized signals are to be utilized by the controller or a plant resource administration framework, a HART card must be utilized or the 4–20 mA circles must be tapped and the signs read through a multiplexer. It is additionally conceivable to utilize HART as an absolutely advanced multidrop transport, however, this is occasionally found in practice. HART gadgets are coordinated into the controller and handheld by methods for device description files (DDs). For resource administration with a FDT outline, device type managers (DTMs) are required.

3.5. SENSING TECHNOLOGIES

The various sensing technologies in use today correlate a specific property or molecular signature of the measured system with a fundamental property of the sensing technology or sensing material. For example, if one were measuring the amount of moisture in a food processor with moisture absorbing thin polymer film, electrical impedance of the film might be measured with no water present as a baseline. Moisture absorbed by the polymer would change its impedance. A good sensing material would come close to showing a linear change with the moisture. Such new materials are emerging from research laboratories. Many of these are not yet patented. A great need for further understanding of new sensor materials remains because the sensing properties are extremely dependent upon their surrounding environment. For instance, some good polymer-based moisture sensing materials are seriously affected by other chemicals, particularly alcohols, and they can temporarily or permanently lose their measuring capabilities in the presence of such chemicals.

3.5.1. Optical Sensors

Optically-based sensing was identified at the beginning of the nineteenth century. The scientific principle has taken a long time to be adapted to a viable sensing technology. Considerable information about the optical properties of the measurement systems is required to develop such technologies; much of this is still the subject of research studies. A light source is focused into a fiber optic cable which transmits the light with very little loss of strength to a sensor. The sensor interacts with the environment or food system to be measured either physically by absorption, or chemically with altered properties. This change in the sensor usually absorbs and reduces the light signal. After continuing through the fiber to a detector, the incoming and detected signals are compared and the change in signal correlated with the attribute environment being measured.

Optical sensing technology became recognized as a valuable measurement tool in the flour, grain and forage industry in the mid-1960's. Its background and principles are well described by Scotter. Improved spectral discrimination of instrument output, as well as the development of new optical performing materials, have led to online application in determining fat, protein, alcohol, and moisture. The technology has continued to develop and more specific capabilities have begun to emerge in practical sensing systems. For example, specificity of optical sensing has been expanded using the electromagnetic spectrum from the visible light range, to near infrared, to far infrared. Also, new optical materials have been developed that reduce signal loss and extend the useful signal transmission range making on-line use very attractive. For example, the Fiber Optic Materials Research Program at Rutgers University has developed porous fiber optic materials which can be used in conjunction with specific chemically sensitive compounds to sense a range of properties including moisture, pH, NO_x, $S0_2$, and H_2S. Others have investigated temperature and thermal profiling.

One of the more significant areas of development that has aided in new use of optical sensing is the source of light. The light emitting diode (LED) and laser diode (LD) have made it possible to obtain much stronger light signals in narrow frequency ranges. This has helped overcome selectivity and signal to noise issues which affect the ability to effectively correlate measurements. The pre-requisites for control of a process are sensors and instruments which measure specified process variables and transmit the information to a process controller.

3.5.2. Electrical Sensors

Electrically-based sensing often uses ceramic or polymer materials that specifically correlate emf (voltage), impedance, or dielectric constant with the desired measurement property. A sensor, placed into an electrical circuit, reacts with moisture, heat, pressure, or some other property in the food environment, or by chemical reaction with the food system, causing a perturbation in the electrical properties of the sensing material. This change can be measured and correlated with the attribute of interest. There are three major types of electrochemical sensors being used for process measurement: ion-selective electrodes (ISE), ion-selective field effect (ISFET) transistors and metal oxide gas sensors. The most commonly used ion-selective electrode is the glass pH sensor, an off-line device. This device cannot be used on-line in food processing operations due to its fragile glass structure.

To overcome this problem, ISFET sensors became important. The development of an ISFET sensor, particularly for pH measurement, was aimed at overcoming the hazard of the fragile membrane of the ion-selective electrode. Various insulating oxide films such as SiO_2 and $A1_2O_3$ have been investigated to optimize performance of the ISFET over a wide range of processing conditions since this type of sensor is influenced by a variety of ionic materials (e.g., Na^+, Ca^+), and temperature. The challenges involved for online use is that the adhesives used in their manufacture generally do not withstand the higher temperatures used in some food manufacturing processes (figure).

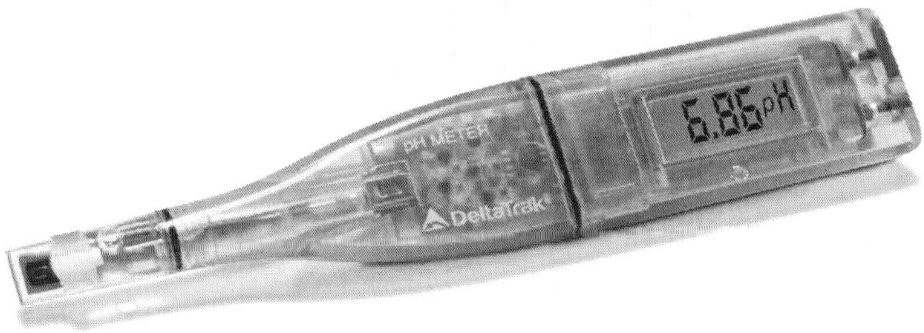

Figure 3.9: ISFET pH Meter, (*Source: http://deltatrakeurope.be/index.php/yt-products/shelf-life/isfet-ph-meters*).

3.5.3. Physical Sensors

Clearly, beginning with the receiving and storage of ingredients, through blending, processing and packaging, measurements of temperature, pressure, weight (continuous and discrete), size, shape, fluid flow of gas and liquids, viscosity, and liquid level, are important for process and product quality assurance. Most of these sensors are currently in use by the food industry. New developments in computer-automated control will assist in fine-tuning their use; in addition, food industry application of on-line rheometry for viscous materials like doughs is also emerging.

3.5.4. Chemical Sensors

Chemical sensors are used to measure a variety of chemical and biochemical components in food systems. Physicochemical sensors are emerging as important to address the measurements of water (bulk moisture, gaseous moisture, and humidity); fat; protein; carbohydrate; chemical composition; soluble solids; and color. Measurement of some combination of these parameters often opens up a new control strategy toward optimizing product quality.

Companion to physical measurements, there are number of chemical parameters of significant influence to food quality. Their measurement will provide the specificity for good control of product manufacture. Chemical measurements include pH, ionic species (e.g., chloride), odors, flavors, dissolved oxygen, developed gases, and alcohols, as well as measures of freshness and ripeness. While most of these parameters can be measured off-line, new on-line sensors are emerging.

3.5.5. Biosensors

Biosensors are often defined as sensors which use a biologically sensitive material producing a biochemical signal that is converted by a transducer into an electrical response. A biosensor is a device which responds to the presence of a specific analyte (a chemical substance that needs to be measured) by producing an electrical signal proportional to the concentration of the analyte. It is usually constructed from three components: receptor, transducer, and electronics (amplification and display. The biomaterial is generally placed on or into a membrane. An important advantage of a biosensor is selectivity and the capability to detect very small amounts of such compounds as glucose. Sterilization or exposures to harsh conditions in a food processor are often fatal to a biosensor. Enzyme-based biosensors used in food measurements generate signals from chemical reactions. Immunosensors are based on antigen-antibody reactions. Currently, the main application of these latter biosensors is in the pharmaceutical industry.

Enzyme-based biosensors are commercially available; glucose and lactate are the two main analytes for which instruments are available and routinely used.

Biosensors can be classified by either their receptor system (e.g., enzymatic, immunological, cellular) or by their transducing system (e.g., potentiometric, amperometric, optical, thermal or mechanical).

3.5.5.1. Potentiometric Sensors

This sensor was the first type available and can detections such as Na^+, H^+, and NH^+. In its simplest form, the potentiometric biosensor uses a standard ion-selective electrode (ISE) in conjunction with a biological element surrounded by a membrane. Changes in potential caused by the binding of selective ions to the sensing membrane are measured by the ISE. The best example of a potentiometric biosensor is the enzyme electrode in which an immobilized enzyme is coated over a conventional ISE. Potentiometric sensors are very simple and reliable but they tend to have relatively slow response and recovery times (30 s to 5 mins).

3.5.5.2. Amperometric Sensors

These sensors rely on measurement of current rather than voltage. The current generated is usually directly proportional to the specific binding reaction, unlike potentiometric measurements where the relationship is logarithmic. Although any redox enzyme can be used, the best known amperometric sensor employs glucose oxidase as catalyst to produce H_2O_2 according to:

$$\beta - D - glu\cos e + O_2 + H_2O \rightarrow gluconicacid + H_2O_2 \quad (7.4)$$

The biosensor monitors either O_2 consumption or H_2O_2 production. The reoxidation current at a platinum electrode is measured. This device can be used to quantify glucose in a concentration range of 107 to 103 mol/L. Electrochemical-sensitive transistors. The output signal from a sensor employing an ISE tends to be noisy due to the connecting wires between receptor and amplifier which act as antennae for the local electromagnetic field. Potentiometric devices have recently been constructed using thin film semiconductor techniques. The noise problem is solved by abolishing wires and placing the membrane directly on the gate of the transistor. One type of field effect transistor (FET) has been modified by the binding of biocatalysts to an ion sensitive membrane that overlays the gate. The membrane is simply an overlay of an electrically conductive chemical. In ion sensitive

FETs (ISFETs), the membrane allows specific ions in the sample to pass through the membrane for detection. Material having an electrical potential will flow through the membrane and release its electrons, thus creating a current flow between the source and the drain of the gate. An enzyme FET (ENFET) is created by the application of an immobilized enzyme gel layer over the ion-selective membrane of an ISFET.

3.5.5.3. Optoelectric Sensors

Light emitting diodes can monitor changes in optical parameters at the biological level as a basis for a biosensor. The photodiode will give different voltage levels, relative to the absorbance levels of a specific enzyme, which reflect the concentration of the enzyme in the measurand. Fiber optic technology can be used to devise reflectance, transmittance or fluorescence-based biosensors by immobilization of reactive chemicals at the terminal of a small fiber optic probe. pH changes have been monitored optoelectronically by using pH sensitive dyes immobilized on a fiber optic/ photodiode combination.

3.5.6. Other Sensing Technologies

The use of microwave sensing to date has been primarily associated with bulk moisture determination. Dielectric constant and loss factor are properties of food materials ideally suited for microwave determination of moisture content. Since microwave instruments are dependent on sample density; their broad application is limited due to the difficulty of measuring density on-line. More recently, a new two-variable technique has permitted development of a more compact and inexpensive instrument. In addition, onboard microprocessors and data storage have enhanced data acquisition and processing capabilities. A better understanding of food dielectric properties coupled with microwave advances provides opportunities for further on-line exploitation. The principal advantage of this instrument is its non-intrusive capability. The cost/benefit ratio may still be too high relative to other developing moisture sensors to see utilization broadly increased.

Nuclear magnetic resonance (NMR) must be mentioned as a potential online process sensing tool. While NMR has been extensively used in the laboratory to study the relaxation time of food matrices leading to determination of bound and unbound water, its cost is prohibitive for on-line use. However, new magnet designs that are far less costly, as well as development of innovative food transfer systems through the magnet, are under study. The use of thermal diffusivity/thermal conductivity to correlate with food moisture content has been developed into a new sensor for highly viscous materials such as doughs, cheeses, and emulsions and for measurements in powder materials. The sensing system, measuring the rate of heat absorption to the food sample, provides a correlation with moisture content and thermal conductivity of the sample from its calibrated database.

Another sensor technology receiving increasing attention and investigation is ultrasonics. Ultrasonic measurement in food systems is of high interest, like NMR, because of its non-intrusive nature. Ultrasonic technology is well established for determining fill levels and for in-process metering and has been adapted to determination of solids content in simple solutions. These sensing techniques use the principle of determining sound absorption, reflection, and phase shift to correlate with properties of the system attenuation. As more complex food systems are investigated, particularly non-homogenous media such as doughs or cheeses signal interpretation becomes limiting.

3.6. REQUIRED CHARACTERISTICS OF SENSORS IN THE FOOD INDUSTRY

The food industry has a general need for methods which are simple; rapid; inexpensive; use readily available, stable reagents; require minimal labor input; and are, ideally, automated. Industry also has specific needs for the QC testing of both ingredients and finished products using rapid methods rather than the traditional, conventional techniques. Highly sensitive and specific methods are needed to detect (relatively) low levels of organisms while maintaining a low incidence of false-positive and false-negative results. Process control in the food industry is more difficult than in other industries due to the utilization of complex and variable raw materials having measured parameters which are often difficult to define. Wren generated a list of desirable characteristics of biosensors, with respect to parameters commonly measured by sensors and are classified into:

(a) Primary measurements (for example temperature, weight, flow rate and pressure).

(b) Comparative measurements, obtained from comparison of primary measurements (for example specific gravity).

(c) Inferred measurements, where the value of an easily measured variable is assumed to be proportional to a phenomenon that is difficult to measure (for example hardness as a measure of texture).

(d) Calculated measurements, found using qualitative and quantitative data from analytical instruments or mathematical models (for example biomass in a fermenter).

It is important to recognize that process variables that are measured and controlled are often only used indirectly as indicators of complex biochemical reactions that take place during processing. Examples include the combination of time and temperature needed to destroy microorganisms, temperature, and pressure as measures of the changes that take place during extrusion or the time required to remove a specified amount of moisture by dehydration. It is, therefore, necessary to know the relationship between the measured variable and the changes that take place in order to be able to exercise effective control. Solid-state electronic sensors have largely replaced older mechanical or chemical types, due to their greater reliability, greater accuracy and precision, and faster response times.

On-line and in-line sensors are widely used because of their rapid response time and accurate positioning. Their main requirements are as follows:

(a) A hygienic sensing head.

(b) Contaminant-free (contains no reagents or microorganisms that could contaminate foods).

(c) No potential hazard from foreign bodies (e.g., no glass components).

(d) Robust to withstand processing temperatures or pressures (for example 200 °C and 10 MPa in cooker-extruders.

(e) Able to withstand chemicals in food components or effluents tolerant to cleaning-in-place or having cheap, easily replaced, disposable sensing heads.

(f) Reliable with good reproducibility, even when exposed to moisture, steam, dust, food volatiles or fouling by fats, proteins or starches.

(g) Resistance to electromagnetic interference in some applications. (e.g., microwaves, ohmic heaters)

(h) Resistant to damage from mechanical vibration.

(i) Low maintenance requirement.

(j) Total cost (capital, operating and maintenance costs) in proportion to benefits gained.

3.7. ADVANTAGES AND DISADVANTAGES OF COMPUTER CONTROL

There are many potential benefits from the application of computer-based process control but the magnitude of potential improvements is application-specific and will vary from plant to plant. The specific advantages of computer control are summarized below (Smallwood, 1985):

(a) *Yields and material consumption.* Precisely programmed operating conditions result in more uniform characteristics in each basic material produced. The consequences of improved uniformity are more predictable product blending (mixing) results; less correction requirements and; a possible reduction in basic material inventory. More precise, responsive control results in minimum usage of materials and maximum yield.

(b) *Yield and quality performance.* Tighter and more regular process control obviously provides the opportunity for improved yield and product quality performance.

(c) *Depletion in energy consumption.* Numerous opportunities are available for energy reduction, including improvements in combustion efficiency and load management. Load management involves also the control of energy loads to minimize use and avoid excessively high peak demands.

(d) Operating faults. Once the process control system is reliably programmed, mistakes are essentially limited to equipment failures. For example, a number of pumping errors and spills may be reduced, and hence the expense associated with the resulting additional loss, reprocessing, and cleanup could be reduced.

(e) *Diagnosis and troubleshooting.* Computer control is invaluable for process diagnostics and troubleshooting. Problems can be quickly identified and resolved.

(f) *Capacity.* Design capacity of plants can be consistently achieved with the computer–supervised operations.

(g) *Safety.* By means of computer control, process safety is improved in terms of alarms, emergency shutdowns, interlocking, and fail-safe modes.

(h) *Dependability.* Combining proper hardware and a well-trained, effective team, statistically greater than 99.0% reliability can be achieved.

(i) *Reports.* Computer systems are able to provide any type of performance and accounting report within the scope of the information collected. New formats for other reports and logs can be easily programmed.

(j) *Inventory management and reduction.* Continuous reporting of stocks (raw materials, intermediates, tank levels, and finished products) provides a database for both operation and management functions. Hence, inventory can be significantly reduced.

(k) *Personnel.* Compared with manually operated plants, a reduction of 30–67% in the total number of personnel (e.g., operating, maintenance, clerical, supervisory, and management combined) can be attained processing operations if effectively managed.

The advantages listed above are convincing, and most plants could justify the cost of retrofitting with computer process control. However, the picture is not complete without exposing the shortcomings, such as:

(a) *Personnel.* With computer control, fewer total personnel are required, but the skill requirements are significantly greater than with conventional operation. For proper operation, every member of the operating team must be capable of understanding, operating, troubleshooting, and maintaining all but the most complex aspects. If not, either an increase in staffing is needed or the potential benefits will not be achieved.

(b) *Start-ups and shutdowns.* The initial start-up with computer process control is much more difficult than with conventional control. During pre-operational checks, defects in hardware, software, and control devices will be found and corrections required before the system could start. Nothing operates under computer process control unless every element is working properly. After resolving all initial problems and after developing the necessary skills in the operating team, computerized control permits greater reliability than conventional control.

(c) *Scheduled start-ups and shutdowns.* Usually, the computer hardware and software is not able to control the startup and shutdown process. Thus, a sequence of commands has to be initiated through the computer system. For this, the operators need the necessary understanding and skills.

(d) *Breakdowns*. Problems such as electric power outages may cause unscheduled shutdowns and subsequent restarts. These interruptions may be detrimental if not taken into consideration during system planning. The time of restart depends on the skills of the operating team. Under optimal conditions, the team can recover the system within minutes, but lack of understanding of functions may result in several lost hours.

3.8. NEURAL NETWORKS

Neural networks can be taught to perform complex tasks and do not require programming as conventional computers. They are massively parallel, extremely fast and intrinsically fault-tolerant. They learn from experience, generalize from examples, and are able to extract essential characteristics from noisy data. They require significantly less development time and can respond to situations unspecified or not previously envisaged. They are ideally suited to real-world applications and can provide solutions to a hos' of currently impossible or commercially impractical problems.

In simple terms, a neural network is made up of a number of processing elements called neurons, whose interconnections are called synapses. Each neuron accepts inputs from either the external world or from the outputs of other neurons. Output signals from all neurons eventually propagate their effect across the entire network to the final layer where the results can be output to the real world. The synapses have a processing value or weight, which is learned during training of the network. The functionality and power of the network primarily depends on the number of neurons in the network, the interconnectivity patterns or topology, and the value of the weights assigned to each synapse.

3.8.1. Artificial Neural Networks

Quality of food is complex term and is assessed by suitable combination of physical, chemical and organoleptic tests. Physical and chemical parameters do not always have straightforward correlations with the sensory evaluation results. However, frequent sensory evaluation is restricted due to the availability of trained judges, and proper ambiance. Several investigators have attempted to apply Artificial Neural Networks (ANN) models for prediction of food properties, and changes during processing and storage of foods. Zhang and Chen introduced a method of food sensory evaluation employing artificial neural networks (1997).

The process of food sensory evaluation can be viewed as a multi-input and multi-output (MIMO) system in which food composition serves as the input and human food evaluation as the output. It has proved to be very difficult to establish a mathematical model of this system; however, a series of samples have been obtained through experiments, each of which comprises input and output data. On the basis of these sample data, the back-propagation algorithm (BP algorithm) is applied to "train" a three layer feed-forward network. The result is a neural network that can successfully imitate the food sensory evaluation of the evaluation panel. This method can also be applied in other fields such as food composition optimizing, new product development and market evaluation and investigation. The human sense of smell is the faculty which has very important role to play in industries such as beverages, food, and perfumes (Figure 3.10).

Figure 3.10: Application of artificial neural network in agriculture, (*Source: https://medium.com/futurefood/artificial-intelligence-in-the-food-industry-empowering-farmers-decision-making-d2fdebb6debe*).

3.9. FUZZY SET THEORY

Zadeh first reported on a novel way of characterizing no probabilistic uncertainties, which he called "fuzzy sets" (1965). The applications, which are multidisciplinary in nature, include automatic control, data classification, chemical processes, and so on. Fuzzy logic embodies the nature of human's mind in some sense, as the conception of possibility and probability is truly underlined in this logic in contrast with the traditional Boolean logic. Hence, any studied object, in fact, can be described as a fuzzy system. The three basic concepts of fuzzy logic are fuzzy set, fuzzy membership, and fuzzy variable and these are the underlying factors of a fuzzy system.

A fuzzy system consisting of a fuzzifier, a knowledge base (rule base), a fuzzy interference engine and defuzzifier has been considered (see figure below). The knowledge base of the fuzzy system is a collection of fuzzy IF-THEN rules. The term fuzzy logic denotes a modeling approach, where functional dependencies between the input and output variables are described by means of a set of IF-THEN rules following the reasoning with the operators AND, OR and NOT. The domain of both the input and output values is subdivided in the fuzzy sets according to technologically reasonable estimates. Fuzzification means that the degree of membership of a given crisp input value to its fuzzy sets is calculated. The fulfillment of the IF part, the premise, is calculated by fuzzy inference. The fuzzy sets and their combination by the operators AND, OR are used the standard models of minimum, maximum (see equation 7.1) and finally, the fulfillment of the THEN part is calculated. Defuzzification ponders the several IF-THEN rules according to their degree of fulfillment, for instance, the degree of membership to the output sets and a crisp value is assigned to the output variable.

$$AND : \mu(xANDy) = \min(\mu(x), \mu(y))$$
$$OR : \mu(xORy) = \max(\mu(x), \mu(y)) \quad (1)$$

Use of fuzzy expert systems in the food processing industry has many advantages. It may reduce labor costs by eliminating labor-intensive tasks, provides a rapid, consistent and more accurate and objective results, may help when expert advice is needed but an expert may not be available and can help to support and verify human expert's opinion. Fuzzy system is independent of human errors and can be used throughout the day without a labor needed. It can also operate in risky situations and under conditions unfit for human experts and can act quickly on the basis of huge databases and knowledge banks.

3.10. FUTURE TRENDS

Despite the fact that the detecting components and related hardware of an estimating gadget are constantly subject to persistent change, no radical changes are not out of the ordinary in the following couple of years. The same is valid for advanced correspondence: the Fieldbus has set up itself as a solid other option to the 4– 20 mA current circle yet it will be the market instead of new improvements that decides the achievement or disappointment of a specific convention. Plant Asset Management, then again, is drawing in a lot of consideration right now: since productive task likewise depends to some degree on the capacity to design and oversee upkeep exercises. Advancements can be seen at a few levels.

At the sensor level, NAMUR suggestion NE107 proposes an improved gadget determination plot that sorts blunder messages into four classes: 'support required,' 'failure,' 'utilitarian check' and 'out of detail.' This enables the client to evaluate the seriousness of the alert before choosing whether the nitty gritty data held by the gadget is required. The proposal has been consolidated into both PROFIBUS and FOUNDATION Fieldbus details and the primary gadgets with this type of finding areas of now available.

As talked about quickly in Section 3.4 there are two open determinations that help gadget arrangement and observing, EDDL and FDT. FDT was considered because of benefit administration and has considerably more extensive help, which reaches out past estimating gadgets and valves. EDDs are a vital part of each HART, PROFIBUS and FOUNDATION Fieldbus instrument. In spite of the fact that FDT will draw out an upgraded detail soon, they are additionally co-working with the EDDL gathering to indicate a joint FDI (Field Device Integration) standard. This should open the best approach to less demanding joining of gadgets in plant administration instruments and better help of the arrangements as of now offered, for instance, condition checking.

Above and beyond is the utilization of a plant resource administration apparatus to associate with the life-cycle information of a gadget: testaments, adjustments, arrange codes, save parts or, on account of suspended lines, pointers to on substitution gadgets. This is data which, obviously, can be physically gone into any database; in any case, it requires a lot of push to do this and is essentially copying information that is held by producer. Opening a web entry with life-cycle information by essentially tapping on a gadget tag or serial number is a probability which as of now exists for a few gadgets. A further click permits the substitution gadget or extra part to be requested. This administration is probably going to be offered by most instrument producers in the following couple of years.

At the time of composing, an idea for remote administration is being tried inside in the purported Future Factory Initiative, whereby estimating instruments are empowered as shrewd gadgets inside an ERP device. These would then be able to make mistake tickets, warnings or cautions and also exchange handling information to the administration application for introduced base setup, administration, and support. On distinguishing an issue, a gadget sends a warning to the administration application which advances it to the administrative division of the client. Here the choice is taken whether the issue can be settled locally, or whether the gadget maker ought to be included. In the last case, the message is again naturally sent to the fabricate's administration focus, which at that point gets to the gadget remotely and proposes suitable activities, if the blame can't be cured by reconfiguration. The outcome is a quickened support process which guarantees that the best-qualified individual tackles the issue.

REFERENCES

1. Avesta Sheffield Corrosion Handbook, (1994). Stainless steels for the Food Processing Industries. ISBN 91-630-2122-6 [Accessed on 11th April 2018]
2. Bentley, (1998). Handbook of Temperature Measurement Vol. 1: Temperature and Humidity Measurement, Springer; ISBN-10: 9814021091; ISBN-13: 978-9814021098.
3. Berrie, (2001). Instrumentation and Sensors for the Food Industry (2nd edition), ed. Kress-Rogers Brimelow, Chapters 13 and 14, Woodhead Publishing, ISBN 1 85573560 1.
4. Bhuyan, M., (2007). Measurement and Control in Food Processing. Taylor & Francis Group, Boca Raton.
5. Bimbenet, J., Dumoulin, E., & Trystram, G. (eds)., (1994). Automatic Control of Food and Biological Processes. Elsevier, Amsterdam.
6. Burns, J., (1999). Resistive thermometers. In The Measurement, Instrumentation and Sensor Handbook (Webster, J.G., ed.). Springer, New York.
7. Davidson, V., Brown, R., & Landman, J., (1999). Fuzzy control system for peanut roasting. *J Food Eng 41,* 141–146.
8. Figura, L., & Teixeira, A., (2007). Food Physics. Springer-Verlag, Berlin.
9. Fraden, J., (1999). Infrared thermometers. In The Measurement, Instrumentation and Sensor Handbook (Webster, J.G., ed.). Springer, New York.
10. Caldwell, G., & Darwin., (2013). Robotics and Automation in the Food Industry: Current and Future Technologies. Woodhead Publishing. 1st edition
11. Kress-Rogers, E., & Brimelow, C., (eds) (2001). Instrumentation and Sensors for the Food Industry, 2nd edn. Woodhead Publishing, Cambridge.
12. Mittal, G. S., (1997). Computerized Control Systems in the Food Industry. CRC Press, New York.
13. Perrot, N., Agioux, L., & Ionnou, I., (2000). Feedback quality control in the baking industry using fuzzy sets. *J Food Proc Eng 23,* 249–280.

CHAPTER 4

MACHINE VISION IN THE FOOD INDUSTRY

"Without food, we cannot survive, and that is why issues that affect the food industry are so important."

—Marcus Samuelsson

CONTENTS

Over the previous decade, machine vision has been connected substantially more generally, consistently and methodically in the food business. This has been guaranteed by consistent improvements in the constituent strategies, to be specific picture preparing and design acknowledgment. In the meantime, propels in PC innovation have allowed reasonable executions to be accomplished at bring down cost. To some degree, advance is presently being held up by the requirement for custom-made improvement in every application: subsequently future calculations should be made trainable to a considerably more noteworthy degree than is as of now conceivable. Likewise, late improvements, for example, hyperspectral imaging ought to end up acknowledged in various specialty regions.

4.1. INTRODUCTION

Machine vision has for some time been viewed as the path forward to help numerous parts of industry–including medication, transport and a decent number of other application territories–to screen and even deal with its shifted tasks. Specifically, it has been connected to the review of items amid make, and on account of food items it has likewise been vital for assessing the crude food materials touching base from ranches and abattoirs. Most importantly, its utilization is viewed as a method for keeping up control of value, which is all the additionally demanding when the quantities of individual items touching base at a handling plant add up to millions every week, speaking to a run of the mill stream of result of 20 things for each second. Obviously, for seeds or oat grains, the relating figures are far more noteworthy, and it is more down to earth to quantify stream rates in kg every second.

To apply machine vision, pictures of the different things initially must be gotten, utilizing reasonable cameras: these can be ordinary 'zone' cameras, not unlike the advanced cameras found on the residential market, however better adjusted for consistent PC interfacing. Notwithstanding, with the relentlessly moving transport line framework found in numerous plants and industrial facilities, it is regularly valuable to utilize a 'line-filter' camera to get thus every vertical line of pixels for (successfully) a boundlessly long level picture of the belt, from which typical 2D pictures of the items and bundles can be acquired.

Once reasonable 2D pictures have been procured, they can be broke down to acquire the required visual data about the items. A significant part of the handling can be accomplished by taking each picture and changing over it into a progression of different pictures that refine the information and consistently separate important subtle elements and estimations. Without a doubt, picture handling is characterized as the transformation of one picture into another, in which similar articles happen at a similar relative positions, however in progressive pictures they may have commotion evacuated, edges improved, edges recognized, edges estimated, and afterward entire items distinguished and surveyed. On the other hand, dim scale pictures might be threshold to get twofold pictures speaking to shapes, and these may then be 'diminished' and broke down further. These procedures will be depicted in more detail in later areas of this section. Here we take note of that moderately basic picture handling methods can separate very abnormal state data while working just in picture space. Be that as it may, if machine vision were constrained to picture handling, it would have extremely limited power, and it is important to add digest design acknowledgment procedures to understand its maximum capacity. Theoretical example acknowledgment depends on the possibility that the PC ought to retain both the attributes of articles, (for example, their shapes) and their arrangements, so it can perform programmed elucidation at whatever point it sees such protests once more. As we should see, negligible memorization is for all intents and purposes unthinkable as a result of the immense varieties between various occurrences of any individual sort of question, so extraordinary strategies must be conjured to perform dynamic recognition. When this has been accomplished, consolidating the capacities of picture preparing and unique example acknowledgment gives the premise to practical machine vision frameworks.

Once appropriate vision frameworks have been conceived, they can be utilized for examination, quality control, and furthermore to direct and control 'gathering' of more intricate foods, for example, pizzas and layer cakes. Certain items require X-beam assessment (e.g., to recognize bones in chicken filets), and when all is said in done it is important to choose the picture modalities that can give the most proper data about items. At long last, there is a requirement for continuous preparing so any vision framework can stay aware of the stream rate on the product offering: while it is typical for committed equipment quickening agents to help with the PC handling, we are finally moving into a period where PCs will have the capacity to do all the essential preparing unaided.

In the accompanying area we audit the standards and strategies for machine vision that emerge from different picture handling and example acknowledgment methods; we likewise take a gander at accessible picture modalities. In Section 4.3 we display an arrangement of contextual investigations that exhibit the estimation of a decent extent of these strategies. In Section 4.4 the talk is widened to cover a substantially more extensive scope of utilizations and late advances in the sustenance business. Segment 4.5 evaluates the requirement for exceptional equipment for sustenance investigation applications. The chapter with a conclusion delineating likely future patterns.

4.2. MACHINE VISION: PRINCIPLES AND METHODS

Machine vision is a composite subject whose methodology is based in two major disciplines, namely image processing and statistical pattern recognition. In fact, the first of these disciplines is very wide in scope and it is helpful to divide it into a number of smaller but closely linked areas including shape analysis, morphology, object location and texture analysis. Also of concern is the means by which images are acquired before analysis– as additional information is clearly available from X-ray, ultra-violet and infra-red images, as well as from color. All the above aspects are considered in the following subsections.

4.2.1. Basic-Image Processing Techniques

A basic method for performing picture preparing is to take the pixel esteem at every pixel area and to put an adjusted an incentive at the comparing area in a yield picture space. For instance, in the process known as thresholding, which is broadly utilized for finding obviously characterized protests in computerized pictures, any pixel whose force is darker than a given limit prompts a twofold estimation of 1 in the yield picture space, while different pixels are given the esteem 0 (Figure 4.1b). Such a procedure is known as a 'pixel– pixel' task. 'Window– pixel' activities are much more intense and are accomplished by looking at the pixel esteems inside a window around every pixel area and registering an incentive to be put at the relating area in the yield picture space.

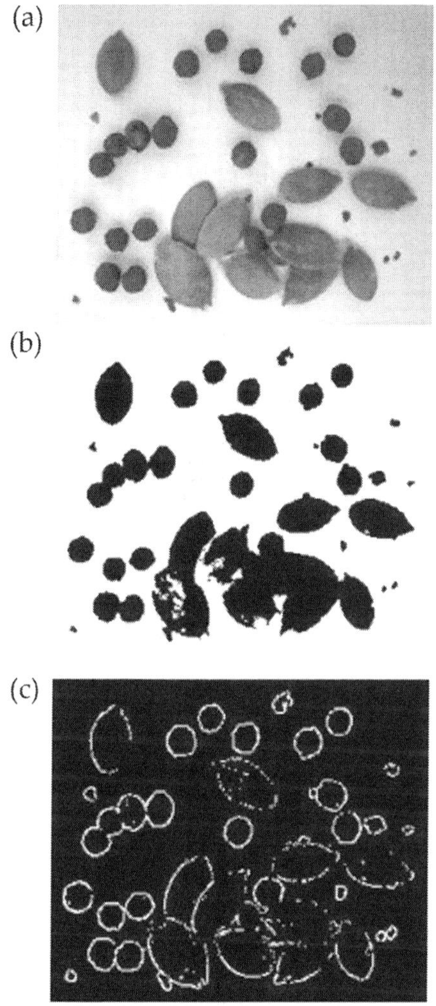

Figure 4.1: Thresholding and edge recognition. (a) Original seed picture (b) Threshold rendition, showing that few seeds are not well pleasing. (c) Edge-location variant, demonstrating that edges are frequently broken and might be a few pixels wide in a few spots, (*Source: G. Caldwell Darwin (2013), Robotics and Automation in the Food Industry: Current and Future Technologies, Woodhead Publishing*).

Essential properties of these covers are appeared in Figure 4.2. While the second cover tends to give a huge flag simply inside the limit of a question (Figure 4.2c), the flag is helped at corners–and considerably more inside gaps–where the numerous limit commitments collect. At first sight the corner improvement veil (Figure 4.2f) gives a comparable reaction to that of the even edge cover (Figure 4.2d); actually, this is normal as the two 3 × 3 veils are very comparable; be that as it may, the correct hand corner exhibits a more noteworthy flag in Figure 4.2f than in Figure 4.2d. The principle truth to watch is that corners are more sizeable highlights than edges, and in this manner require bigger veils to distinguish them unambiguously (commonly 7 × 7 pixels as opposed to 3 × 3).

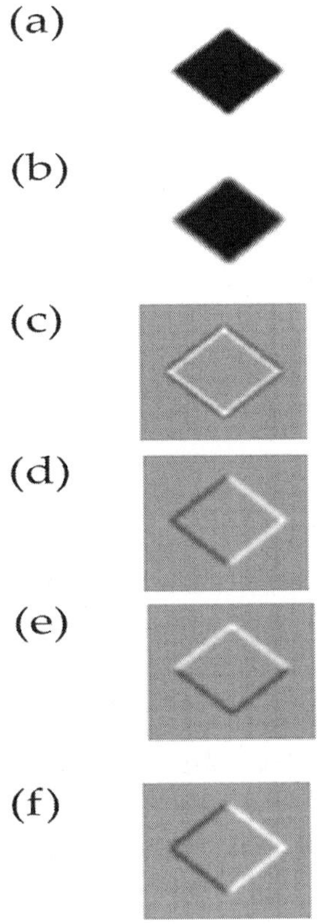

Figure 4.2: (a) Original figure of a square shape. (b)-(f) Effect of applying the five covers depicted in the content. Panel (b) demonstrates a slight burring impact that can smooth out clamor; (c) indicates how a Laplacian veil gives a positive signal just inside an edge and a negative signal just outside it; (d) demonstrates a positive edge signal on the right of the object and negative signal on the left; (e) shows a positive edge signal on the top of the object and a negative signal at the bottom; (f) demonstrates a comparative impact to (d) but there is a more grounded motion on the right-hand corner. In (c)–(e) the signs are appeared on a dim platform with the goal that positive and negative signs

are effectively identifiable, (*Source: G. Caldwell Darwin (2013), Robotics and Automation in the Food Industry: Current and Future Technologies, Woodhead Publishing*).

While convolutions have wide utility and extensive power, Equation (1) demonstrates that they are constantly straight. Actually, non-direct channels are expected to perform helpful acknowledgment capacities. Specifically, for really recognizing improved questions and highlights, tasks, for example, thresholding and non-most extreme concealment (i.e. overlooking signs that are littler than the encompassing ones) should be connected. Next, we look all the more deliberately at the commotion concealment cover given above. This disposes of commotion by smoothing it out finished neighboring pixels; truth be told, it additionally soothes out the picture flag. To stay away from this impact, middle channels are regularly utilized. These include taking the pixel esteems inside the window, putting in them in request of size, and taking the center one in the subsequent succession. For a 3 × 3 window, the fifth of the nine esteems is taken as the yield esteem; for instance, if the nine esteems are 5, 3, 6, 2, 1, 18, 4, 6, 7, reordering them as per measure gives the sequence, 2, 3, 4, 5, 6, 6, 7, 18; the center one (the 'middle') is 5, and is taken as the yield esteem. Middle channels are profoundly compelling at taking out commotion without causing picture obscuring; specifically, they are particularly great at evacuating 'drive' clamor, comparing to sudden extraordinary esteems, for example, 18 in the above arrangement. Regardless of these great characteristics, this kind of channel can present slight contortions by minutely moving edges, and ought to in this way not be connected unnecessarily when exact estimations are going to be made.

We now swing to the issue of edge location. To accomplish this we utilize the yields from the vertical and level edge improvement covers. Truth be told, these separate the picture locally in the x and y headings, allowing the x and y segments of power inclination, gx , gy , to be resolved. As g x , and g y demonstrate how quickly the power is changing along the x and y tomahawks, it isn't amazing that this data can be utilized to discover the extent and bearing of most extreme rate of progress of force. Truth be told, the extent and bearing can be ascertained from the accompanying equations:

$$g = [g^2_x + g^2_y]^{1/2} \qquad (2)$$

$$\Theta = \arctan (g_y/g_x) \qquad (3)$$

The final process of edge detection is now achieved by thresholding *g* (Figure 4.1c).

Next, we look all the more carefully at the corner improvement veil, which just upgrades corners of a specific introduction. (The principal purpose behind this is the veil is a model corner, and moving a model corner over the entire picture will just give a nearby counterpart for corners of comparable introduction.) This implies discovery of corners of subjective introduction will require eight covers of this write, each fittingly orientated. As a rule this circumstance will apply when planning finders for any subjective component, however with bigger highlights, bigger more unpredictable veils will be required, and more prominent quantities of covers with various introductions will on a basic level must be formulated. Actually, outlining general component indicators represents the rule, surely understood in radar, that a 'coordinated' channel must be indistinguishable in frame with the profile it is to be utilized to identify, so as to accomplish ideal flag to-commotion proportion. Notwithstanding, in picture handling there is the admonition this would offer ascent to affectability to the level of brightening of the foundation. Henceforth coordinated channels utilized as a part of picture preparing need to utilize covers whose coefficients entirety to zero. This can promptly be comprehended as takes after: if the level of light ascents by a similar sum δ wherever in the picture, at that point the expansion accordingly from the entire veil will be δ times the whole of the cover coefficients: and for a zero-entirety cover the general change accordingly will fundamentally be zero. This applies for the last four of the convolution covers recorded previously. (The main veil is for smothering clamor and isn't expected to copy a coordinated channel.)

To condense, convolutions are very vital picture handling activities as they give ideal location execution to an assortment of picture highlights, yet to understand the identification itself, which brings about the real acknowledgment of the highlights, their yields should be liable to non-straight channels including thresholding and non-greatest concealment. Actually, non-most extreme concealment is more critical than may be thought, on the grounds that thresholding relies upon the unwavering quality with which worldwide limits can be chosen: by and large choice is troublesome on the grounds that it depends to a great extent on there being a reasonable outline between obvious highlights and clamor; furthermore, while viable nearby edges can frequently be chosen, worldwide edges will change with the level of enlightenment and the difference, both of which are at risk to shift altogether starting with one a player in the picture then onto the next. Much exertion has been committed to these issues, not simply previously, but rather likewise up to the present day (Davies, 2007). This is as yet a hot issue, when the eye can observe a wide range of highlight and protest without inconvenience, is down to the very idea of picture information, which is fantastically variegated, with numerous angles that are mistaking for the PC translation framework. It will serve to feature this circumstance on the off chance that we here allude to one such issue–that of auxiliary (intelligent) lighting which changes the enlightenment of a given question when another protest is put close-by.

4.2.2. Shape Analysis

Shape investigation is one of the prime means by which people can recognize and perceive objects: it is no less imperative for PC acknowledgment. Here we accept that any information pictures have been threshold to deliver parallel pictures in which the articles show up as 1s of every a foundation of 0s, as demonstrated in Section 4.2. At this stage the PC should decide what number of items there are in a picture and precisely what areas they cover. While this is 'self-evident' to a human looking at the picture, the PC needs to decide it by 'associated parts investigation.' This isn't an insignificant procedure. When looking over the picture in a forward raster examine (that utilized on a simple TV screen), initial one question is met, at that point another, at that point the first once more, so naming the pixels in grouping won't give objects extraordinary arrangements of names (Figure 4.3). While it is frequently discovered that mixes of sweeps (e.g., forward raster checks took after by turn around raster filters) can land at an exceptional marking, it is more solid and more productive to look at the primary arrangement of names and to make a 'conflict table' indicating which names coincide. Watchful iterative investigation of the conflict table will at that point demonstrate to accomplish a perfect marking. When this has been done, the items have essentially been checked, and their regions, outlines and straight estimations can be classified without trouble. Obviously, associated segments investigation makes it feasible for some further size and shape estimations to be made and recorded.

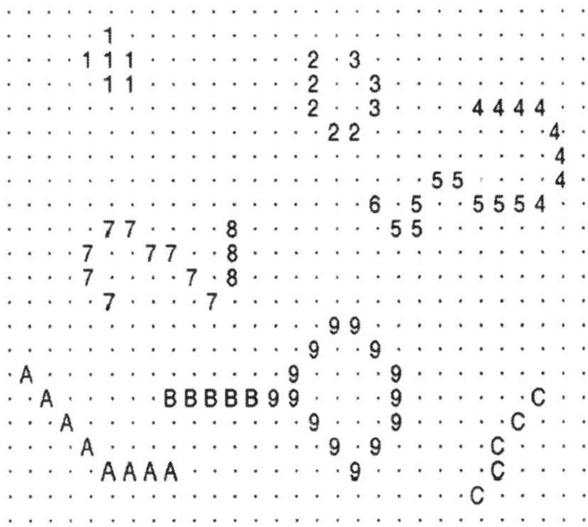

Figure 4.3: Aftereffect of a fundamental calculation for associated parts investigation. In few cases objects have in excess of one mark: examination for name conflicts is required to give a remarkable naming, (*Source: G. Caldwell Darwin (2013), Robotics and Automation in the Food Industry: Current and Future Technologies, Woodhead Publishing*).

On the off chance that articles are to be arranged rapidly, basic shape estimations that are invariant to size can be valuable. Prime among these is the 'circularity' measure, $C = A/P\ 2$, where A is the territory and P is the border of a blob: C is biggest for a circle and littlest for a thin stick. C is broadly utilized for cell checking in natural examples, but on the other hand is helpful for arranging little questions, for example, seeds. While not a modern measure, it is valuable for making a preparatory distinguishing proof of the items in a picture, preliminary to a full examine of any suspicious cases, utilizing further developed and precise devices. One such instrument is that of minutes, in which the main, second, third and higher snapshots of any unpredictable shape are registered: as this should be possible to a high level of exactness, it can leave little uncertainty with regards to the kind of protest that is being tried–insofar as a sufficiently expansive database of qualities is accessible. Be that as it may, this technique is to some degree computationally serious, so limit following strategies that emphasis on protest limit focuses as opposed to every one of the focuses in the question shape are for the most part favored.

The original limit following portrayal is the 'centroidal-profile' strategy, which depicts the protest limit as far as a (r, θ) polar diagram in respect to the centroid area (Figure 4.4). For a circle this is a straightforward diagram comprising of an even straight line at r = a, where an is the circle sweep. For a square it is an arrangement of four bends meeting at four cusps, which speaks to the four corners of the square; the introduction of the square is given by the points θ at which the cusps happen. Plainly, such a chart will show the question's size and introduction, and will likewise allow any shape mutilations to be distinguished and estimated. In like manner it will allow the protest be perceived. A similar circumstance applies for any of the items that have been incorporated into an appropriate database. Shockingly, this strategy isn't extremely hearty: for instance, if two items touch or cover or if there is one question that is genuinely deformed, the centroid will be in a self-assertive area; accordingly, the centroidal profile will be terribly misshaped and won't be conspicuous (Figure 4.4). While there are elective kinds of limit profile, for example, curvature– limit remove (κ , s) bends, which are not dependent on exact area of the centroid, these are clumsier to deal with, and less strong, than the Hough change approach, which we treat in Section 4.4.

Skeletonisation is a further way to deal with shape acknowledgment. The thought is at its most regular when managing objects that are made essentially out of thin strips and bends–especially alphanumeric characters, asbestos filaments, branches in a tree, et cetera. In such cases, by disintegrating the limits of the items more than once, the point is to come to the deepest 'average' lines, which on account of hand-attracted characters would compare to the ways took after by the pen (Figure 4.5). In any case, the skeleton idea is firmly characterized. It is intended to speak to the state of the associated question, and should never end up detached amid the disintegration procedure. This influences calculation to outline very mind boggling and demanding. Once framed, the skeleton holds every one of the appendages and openings of the first shape; without a doubt, it is proposed to be a topologically correct portrayal, with every one of the intersections and line closes giving a helpful, glorified width-autonomous depiction of the shape. It is a broadly utilized procedure, yet has the disservice that if clamor in the first picture prompts extra openings in the protest shape, a deceptive skeleton will come about. Be that as it may, if intemperate clamor can be disposed of, mistakes from this reason will be significantly lessened.

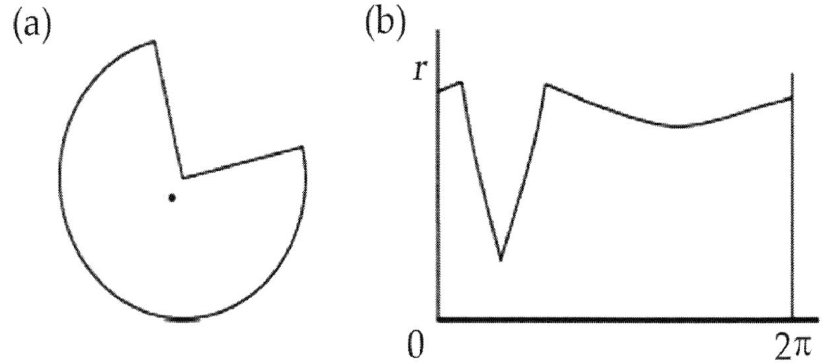

Figure 4.4: Centroidal-profile technique for protest-shape investigation. (a) An original object. (b) Centroidal profile, which postures undue issues of elucidation, (*Source: G. Caldwell Darwin (2013), Robotics and Automation in the Food Industry: Current and Future Technologies, Woodhead Publishing*).

Figure 4.5: The skeleton idea. Here the skeleton is planned to demonstrate the path the pen travelled while drawing the shapes, (*Source: G. Caldwell Darwin (2013), Robotics and Automation in the Food Industry: Current and Future Technologies, Woodhead Publishing*).

4.2.3. Morphology

While inherently morphology implies much the same as shape examination, it became out of an alternate approach–that of the utilization of basic window activities to change shapes with the goal that they could be separated to find specific structures. The beginning stage was that of disintegrating objects so the peripheral layer of pixels was killed, or widening them, so an extra limit layer was included (Figure 4.6): the two strategies could be connected isotopically or directionally, for instance, parallel to the picture x – hub. The reason for these activities turns out to be clear if an arrangement of vertical striations on a smooth surface is to be found. Disintegrating the picture evenly would take out the striations, and after that enlarging the picture would restore the picture to its unique state, however without the striations. At that point, subtracting the last picture from the first picture would uncover the striations that had been wiped out.

Such a technique would for instance uncover any scratches on a PC plate or cleaned metal surface. Figure 4.7 shows brings about a comparative however additionally demanding case–that of scratches on a wood surface. In spite of the fact that the calculation works by wiping out the scratches and afterward subtracting in order to recoup them specifically, the general impact is to sift through the majority of the (even) wood-grain markings, along these lines disengaging the vertical scratches.

Morphology is detailed scientifically, in light of the fact that doing morphological activities effectively expects them to be separated into essential tasks and after that examined to decide the most ideal method for collecting them into valuable methodology. In this way the way toward enlarging picture An utilizing a (perhaps directional) cover B is composed as $A \oplus B$, while the way toward dissolving An utilizing veil B is composed as $A \ominus B$. As per the sorts of handling demonstrated above, enlargement took after by disintegration is depicted as 'shutting,' spoke to by '•,' and disintegration took after by widening is portrayed as 'opening,' spoke to by '∘.' These tasks are composed separately as:

$$A \bullet B = (A \oplus B) \ominus B \quad (4)$$

$$A \circ B = (A \ominus B) \oplus B \quad (5)$$

With this documentation, we can immediately write the process of identifying the striations described above as the following:

$$S = A – A \circ B \quad (6)$$

While this approach demonstrates how one kind of blame can be found, discovering splits in an egg may well require a task of the accompanying sort:

$$T = A \bullet B – A \quad (7)$$

This adds up to filling in the splits and then comparing with the original image in order to accurately locate them.

Space does not allow a full investigation of the science basic morphology. Nonetheless, we can record two decides that give facilitate knowledge into the thoroughness it forces:

$$(A \oplus B) \oplus C = (A \oplus C) \oplus B \quad (8)$$

$$(A \bullet B) \bullet B = A \bullet B \quad (9)$$

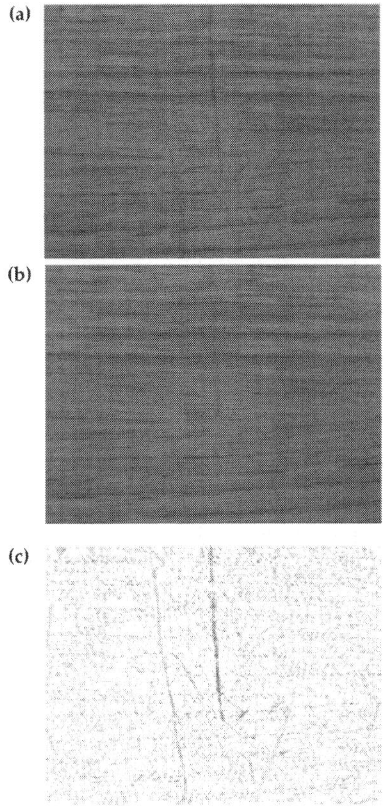

Figure 4.6: Use of morphology for scratch detection (a) Original picture of a scratched wood surface (b) Removal of scratches by a level shutting algorithm. (c) Using subtraction to recoup the scratches, (*Source: G. Caldwell Darwin (2013), Robotics and Automation in the Food Industry: Current and Future Technologies, Woodhead Publishing*).

These can be understood as 'it doesn't make a difference in what arrange two expansion tasks are completed' and 'once an opening has been filled in, it stays filled in, so rehashed shutting is inefficient and superfluous.' To make best utilization of morphology, the sizes and directionalities of the essential disintegration and enlargement activities should be chosen precisely. It ought to likewise be noticed that dark scale adaptations of these fundamental tasks have been conceived, some of these being as basic as greatest and least activities inside pixel windows.

By and large, morphology gives us an extra arrangement of devices that are able to do productively finding a scope of highlights in advanced pictures, including especially deformities, for example, gaps, concavities, breaks, spots, prominences and hairs–and furthermore scratches, as said prior.

4.2.4. Efficient Object Location

In Section 4.2.2 we considered the effective area of little highlights, epitomized by edge, gap and corner focuses, utilizing the coordinated channel worldview. Tragically, it is hard proportional up this technique to find vast items, (a) due to varieties in enlightenment over the articles, (b) on the grounds that variegated foundations will make distinctive parts of their limits endure low or even altered complexity, and (c) due to the improved probability of fractional obscuration or bending. Consequently protest area strategies must be intended to be significantly heartier. (The requirement for this was at that point implied in Section 4.2.2 with regards to the centroid-profile method) (Figure 4.7).

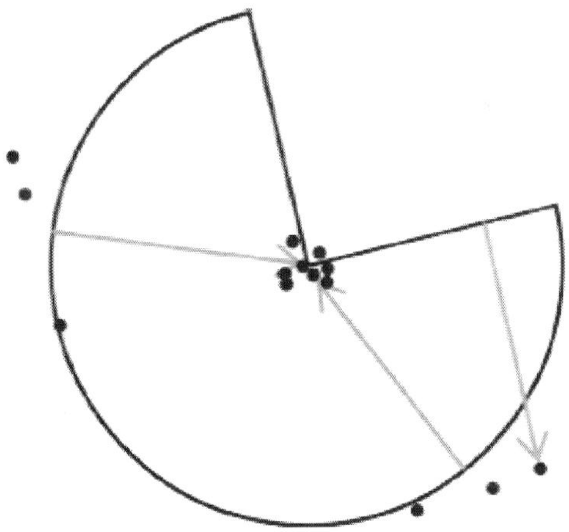

Figure 4.7: Guideline of the Hough change. Here an incomplete circle is found powerfully: insignificant votes are scattered and don't meddle with the protest area, (*Source: G. Caldwell Darwin (2013), Robotics and Automation in the Food Industry: Current and Future Technologies, Woodhead Publishing*).

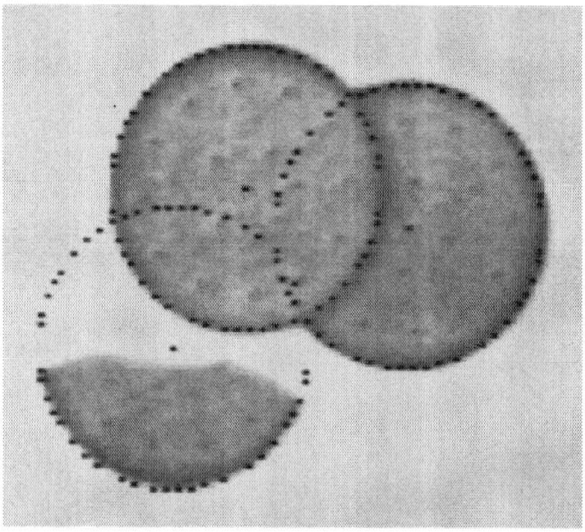

Figure 4.8. Use of Hough change for biscuit area. The broken biscuit and the halfway impediment of another don't keep any of the biscuits from being found dependably, (*Source: Reproduced from Davies (1984) with permission, IFS International Limited, 1984*).

To accomplish an abnormal state of power, a radical new approach is required: it winds up important to induce the nearness of items from fractional confirmation. The Hough change approach accomplishes this by a voting plan, in which just positive proof for objects is tallied. The most straightforward case of this is the discovery of roundabout articles. All edge focuses in the picture are made to vote in favor of a hover focus by moving a separation equivalent to the gathered sweep along the edge normal and making a choice in an optional picture space called a parameter space (Figure 4.8). On the off chance that exclusive p pixels are found on the circle limit, p votes will at present be thrown at the circle focus, and a pinnacle of this weight will be found at that situation in the parameter space. Consequently the circle can be discovered vigorously, disregarding the way that a lot of its limit might be lost (Figure 4.9) (Davies, 1984). Likewise, the issue of obscure circle span might be illuminated in different routes, for instance, by attempting an arrangement of sweep esteems.

The strategy can be reached out to enable straight lines and circles to be found, or different shapes made out of these shapes. It can likewise be utilized to find discretionary shapes as long as these can be indicated precisely. As demonstrated over, the energy of the approach relies upon amassing positive confirmation: mistaken proof should be overlooked as it would just serve to inclination the outcomes.

The technique known as RANSAC works rather in an unexpected way. Sets of information indicates are utilized produce speculations and these are each tried to figure out which gives most noteworthy concurrence with the picture information. In spite of its distinctive methodology, it really encapsulates a voting plan, looking for the arrangement giving the best quantities of votes.

A very unique approach called diagram coordinating has as often as possible been utilized when the articles being looked for are described by sets of point features, for example, corners or openings–as on account of pivots, sections and certain kinds of scone. In such cases it is important to scan for subsets of the watched point includes that precisely coordinate with subsets in the protest layout. The arrangement that is taken to be the most plausible one is the biggest subset that gives a correct match. Here once more, we see this is a sort of voting plan, in which the best arrangement is the one for which the best number of votes are recorded. Sadly, applying this thought thoroughly acquires a computational load that is exponential in the quantity of highlights to be coordinated. This is on the grounds that the meaning of a subset is very broad thus all subsets of all sizes and pieces must be attempted. Exponential issues, for example, this have the property of requiring paltry measures of processing for say three model highlights being looked for in a picture containing five highlights (in which case there are only ten conceivable outcomes to be checked), yet the measure of figuring heightens lopsidedly when six model highlights must be looked for in a picture containing ten highlights (in which case there are 210 potential outcomes to be checked), while for nine model highlights in a picture containing 15 includes there are an enormous 5005 potential outcomes to be checked. Luckily, for 2D understanding of roll pictures, this issue can be unravelled carefully utilizing a unique type of the Hough change, as appeared in Section 4.3.2.

4.2.5. Texture Analysis

Object surfaces are said to have texture when they are unpleasant, woven or made out of numerous little particles or strands. In such cases they gain trademark power designs that have fluctuating degrees of normality and arbitrariness. For the more consistent examples, periodicities will be quantifiable, however they may differ significantly finished the surface; they will likewise display solid directionality, which might be show in a few headings–with respect to the weave of a texture. A heap or sand or seeds will have a significantly more arbitrary surface than a texture, and won't be directional.

In view of the level of haphazardness that is available in many surfaces, estimating surface examples presents issues of measurements, so it ends up important to normal over an adequate territory with a specific end goal to touch base at a precise appraisal of the surface. This is vital with the end goal of acknowledgment, and for outline of question limits–and furthermore for the area of any flaws, absconds or remote bodies. Getting a remarkable characterization of a surface can include impressive calculation, as every pixel must be gotten to ordinarily with a specific end goal to quantify textural cognizance over various separations and headings. In any case, alternate ways might be worthy in specific conditions. For instance, on a theoretical 'bixit' line, just bixits ought to be available, so basic tests for (a) nearness of a bixit, (b) nearness of a non-bixit, (c) nearness of a flawed bixit might be adequate. Maybe the easiest test is that for 'hecticness,' which might be estimated as the quantity of edge focuses, or the quantity of minor spots per unit territory (spots can be identified utilizing the second veil in Section 4.2.2) (Figure 4.10).

(a)

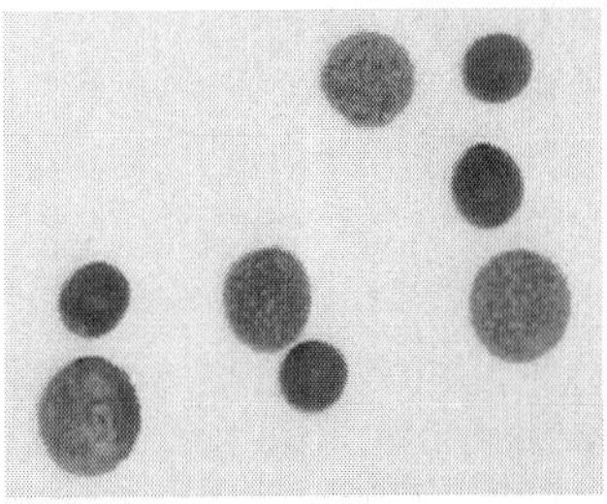

(b)

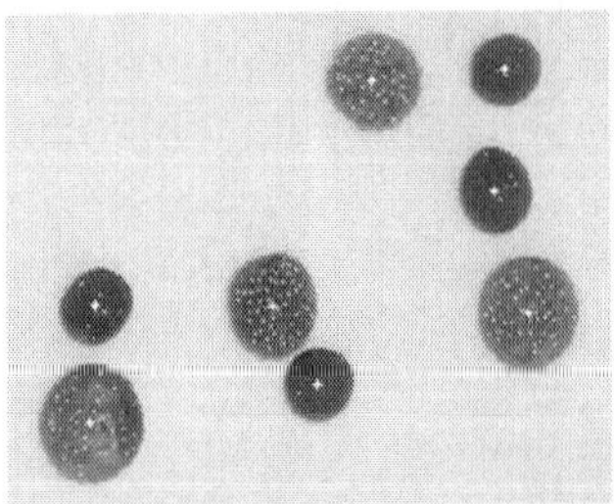

Figure 4.9: Basic surface discovery utilizing the 'hectiness' idea (a) An unique picture of assult and charlock seeds (b) Light features in the seeds have been found and are shown by white spots. In the event that more than about 20 white dots appear in the vicinity of the object centers (shown by white crosses), the relating seed is dependably named an assault seed, (*Source: G. Caldwell Dar-*

win (2013), Robotics and Automation in the Food Industry: Current and Future Technologies, Woodhead Publishing).

4.2.6. Statistical Pattern Recognition

As showed in Section 4.1, picture handling alone won't allow solid judgments to be made about classes of articles. In any case, it will allow the collection of information that would then be able to be utilized as a part of an entire machine vision framework that is equipped for settling on valuable choices. Here it will be critical not exclusively to arrange and sort objects into one of numerous classifications, or to put items into dismissal containers, yet additionally to know the 'false positive' (FP) and the 'false negative' (FN) rates, with the goal that the classifier framework can be enhanced. It ought to be noticed that there will be costs related with incorrect choices and it will for the most part be important to limit these expenses instead of the blunders themselves. Such contemplations are critical when endeavoring to dispense with noxious molds, for example, ergot for instance.

To accomplish this, a theoretical example acknowledgment plan will be required. One of the least difficult such plans is the supposed 'closest neighbor' strategy, which might be executed by setting up a 'preparation set,' comprising of a database of examples of each class, including the element estimations that have been made on them. At that point, amid testing, the highlights of a test design are contrasted and those of each preparation design, and the class of the example giving the nearest coordinate is relegated to that test design. This strategy is computationally concentrated, in light of the need to make correlations with an expansive number of put away preparing set examples when testing. In any case, if an adequately huge preparing set is utilized, the technique has the benefit of a near perfect (Bayes) blunder rate.

There are numerous techniques giving preferable execution over the closest neighbor strategy delineated previously. Be that as it may, whichever is utilized, it will be important to mastermind the best tradeoff between the quantities of FPs and FNs?

To guarantee setting up the framework accurately, it is helpful to plot the 'collector working trademark (ROC), which is a plot of the genuine positive versus the FP rate (Figure 4.11). Be that as it may, it can be hard to do this thoroughly, if by any means, while endeavoring errands, for example, the ideal identification of uncommon imperfections.

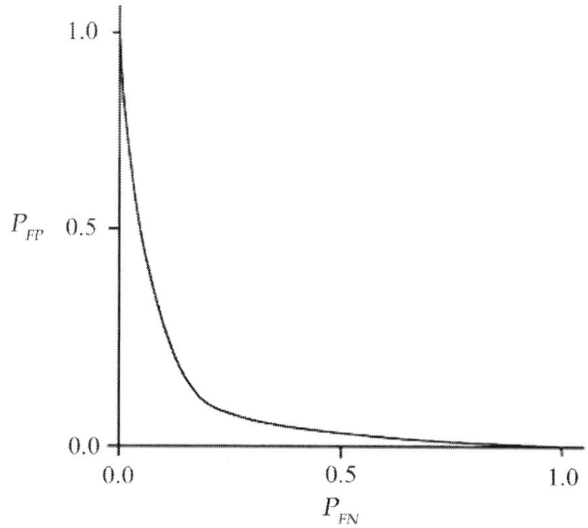

Figure 4.10: ROC. This diagram of likelihood of false positives (P_{FP}) against false negatives (P_{FN}) is a case of a ROC bend. A common working point is where $P_{FP} = P_{FN}$, (*Source: G. Caldwell Darwin (2013), Robotics and Automation in the Food Industry: Current and Future Technologies, Woodhead Publishing*).

4.2.7. The Variety of Image Modalities

Up until now, the talk has focused on the examination of dark scale and double pictures. Truth be told, present day cameras ordinarily have shading securing abilities and this can be gigantically valuable, however it has a tendency to include more mind boggling preparing systems. Here RGB input is regular, yet it is as often as possible helpful to change over RGB pictures into different arrangements, for example, HSI (H is the 'shade,' which is the principle shading part, S is the 'immersion' or level of shading versus whiteness, and I is the force) before continuing with nitty gritty picture examination. The specific favorable position offered by HSI is that the impacts of any arbitrary varieties in light will be packed in the I divert and scarcely resounded in the S and H channels. This implies while watching apples, for instance, the H channel ought to demonstrate unambiguously whether green or red patches are being watched.

For a few purposes, infrared imaging or warm imaging can be helpful. The last applies to the location of live protests, for example, creepy crawlies or slugs, while the previous can be helpful for distinguishing the emanation groups of specific particles in vegetation or substance. Bright light can likewise demonstrate follows from live issue, for instance, fingerprints, rodent pee, ergot and molds that create aflatoxin. While multispectral picture information, covering unmistakable light, infrared and the microwave locales, is as often as possible accessible from satellite information, this methodology might be excessively mind boggling for most sustenance applications, however methods for joining the yields of the different stations to acquire sound data, for instance, utilizing main segments investigation (PCA), are profoundly powerful.

The other methodology that is of huge incentive in the sustenance business is that of X-beams. Its entering powers are notable, yet the possibility of radiation harm to the human body carries with it wellbeing issues and the prerequisite for overwhelming screening. The outcome is devoted apparatus that is substantial, expensive and non-compact. Be that as it may, as of late huge changes have been made to semiconductor direct cluster identifiers and in the meantime double vitality recognition (DEXA) has been accomplished and is in wide utilize. The estimation of DEXA innovation is that by taking the proportions of the double vitality indicator yields, the thickness of an example can to a great extent be counteracted. Consequently cautious investigation can uncover more nitty gritty structure in the X-beam pictures; therefore location of outside bodies turns out to be significantly more solid. Creating and preparing X-beam pictures are somewhat specific subjects and can't be canvassed in full here. For additionally subtle elements and direction the peruser is alluded to Batchelor et al. (2004) and Kroger et al. (2006).

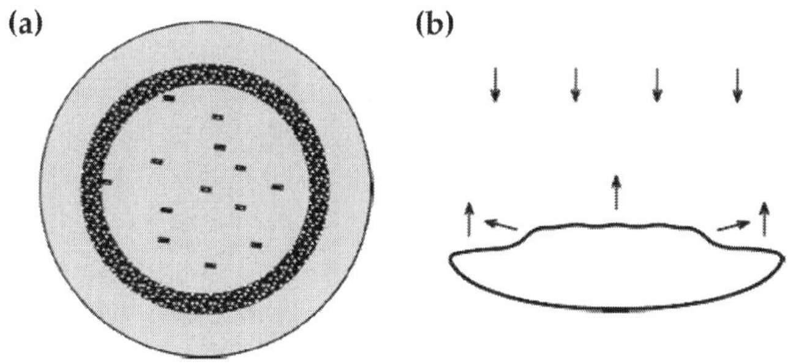

Figure 4.11: Chocolate cakelets and methods of enlightenment. With this mode of brightening, a dark ring shows up over the edge of jam position (a) Top perspective of cakelet (b) Side perspective of cakelet demonstrating profile and bearings of enlightenment and reflected light, (*Source: Reproduced from Davies (2001) with permission; Editions Cepaudues, 2001*).

4.3. APPLICATIONS OF MACHINE VISION TECHNOLOGY

Computer vision systems are being used increasingly in the food industry for quality assurance purposes. It offers the potential to automate manual grading practices thus to standardizing techniques and eliminating tedious inspection tasks. Computer vision has proven successful for the objective; online measurement of several food products with applications ranging from routine inspection to the complex vision guided robotic control (Gunasekaran, 1996). Kanali et al. reported that the automated inspection of produce using machine vision not only results in labor savings, but can also improve quality assurance objective of food industry (1998).

4.3.1. Meat and Meat Products

Visual discernible characteristics are used extensively for the quality appraisal of meat and meat products applied to processes from the initial grading through to consumer influences. The correct appraisal of meat quality (i.e., to execute the consumer's needs) is crucial element with in the meat industry. Although there are several factors that affect the perception of taste, tenderness is considered the most important characteristic. Locht et al. examined the development of full color machine vision analysis in the food industry (1997). This Win Grain system analyzed sirloin steak for fat content and chives for meat for component percentage composition and findings indicated that the system provides objective and quantifiable in-line visual information. Cortez and Portelinha exhibited a feature selection procedure, based on a Sensitivity Analysis are combined with a Support Vector Machine, in order to predict lamb meat tenderness (2006). This real-world problem is defined in terms of two difficult regression tasks, by modeling objective (e.g., Warner-Bratzler Shear force) and subjective (e.g., human taste panel) measurements. In both cases, the proposed solution is competitive when compared with other neural (e.g., Multilayer Perceptron) and Multiple Regression approaches. McDonald and Chen explored betimes work in the area of image based beef grading (1990). Based on reflectance characteristics, they discriminated between fat and lean in the Longisimus muscle and generated binary muscle images. Li et al. measured image texture as a means of predicting beef tenderness (1997). Color, marbling and textural features were extracted from beef images and analyzed using statistical regression and neural networks. Their findings indicated that textural features were a good indicator of tenderness. Image analysis was also used for the classification of muscle type, breed and age of bovine meat (Basset et al., 2000). It investigated as an effective tool for evaluating fresh pork color and pork quality (Tan et al., 2000; Lu et al., 2000). Color image skins were obtained from segmented images. Both statistical and neural network models were employed to predict color scores by using the image features as inputs and then compared to the sensory scores of a trained panel. More than 90% agreement was found between the vision system and the panelists (Tan et al., 2000).

4.3.2. Milk and Milk Products

Presently, inspections are tedious and subjective based on empirical and sensory assessments, hence, the use of computer vision has been implemented for these tasks. With numerous foods containing cheese available on the market the necessity for the evaluation of cheese functional properties continues to grow. The evaluation of the functional properties of cheese is assessed to ensure the necessary quality is achieved, especially for specialized applications such as consumer food toppings or ingredients. Wang and Sun developed a computer vision method to evaluate the melting and browning of cheese (2001). This unique non-contact method was employed to analyze the characteristics of cheddar and mozzarella cheeses during cooking and the results showed that the method provided an objective and easy approach for analyzing cheese functional properties (Wang and Sun, 2002ab).

Ni and Gunasekaran developed an image-processing algorithm to recognize individual cheese shred and automatically measure the shred length (1995). It was found that the algorithm recognized shreds well, even when they were overlapping. It was also reported that the shred length measurement errors were as low as 0.2% with a high of 10% in the worst case. Gunasekaran and Ding obtained 3-D images of fat globules in cheddar cheese from 2-D images (1999). This enabled the in situ 3-D evaluation of fat globule characteristics so as the process parameters and fat levels may be changed to achieve the required textural qualities. Jelinski et al. used computer vision to determine the distribution and quantity of spices in pasteurized cheese (2007). For localization of the spices they used threshold that is appropriate in this case since the color of spices differs from the cheese color (paprika, pepper, etc.). The results show that computer vision can be successfully applied to monitor the production of cheese with spices included. Wang and Sun studied oiling off property of cheese with computer vision and implementation of the traditional method with the formation of ring of oil released onto filter paper (2004).

4.3.3. Aquatic Food and Food Products

Computer vision systems promise faster, cheaper and more consistent operations than manual methods. Since the aquatic foods industry requires skilled seasonal workers to perform repetitive and arduous tasks, automation is desirable. Considering the limited existing labor pools, the current congressional reduction in foreign temporary labor supply, a computer vision system could be an alternative for high speed grading. Visual quality of aquatic foods (size, shape, and color) has a direct influence on their value and acceptance. Computer vision can evaluate all these attributes.

4.4. APPLICATION IN EVALUATION OF QUALITY AND SAFETY

The system offers the potential to automate manual grading practices thus standardizing techniques and eliminating tedious human inspection tasks. Machine vision has proven successful for the objective; online measurement of several food products with applications ranging from routine inspection to the complex vision guided robotic control (Gunasekaran, 1996).

4.4.1. Fruit and Vegetables

Shape, size, color, blemishes and diseases are important aspects, which need to be considered when grading fruits and vegetables (Kanali et al., 1998). Color provides valuable information in estimating the maturity and examining the freshness of fruits and vegetables. The automated inspection of produce using machine vision not only results in labour savings, but can also improve quality inspection objective (Kanali et al., 1998). Machine vision is being implemented for the automated inspection and grading of horticulture produce to increase product throughput and to improve objectivity of the industry (Brosnan and Sun, 2002). Narendra and Hareesh reported variety classification, defects detection and segmentation, identification of stems and calyxes and sugar content prediction in apples (2010). They also investigated the use of computer vision in sorting fresh strawberries based on size and shape, shape characteristics of papaya, tomato color, size, shape and abnormalities with inner quality, inspect and grading of mushrooms, separation method for grading of potatoes, color of lemon. Marakeby et al. also studied on fast quality inspection of food products using computer vision (2013).

Golmohammadi developed a machine vision system for online evaluation of potato sorting (2013). To separate potatoes, an accelerator system based on Pneumatic control valves and air pressure was used. As the system is online, all of timings related to image processing, product transferring and accelerators operation were calculated and applied based on the speed of the potatoes being fed to processing and sorting system. For the primary evaluation of the system, 100 kg of potatoes were manually sorted into three categories. Then the same potatoes were fed to the system, which sorted them out with an accuracy of 97.4% with the speed of two potatoes per second.

External quality is considered of paramount importance in the marketing and sale of fruits. The appearance i.e., size, shape, color and presence of blemishes influences consumer perceptions and therefore determines the level of acceptability prior to purchase. The consumer also associates desirable internal quality characteristics with a certain external appearance. This relationship of internal quality to external quality has a considerable impact on future purchases. To meet the quality requirements of customer, computer vision is being implemented for the automated inspection and grading of fruit to increase product throughput and improve objectivity of the industry. Computer vision has been used for such tasks as shape classification, defect detection, quality grading and variety classification.

4.4.2. Beverages

The classification of beverages was conducted using three approaches: by using the electronic nose only, by using machine vision alone and by using the combination of electronic nose and machine vision. A supervised support vector machine was used to classify beverage according to their brands. Results show that by using the electronic nose alone and machine vision alone were able to respectively classify 73.7% and 92.9% of beverage correctly. When combined the electronic nose and machine vision, the classification accuracy increased to 96.6%. Based on results, it can be concluded that combination of both is able to extract more information from sample, hence improving the classification accuracy.

Yin and Panigrahi used computer vision to evaluate the internal texture of French fries (1997). A set of 150 sample images were categorized into normal and hollow classes by three algorithms. The co-occurrence matrix gave the best results with 100% accuracy.

Kondo, Ahmada, Montaa, and Muraseb evaluated the sugar content and acid content of Iyokan orange fruit using a machine vision system (2000). Images of 30 Iyokan orange fruits were acquired by a color TV camera. Features representing fruit color, shape and roughness of fruit surface were extracted from the images. The correlation coefficient between measured sugar content values and predicted sugar content values was 0.84 while the correlation coefficients between measured pH values and predicted pH values was 0.83 for the developed system.

Ruiz, Molto, Juste, Pla, and Valiente also used computer vision for the location and characterization of stem calyx area on mechanically harvested oranges (1996). Color segmentation, contour curvature analysis and a thinning process were the image analysis techniques used in this study, yielding accuracies of 93%, 90% and 98% respectively, for stem absence or presence.

4.4.3. Confectionaries and Other Processed Food Products

Davidson et al. measured the physical features of chocolate chip biscuits, including size, shape baked dough color, and fraction of top surface area that was chocolate chip using image analysis. Four fuzzy models were developed to predict consumer ratings based on three of the features (2001). Vision systems have been used in inspection applications ensuring that molds used in the production of confectionery are empty and properly cleaned. The internal and external appearances contribute to the overall impression of the products quality consequently such characteristics have been evaluated by computer vision.

Scott described a system which measures the defects in baked loaves of bread, by analyzing its height and slope of the top (1994). The internal structure (crumb grain) of bread and cake was also examined by machine vision (Sapirstein, 1995). The brightness, cell density, cell area and uniformity of the grain analyzed indicated that even the most minor deviations from the required specifications can be identified through machine vision system, allowing corrective measures in the bakery to be taken sooner. In a more recent study, digital images of chocolate chip cookies were used to estimate physical features such as size, shape, baked dough color and fraction of top surface area that was chocolate chip (Davidson, Ryks and Chu, 2001).

Abdullah et al. developed a prototype-automated system for visual inspection of muffins and they reported that it was able to correctly classify 96% of pre-graded and 79% of ungraded muffins with an accuracy of greater than 88% (2000). Visual features such as color and size indicate the quality of many prepared consumer foods. Sun investigated this in a research on pizza in which pizza topping percentage and distribution were extracted from pizza images (2000). Then it was found that the new region-based segmentation technique could effectively group pixels of the same topping together and the topping exposure percentage can be easily determined with accuracy 90%. To avoid the wrong application of quality assessment by visual based human perception, computer vision has been widely used in the assessment of confectionary products so far. Here are some cases where machine vision system was applied.

Machine vision can be used in the food processing industry to:

(a) Classify objectively potato chips according to their color in different categories,

(b) Identify broken crackers.

(c) Determine edge.

(d) Ensure uniform baking/cooking and color development.

(e) Detect defects in color, shape, topping and packaging.

(f) Find pizza topping percent and distribution.

(g) Visually inspect chocolate chip cookies and muffins.

4.4.3.1. Cheese

The evaluation of the functional properties of cheese is assessed to ensure the necessary quality is achieved, especially for specialized applications such as consumer food toppings or ingredients. Wang and Sun developed a computer vision method to evaluate the melting and browning of cheese (2001). This novel non-contact method was employed to analyze the characteristics of cheddar and mozzarella cheeses during cooking and the results showed that the method provided an objective and easy approach for analyzing cheese functional properties. Ni and Gunasekaran (1998) developed an image-processing algorithm to recognize individual cheese shred and automatically measure the shred length. It was found that the algorithm recognized shreds well, even when they were overlapping. It was also reported that the shred length measurement errors were as low as 0.2% with a high of 10% in the worst case.

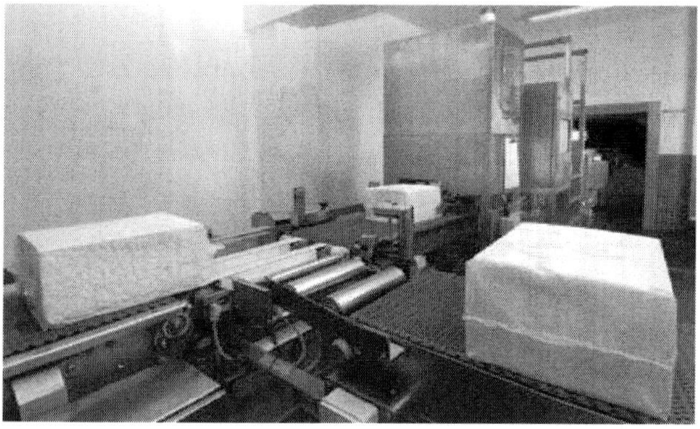

Figure 4.12: Machine vision system inspecting a cheese with faulty seal, (*Source: https://www.vision-systems.com/articles/print/volume-18/issue-9/features/vision-helps-dairy-spot-slack-cheese-bags.html*

In the figure above, as the cheese with the faulty seal passes and exits the vision inspection station, its presence is detected by a laser sensor).

4.4.3.2. Noodles

The important of noodles in the Asian diet is very significant, as currently they account for 30–40% of most countries' wheat consumption (Miskelly, 1993). The majority of Asian noodle manufacturers noodle products to be derived from white seed-coated wheat as compared to red seed-coat material. Initial research had used image analysis to characterize the white wheat products (Hatcher and Symons, 2006). As noodles appearance and color are the consumer's initial quality parameters, there is a preference for manufacturers to use high quality, low yield patent flours and free from various contaminants of the flour (i.e. bran and germ), as they result in brighter noodles. The standard methodology to assess bran contamination has been ash content, the use of auto fluorescence by different wheat seed tissue (Munck et al., 1979). Hatcher demonstrated that the image analysis could be effectively used to quantify, measure, and discriminate varietal differences in white seed coated wheat varieties in preparation of yellow alkaline noodles prepared from high-quality patent flour (2001).

4.4.3.3. Potato Chips and French Fries

The images of commercial potato chips were evaluated for various color and textural features to characterize and classify the appearance and to model the quality preferences of a group of consumers. The most frequently used color model is the RGB model, in which each sensor captures the intensity of the light in the red (R), green (G) and blue (B) spectra respectively, for food quality evaluation (Du and Sun, 2004). However, the features derived from the image texture contained better information than color features to discriminate both the quality categories of chips and consumers preferences.

Pedreschi et al. recently designed and implemented a computer vision system to measure representatively and precisely the color of highly heterogeneous food materials, such as potato chips (2006). Leon et al. developed a computation color conversion procedure that allows the obtaining of digital images from the RGB images by testing five models: linear, quadratic, gamma, direct and neural network (2006). After the evaluation of the performance of the models, the neural network model was found to perform the best. Mendoza et al. developed a stepwise logistic regression model in another study and they reported that that it was able to explain 86.2% of the preferences variability when classified into acceptable and non-acceptable chips (2007).

4.5. DETECTION OF MICROBIAL AND FECAL CONTAMINATION IN FOODS

4.5.1. Cereal Grains

Fungal infections and insect infestations are the two major causes of reduced market value of cereal grains. *Aspergillus, Penicillium, and Fusarium* are the main toxigenic species causing contamination of cereal grains. If fungal growth is not detected in its early stages, some strains can produce mycotoxins affecting both humans and animals (Shephard, 2008). To decrease health risks and economic losses, cereal grains are regularly monitored for mycotoxin contamination. Though products with surface fungal growth can be removed, mycotoxins already produced in the grains are not destroyed or removed during processing steps. Thus, early detection of microbial infections of cereal grains must be carried out in the field to reduce impact of the disease.

4.5.2. Wheat

Fusarium head blight or scab is a prominent fungal infection in wheat that can result in production of the toxindeoxynivalenol (DON) in wheat kernels. Visual screening is a common, subjective, labor-expensive method with inherent inconsistencies as characteristics of the disease are not always evident to the naked eye. Delwiche and Hareland inspected wheat scab by NIR reflectance (1,000–1,700 nm) and accurately classified 95% of healthy and scab-damaged kernels (2004). DON and ergosterol were also quantified using NIR spectroscopy in single wheat kernels and showed accurate classification as determined by expert inspectors (Dowell et al., 1999). DON levels exceeding 700 ppm could be identified in individual kernels by this ethod. Fusarium-damaged kernels containing DON levels below 60 ppm could not be estimated using NIR absorption spectra (350–2500 nm) (Peiris et al., 2009).

Wheat can develop fungal infections such as *Aspergillus* glaucus, *A. niger*, and *Penicillium* spp. during storage. Chelladurai et al. identified fungal infection in stored grains after heating for 3 min at 90°C using a plate heater in an infrared thermal imaging system (2010). Pairwise linear and quadratic discriminant analysis models were applied and accurately classified all of the healthy samples and 97% of the infected samples. The changes in temperature patterns resulted from higher reducing sugar and protein contents in infected kernels.

4.5.3. Seafood

Foods containing raw seafood such as "sushi" or "sashimi" are prone to pathogenic microorganisms and may cause foodborne illnesses. Fillets of marine fish, molluscs, crustaceans, fish roe, or other seafood are sometimes used in its raw state to make these products. *Vibrio parahaemolyticus* is commonly found in seafood, whereas *Staphylococcus aureus* and Salmonella species may enter the food by cross-contamination or improper handling during food processing (Gram and Huss, 1996).Considering the health implications caused by these pathogenic microorganisms, it is critical to study the applications of non-destructive computer vision techniques to detect the microbial quality of raw fish products.

4.5.4. Corn, Maize, and Barley

Fusarium spp. are frequently associated with destructive diseases in maize plants, particularly ear rots. Presently, effective methods to remove entire corn kernels that are infected with aflatoxin are lacking. Mycotoxin analysis using the most modern chromatographic techniques are accurate but expensive and do not provide real-time responses. ELISA-based methods used to detect aflatoxin in corn samples are also inconvenient and impracticable. Chemical and immunological methods may provide accurate toxin quantification but are expensive, time-consuming, equipment dependent, and skill demanding. Hence, machine vision techniques have been tested for correlating imaging data to predetermined mycotoxin levels in cereals and grains. Corn kernels infected with *A. flavus* exhibit a bright greenish yellow fluorescence (BGYF) when exposed to high intensity UV light (Yao et al. 2006). Targeted corn samples were examined and kernels were classified as BGYF positive or negative by a hyperspectral BGYF technique under ultraviolet-A (UV-A) excitation. Gordon et al. also classified infected corn kernels exhibiting BGYF in the germ using Fourier-transform infrared (FTIR) photo acoustic spectroscopy (1997). Spectral features were selected visually from spectra of infected and uninfected kernels. However, this detector could only accommodate a single kernel at a time. This limitation was overcome by using transient infrared spectroscopy to detect *A. flavus* contamination in a moving bed of corn kernels.

4.5.5. Fruits

Apple and unpasteurized apple juice or cider that can be contaminated with pathogenic Escherichia coli is recognized as an important food safety issue. Apples with cuts and bruises are prone to bacterial invasion and spoilage. The main source of fecal contamination on fruits is when they come into contact with the ground or unclean handlers or containers containing domestic animal feces. Usually, fecal contamination on apples is detected visually on inspection tables. Kim et al. used hyperspectral and multispectral fluorescence imaging to produce reflectance images of apples with varying colors from green to red (Red Delicious, Gala, Fuji, and Golden Delicious) and experimentally contaminated with fresh dairy feces (2002a;b).

Simple thresholding was enough to detect thick smears of fecal contaminants on all apple cultivars. PCA identified 450, 530, 685, and 735 nm as the four optimal multispectral bands to classify contaminated apples. However, the reflectance imaging method was not sensitive enough to detect minor fecal spots though effective identification was obtained when an expensive Ultraviolet (UV) excitation tool and dual band ratio algorithms were applied (Liu et al., 2007). While using laser-induced fluorescence imaging, it was possible to detect dairy feces (20 μl of 1:20 dilution) applied to the Red Delicious apple calyx even after washing (Lefcourtet al., 2005).

Early detection of fungal infections in citrus fruits is extremely important in packing lines to reduce economic losses and to maintain fruit quality.

Generally, UV illumination is employed to detect and remove rotten fruits manually that can cause serious health risk to humans. Mandarin rottenness caused by *Penicillium digitatum* was detected using a hyperspectral imaging system applied in a closed space (Gómez-Sanchis et al., 2008). A number of band selection and pixel classification methods were tested and the highest success rate of 91 % was achieved using a genetic algorithm linear discriminant analysis band selection method with the classification and regression tree pixel classification method and a number of liquid crystal filters.

4.6. ADVANTAGES AND DISADVANTAGES OF MACHINE VISION TECHNOLOGY

Machine vision is seen as an easy and quick way to obtain data that would be otherwise difficult to obtain manually. Various advantages indicated were the generation of precise descriptive data, objective operation, reduction of tedious human contribution and automation of many labor intensive processes, consistency, efficiency and cost effectiveness. Gerrard et al. recognized that this technology provides a rapid, alternative means for measuring quality consistently (1996). Another benefit of computer vision systems is the non-destructive, least intrusive natures and less interruptive manner in which information is attained, making inspection unique with the potential to assist humans involving visually intensive work. Tarbell and Reid noted that an amazing feature of a vision system is that it can be used to create a permanent record of any measurement at any point in time (1991). Hence archived images can be recalled to look at attributes that were missed or previously not of interest.

Quality control in combination with the increasing automation in all fields of production has led to the increased demand for automatic and objective evaluation of different products. Sistler confirms that computer vision meets these criteria and states that the technique provides a quick and objective means for measuring visual features of products (1991). Lu et al. found computer vision inspection of food products to be consistent, efficient and cost effective (2000). Gunasekaran and Ding agreed that vision system was becoming increasingly popular in the food industry, and pointed out that its development was at a level where it is a robust and competitively priced sensing technique (1993). Disadvantages include the need for defined and consistent lighting, calibration requirements, the difficulties encountered with overlapping objects or objects that are difficult to separate from the background, speed of operations, and when both sides of a food need to be evaluated. An ambiguity of computer vision is that its results are influenced by the quality of the captured images. Often due to the unstructured nature of typical agricultural settings and biological variation of plants within them, object identification in these applications is considerably more difficult. Also if the research or operation in being conducted in dim or night conditions, artificial lighting is needed.

4.7. ELECTROMAGNETIC RADIATION

Electromagnetic radiation has properties in common with other forms of waves such as reflection, refraction, diffraction, and interference. Moreover, it may be characterized by the frequency with which it varies over time or by its wavelength. Electromagnetic radiation, however, has particle-like properties in addition to those associated with wave motion. It is quantized in that for a given frequency, its energy occurs as an integer times h, in which h is a fundamental constant of nature known as Planck's constant. A quantum of electromagnetic energy is called a photon. Visible light and other forms of electromagnetic radiation may be thought of as a stream of photons, with photon energy directly proportional to frequency. The visible part of the electromagnetic spectrum is, by definition, radiation visible to the human eye. Other detection systems reveal radiation beyond the visible regions of the spectrum and these are classified as radio wave, microwave, infrared, ultraviolet, X-ray and γ-ray. These regions are illustrated in the Figure 4.13.

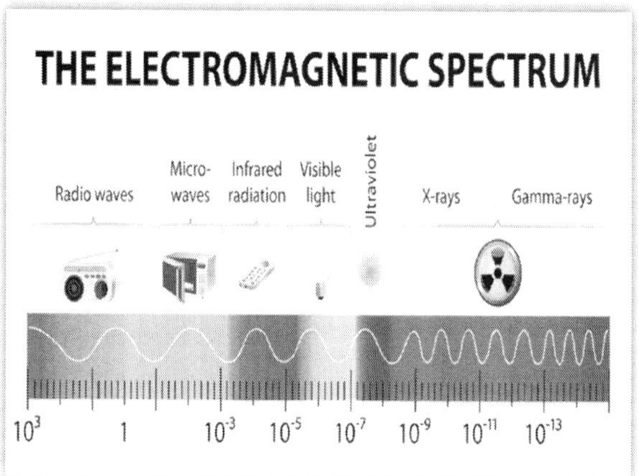

Figure 4.13: Electromagnetric spectrum, (*Source: https://www.tnuda.org.il/en/physics-radiation/what-radiation/types-electromagnetic-radiation*).

The electromagnetic spectrum and the varied interactions between these radiations and many forms of matter can be considered in terms of either classical or quantum theories. The nature of these various radiations has been interpreted by Maxwell's classical theory of electro- and magneto-dynamics–hence, the term electromagnetic radiation. According to this theory, radiation is considered as two mutually perpendicular electric and magnetic fields, oscillating in single planes at right angles to each other. These fields are in phase and are being propagated as a sine wave (see the Figure 4.14).

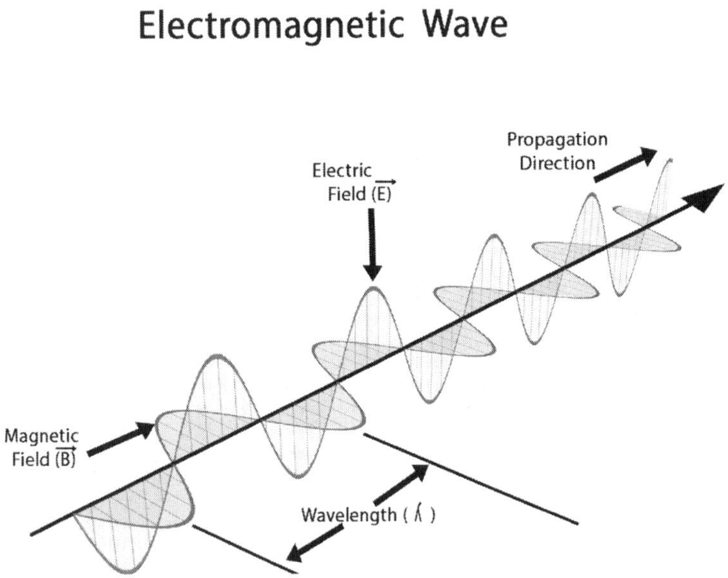

Figure 4.14: Electromagnetic wave, (*Source: https://byjus.com/physics/characteristics-of-em-waves/*).

The magnitudes of the electric and magnetic vectors are represented by E and B, respectively. A significant discovery made about electromagnetic radiation was that the velocity of propagation in a vacuum was constant for all regions of the spectrum. This is known as the velocity of light, c and has the value $2.997\ 925 \times 10^8\ \mathrm{ms}^{-1}$. If one complete wave travelling a fixed distance each cycle is visualized, it may be observed that the velocity of this wave is the product of the wavelength, λ (the distance between adjacent peaks), and the frequency, ν (the number of cycles per second). Therefore:

$$c = \lambda \nu \quad (10)$$

The presentation of spectral regions may be in terms of wavelength as meters or sub-multiples of a meter. The following units are commonly encountered in spectroscope:

1 Å = 10^{-10} m 1 nm = 10^{-9} m 1 μm = 10^{-6} m

Another unit which is widely used in infrared spectroscopy is the wavenumber, $\bar{\nu}$ in cm^{-1}. This is the number of waves in a length of one centimeter and is given by the following relationship:

$$\bar{\nu} = 1/\lambda = \nu/c \quad (11)$$

Work by Einstein, Planck and Bohr indicated that in many ways electromagnetic radiation could be regarded as a stream of particles (or quanta) for which the energy, E, is given by the Bohr equation, as follows:

$$E = h\nu \quad (12)$$

where h is the Planck constant ($h = 6.626 \times 10^{-34}$ J s) and ν is equivalent to the classical frequency.

4.8. PRINCIPLE OF IR SPECTROSCOPY

The investigation of characteristics of near infrared (NIR) region of the electromagnetic spectrum starts with Herschel in 1800 (Herschel, 1800). In the last ten years, near infrared spectroscopy (NIRS) has become a very popular technique for a wide range of analyses in various fields namely agriculture, pharmaceuticals, industries, process control, food processing, remote imaging spectroscopy and other diverse application (Kradiel, 2001).

The advantage of this technique is mainly attributed to the rapid and nondestructive analysis of much thicker samples. Infrared spectroscopy is used to investigate the vibrational properties of a sample. The infrared region of the electromagnetic spectrum (Range: 700–106 nm) is divided into three regions, near infrared (NIR) in the wavelength range 700–2500 nm, mid-infrared (generally referred to as IR) in the range 2500 – 5 × 10^4 and far infrared in the range 5 × 10^4 –10^6 nm. The periodic vibrational motions of atomic nuclei within molecules result in respective molecular spectra. These nuclei move close or apart (stretch) along a straight line vector and they vibrate, rotate, wag, and bend with respect to their centers of gravity. Molecular vibrations give rise to absorption bands generating spectra. When a molecule absorbs infrared radiation, the individual covalent bonds vibrate very similar to the vibrations of a diatomic oscillator.

4.8.1. Molecular Vibrations

In the study of molecular vibrations and rotations we can start with a classical model of the molecule where the nuclei are represented by mathematical points with mass and the internuclear forces holding the molecule together are assumed to be similar to those exerted by massless springs which tend to restore bond lengths or bond angles to certain equilibrium values. We assume also that we can separate the electronic motions from the vibrational and rotational movements of nuclei. The calculations are done in the harmonic approximation.

If there are N atomic nuclei in the molecule, there will be a total of 3N degrees of freedom of motion for all the nuclear masses. After subtracting the translational and rotational degrees of freedom, we are left with 3N−6 internal degrees of freedom for a nonlinear molecule and 3N−5 degrees of freedom for a linear molecule. Rotation of a linear molecule about the molecular axis is not considered as degree of freedom since no displacements of nuclei are involved.

A molecule can absorb IR radiation if it vibrates in such a way that its charge distribution, and therefore its electric dipole moment, changes during the vibration. Although there are many possible vibrations in a polyatomic molecule, the most important vibrations that produce a change in dipole moment are stretching and bending (scissoring, rocking, twisting, wagging) motions (see Figure 4.15). Note that the stretching motions vibrate at higher frequencies than the scissoring motion. Also, asymmetric stretches are more likely to result in a change in dipole moment, with corresponding absorption of IR radiation, than are symmetric stretches.

Given the masses, the molecular geometry and the force constants, complete mathematical procedures exist for calculating the form and frequencies of vibration and rotation of relatively simple molecules. As the molecules become more complex the difficulty of the mathematical treatment increases enormously. Here the diatomic vibrational and rotational calculations serve as the model.

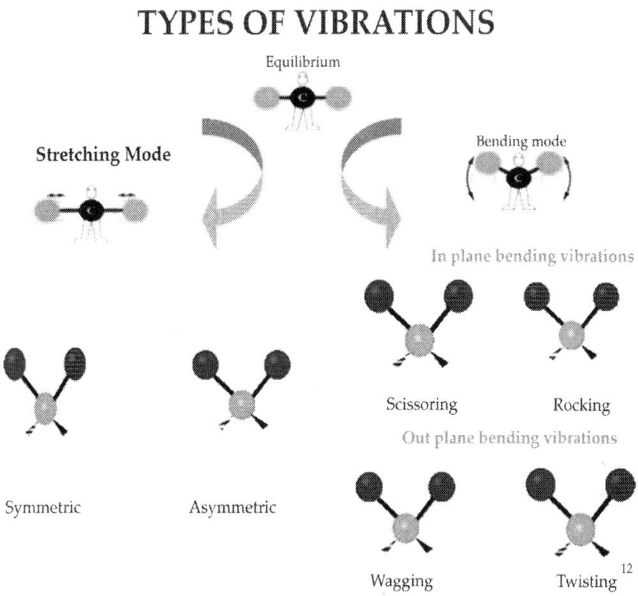

Figure 4.15: Vibrations in Infrared rays, (*Source: https://www.quora.com/What-is-the-basic-principle-of-infrared-spectroscopy*).

4.8.2. Diatomic Oscillator

For an ideal harmonic oscillator, the potential energy V is given by a quadratic expression:

$$V = \frac{1}{2}k(r - r_e) \tag{13}$$

Or $$V = \frac{1}{2}kx^2 \tag{14}$$

where k is the force constant of the bond, $x = (r - r_e)$ is the displacement coordinate, r is the inter-nuclear distance, r_e is the inter-nuclear distance at equilibrium. The frequency at which the ideal harmonic oscillator (atomic or molecular dipole) stretches or bends depends on the bond strength and the masses of the atoms bonded together. When the harmonic oscillator vibrates, the vibrational energy is continuously changing from kinetic to potential and back again. The total energy in the bond is proportional to the frequency of vibration. Considering the elastic properties of a harmonic oscillator, Hooke's law is applied to explain the properties of the two atoms with a spring–like bond between them. The natural vibrational frequency for a bond (or any two masses attached to the ends of a spring) is expressed as:

$$v = \frac{1}{2\pi}\sqrt{\frac{k}{\mu}} \quad (15)$$

where $\mu = m_1m_2 / (m_1 + m_2)$ is the reduced mass of the system; m_1 and m_2 are the masses of the atoms or nuclei. Note that as masses of atoms increase, the frequency of vibrations decreases. Also note that the force constant k varies from one bond to another. The force constant for a single bond is approximately ½ times that of a double bond and $^1/_3$ times that of a triple bond. The ideal or simple harmonic oscillator is able to predict, with relative accuracy, the frequency of energy of fundamental absorption bands such as symmetric and asymmetric stretching, bending, scissoring and wagging. However the simple harmonic oscillator cannot predict the overtone band positions because bonds are not true harmonic oscillators. The quantum mechanical effect on a simple harmonic oscillator indicate that the bond between two atoms cannot be treated quite as simply as two masses connected by a spring. A quantum mechanical treatment has shown that the vibrational energy between atoms in a molecule is quantized and may have only discrete energy levels. Under suitable conditions, the vibrational energy in a molecules jumps from one level to another.

While the absorption frequency depends on the molecular vibrational frequency, the absorption intensity depends on how effectively the infrared photon energy can be transferred to the molecule, which depends on the change in the dipole moment that occurs as a result of molecular vibration. Since the wavelength of infrared radiation is far greater than the size of most molecules, the electric field of the photon can be considered uniform over the whole molecule. The oscillating electric field exerts forces on the molecular charges, tending to change spacing between positive and negative centers of charge, tending to induce the dipole moment of the molecule to oscillate at the frequency of the photon, respectively. In order to absorb infrared radiation, a molecular vibration must cause a change in the dipole moment of the molecule.

4.8.3. Selection Rules for Transitions

The selection rule states that, in the quantum mechanical harmonic oscillator (dipole involving two atoms), a molecule may only absorb (or emit) electromagnetic radiation of energy equal to the spacing between the two levels. The transitions can only occur from one level to the next higher (or lower) level. Hence, transitions are allowed only if the value of the transition moment is non-zero. In essence, we can deduce that photons can be absorbed or emitted, and the harmonic oscillator can go from one vibrational energy state to another. For IR absorption to occur two conditions must be met:

(a) There must be a change in the dipole moment of the molecule as a result of a molecular vibration (or rotation). The change (or oscillation) in the dipole moment allows interaction with the alternating electrical component of the IR radiation wave. Symmetric molecules (or bonds) do not absorb IR radiation since there is no dipole moment.

(b) If the frequency of the radiation v matches the natural frequency of the vibration, the IR photon is absorbed and the amplitude of the vibration increases.

4.9. SOURCE OF IR RADIATION

A source of infrared radiation is usually some solid material heated to incandescence by an electric current. We can approximate its energy distribution with the theoretical black body radiation. It is important to note that energy output falls off very rapidly at long wavelengths. Most IR may be processed to generate a spectrum. There are two basic types, thermal detectors (thermocouples, bolometers, pyroelectric detectors) which measure the heating effects of radiation and respond equally well to all wavelengths, and selective detectors (photoconductive cell) whose response is markedly dependent on the wavelength, but they have a very rapid response and a high sensitivity.

4.9.1. Infrared Detectors

The infrared detector is a device which measures the infrared energy of the source which has passed through the spectrometer. These devices change radiation energy into electrical energy which can be processed to generate a spectrum. There are two basic types, thermal detectors (thermocouples, bolometers, pyroelectric detectors) which measure the heating effects of radiation and respond equally well to all wavelenghts, and selective detectors (photoconductive cell) whose response is markedly dependent on the wavelength, but they have a very rapid response and a high sensitivity. A thermal detector, which is commonly used, is a pyroelectric crystal such as deuterated triglycine sulfate (DTGS). Its residual polarization is sensitive to changes in temperature. An example for selective detectors is the mercury cadmium telluride detector (MCT), which has to be cooled with liquid nitrogen. The photoconductive cells show an increase in electrical conductivity when illuminated by infrared light. It is used for samples with low transmittance or for ATR (attenuated total reflection) spectroscopy.

4.9.2. Interferometers

Between the source and detector, there must be some kind of device to analyze the radiation so that the intensity can be evaluated for each wavelength resolution element. There are two basic types namely monochromators used in dispersive instruments and interferometers used in Fourier transform instruments. In a monochromator, a prism or grating separates the components of polychromatic radiation by bending it by an amount that varies with wavelength. The whole spectrum is measured one wavelength at time, which has some deficiencies compared to interferometers. The biggest advantage that Fourier transform spectrometers have over dispersive spectrometers is the large signal to noise ratio: the intensities of all the resolution elements are measured simultaneously that means the whole time the total radiation is reaching the detector. In addition a complete spectrum can be obtained very rapidly.

The basic principle of the Michelson Interferometer is to split a beam into two and then recombining the two beams after introducing a path difference using fixed and moveable mirror, so an interference pattern is obtained. The interferogram is measured by recording the detector signal as a function of the path difference between two beams. For signal averaging, successive interferograms have to be measured at exactly the same points; this is achieved by using a helium-neon laser as a reference. The customary spectrum, showing energy as a function of frequency, can be obtained by the mathematical process of Fourier Transformation, which analyzes the interferogram as the sum of a series of sine and cosine waves with discrete frequencies.

$$S(k) = \frac{1}{2\pi}\int_{-\infty}^{\infty} f(x)\exp(ikx)dx, \quad k = 2\pi/\lambda \qquad (16)$$

where S(k) is the spectral distribution of light that passed through the sample and f(x) is the difference between energy reaching the detector through the empty machine (the background spectrum) and trough the sample. Usually a number of scans are taken and signal-averaged by the computer, which reduces the noise by the square root of the number of scans. The resolution is constant over the whole spectrum and is increased by increasing the length of travel of moveable mirror.

4.9.3. Infrared Transmitting Materials

Cells for holding samples or windows within the spectrometer must be made of infrared transmitting material. Some of the most useful window materials for mid IR are NaCl (approximate transmission limit around 600 cm^{-1}; it depends on the thickness of material being used), KBr (350 cm^{-1}) and ZnSe (450 cm^{-1}). Because some of them (NaCl, KBr) are quite soluble in water which can occur badly scratches, they have to be polished with alcohol occasionally. For the far IR below 600 cm^{-1} polyethylene is used.

4.10. PRINCIPLES OF NEAR-INFRARED (NIR)

NIR spectrometry is an instrumental method for acquiring spectra of foods and other materials. The acquired spectra are used for determining both qualitative and quantitative characteristics. NIR spectrometry is popular because of its four main advantages: speed (a spectrum can be acquired in as little as a tenth of a second), little or no sample preparation (if any preparation is required, it is usually quite simple), multiple analyses from a single scan (it is not necessary to scan the sample for each constituent), and a nondestructive measurement process (allowing for the return of the analyzed subsample to the original lot).

Currently, NIR spectrometry has three shortcomings. First, it is a technology that must be trained. That is, the instrumentation must be calibrated by scanning a set of sample with known qualitative/quantitative parameters. Second, the known levels often involve expensive and complicated reference methods closely tied to wet chemistry that demand the input of highly skilled personnel. Third, modern-day calibration methods rely on rather sophisticated chemometric techniques, thus calling into play the assistance of personnel who are highly trained in chemometrics. In recent years, costs of wet chemistry have risen at a staggering rate. The need for trained personnel to do the wet chemistry further exacerbates the problems. At the same time, an increased need for more chemical information overloads many laboratories to the point where turnaround of analyses is no longer timely. Little wonder that researchers have worked anxiously to develop short–cut wet chemistry methods, typified by autoanalyzers and autosamplers. Even then, the problems of multiplicative wet chemistry analyses leave modern laboratories in a state of frustration.

4.10.1 NIR Regions

The NIR region range from 720 to 2500 nm. There are three sections of the NIR spectral region namely Transflectance, Transmission and Reflectance.

(a) Transflectance: The range is between 720 to 1100 nm. This section is most suited to transflectance through a thick sample, such as, seeds, slurries, liquids and pastes. The absorption bands are due to 3rd overtones of the fundamental stretching vibrational motions in the Mid IR region.

(b) Transmission: The range is between 1200 to 1850 nm. This section can be used for transmission through liquids and films, as well as for diffuse reflectance measurements off samples with high water contents. The absorption bands are due to the 1st and 2nd overtones of the fundamental stretching vibrational motions in the Mid IR region.

(c) Reflectance: The range is between 1850 to 2500 nm. This section is predominantly used for making diffuse reflectance measurements off ground or solid materials. The absorption bands are due to combination bands, i.e., combinations of C-H stretch and bend vibrational bonds.

The Transflectance region is of particular interest in the analysis of foods because it is suitable for measuring high moisture and high fat content products including meat, dairy products, jams and conserves, dough and batters. The major advantage of working in this region is that longer pathlength sample cells can be used to collect the NIT spectra. Typically, a 10–20 mm path length can be used. This makes sampling easier and allows viscous and non-homogeneous samples to be scanned without further sample processing (Figure 4.16).

On the other hand, the Reflectance region, 1900–2500 nm, should be used for measuring powdered materials or solids because the spectral bands are better resolved and the amount of energy reaching the detector is highest in this region.

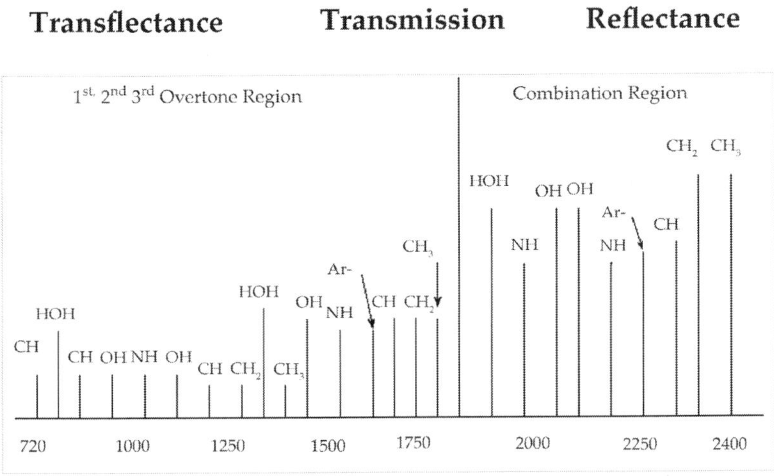

Figure 4.16: NIR Regions, (*Source: https://www.nextinstruments.net/index.php/technical*).

4.10.2. Absorptions in NIR region

If chemical bonds behaved exactly like weak springs then quantum mechanics would restrict their vibration to just two states and there would be very few absorptions in the NIR region. Absorptions in the NIR region (780–2500 nm) are generated from fundamental vibrations by two processes; overtones and combinations. Overtones can be thought of as harmonics. So every fundamental will produce a series of absorptions at (approximately integer) multiples of the frequency (frequency is the reciprocal of wavelength). Combinations are rather more complex. NIR absorptions are at a higher state of excitement so they require more energy than a fundamental absorption. Combinations arise from the sharing of NIR energy between two or more fundamental absorptions. While the number of possible overtones from a group of fundamental absorptions in a molecule is limited to a few, a very large number of combinations will be observed. The effect of all these absorptions combines to create several NIR spectra to look rather uninteresting and to consist of only a few rather broad peaks.

It is an important generalization that NIR spectroscopy is dominated by hydrogen. A spectrum of methanol, CH_3OH, which contains four hydrogen atoms (but three are equivalent) and this spectrum is much more like a typical NIR spectrum with broad peaks while the spectrum of sucrose, $C_{12}H_{24}O_{12}$, which shows very broad areas of absorption but also some quite narrow peaks (see Figures 4.17 and 4.18). It is important to realize that all of these broad absorptions are caused by multiple narrow, over-lapping absorptions. NIR spectra are much more complex than they appear.

The absorption bands in the NIR region tend to be broad and frequently overlap, yielding spectra that are quite complex. However, these broad bands are especially useful for quantitative analysis. Typical NIR spectra of wheat, dried egg white, and cheese are illustrated below. Note that strong absorption bands associated with the −OH groups of water are centered at ca. 1450 and 1940 nm. These bands are the dominant features in the spectrum of cheese, which contains 30–40% moisture, and they are still prominent even in the lower moisture wheat and egg white samples. Bands arising from the −NH groups in protein can be observed at 2060 nm and 2180 nm in the egg white spectrum but are partially obscured by a starch absorption band, centered at 2100 nm, in the wheat sample. Relatively sharp absorption bands arising from −CH groups in lipid can be observed at 2310 and 2350 nm, and another band from these groups is seen around 1730 nm. The 1730 nm band overlaps a weak protein absorption. The lipid bands are distinctly observable in the cheese spectrum.

4.10.3. Principles of Diffuse Reflection Measurements

When radiation strikes a solid or granular material, part of the radiation is reflected from the sample surface. This mirror-like reflection is called specular reflection, and gives little useful information about the sample. Most of the reflected radiation is directed back toward the energy source. Another portion of the radiation will penetrate through the surface of the sample and be reflected off several sample particles before it exits the sample. This is referred to as diffuse reflection, and this diffusely reflected radiation emerges from the surface at random angles through 180°. Each time the radiation interacts with a sample particle, the chemical constituents in the sample can absorb a portion of the radiation. Therefore, the diffusely reflected radiation contains information about the chemical composition of the sample, as indicated by the amount of energy absorbed at specific wavelengths. The amount of radiation penetrating and leaving the sample surface is affected by the size and shape of the sample particles. Compensation for this effect may be achieved by grinding solid or granular materials with a sample preparation mill to a fine, uniform particle size, or by applying mathematical corrections when the instrument is calibrated.

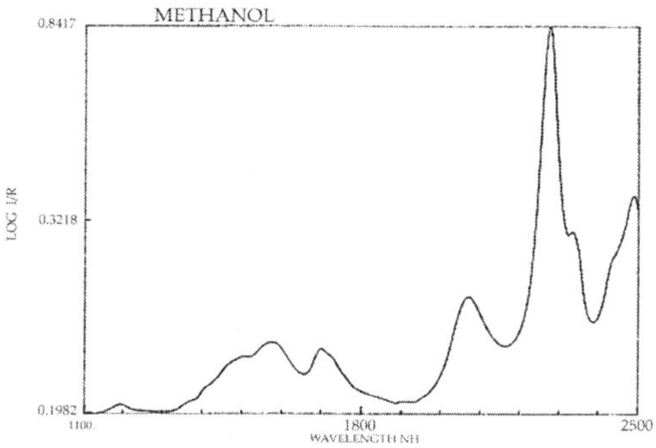

Figure 4.17: NIR spectrum of methanol, (*Source: http://www.impublications.com/content/introduction-near-infrared-nir-spectroscopy*).

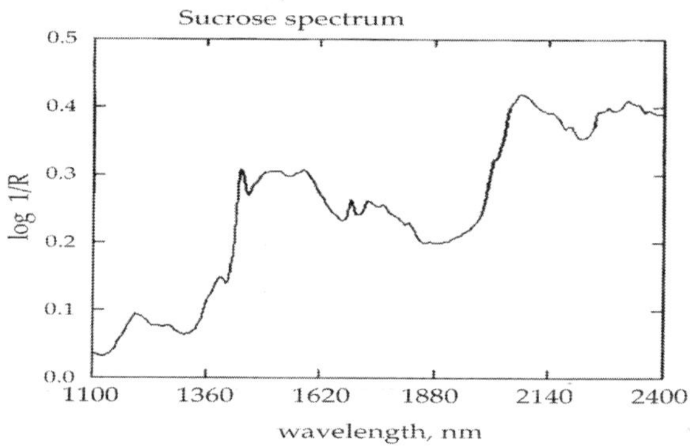

Figure 4.18: NIR spectrum of sucrose, (*Source: http://www.impublications.com/content/introduction-near-infrared-nir-spectroscopy*).

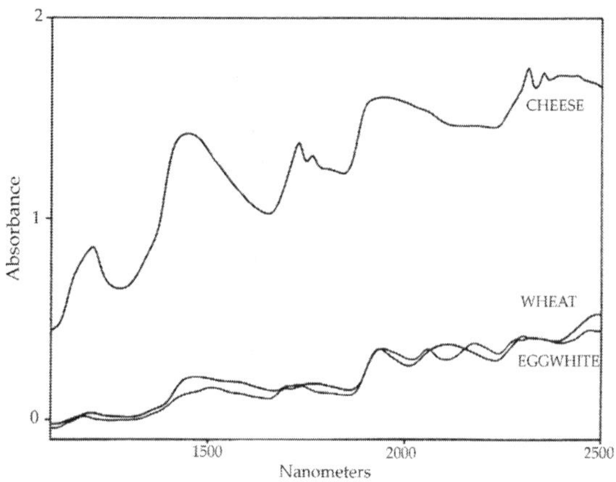

Figure 4.19: NIR spectrum of cheese, wheat, and dried egg white, (*Source: Nielsen Sussane (2017). Food Analysis*).

4.10.4. Applications

Near-infrared (NIR) spectroscopy has been primarily employed in the quantitative analysis of foods. The spectroscopy has been applied to measure moisture, fat, protein and carbohydrate content in wide variety of foods. The most significant advantage is its ability to determine simultaneously several components in a food sample within a short time. The precision of NIR analysis for a wide range of applications is comparable to or better than that of the chemical techniques it replaces.

NIR was used in Australia to predict optimum fertilizer requirements of cereal crops by analysis of total nitrogen and carbohydrate (fructan) in plant tissue samples. Wheat hardness had been measured in both meal and whole grain by using NIR. NIR spectroscopy has also been shown to be useful in the study of changes in starches during processing and storage. This utility is primarily due to the sensitivity of the O–H stretching mode overtone absorptions of starch and of the water bound to starch to changes in hydrogen bonding that accompany changes in starch structure. The spectroscopy combined with chemometrics has been applied to discriminate wheat varieties. It established a discriminant equation, which gave 94% correctly identified varieties.

The study of food authenticity involves establishing whether a sample is genuine in terms of its description, including geographical origin. The application of NIR for authenticity testing of coffee, fruit pulps, milk powders, orange juice, pig carcasses, rice, sausages, sugars, vegetable oils, wheat grain and wheat flour have been reviewed by Downey. This spectroscopy is required to classify within a series of possible classes, to identify a particular kind of adulteration or to quantify adulteration.

4.11. MID INFRARED SPECTROSCOPY

In the Mid Infrared region, i.e. 2500 to 10000 nm, energy is absorbed by molecules at resonant frequencies for each type of vibrational motion. In the Mid IR region, all chemical bonds in organic molecules exhibit strong absorption bands. Typically, the Mid IR region is suitable for characterizing C-C, CO, C-N, C-H, O-H, N-H, N-O and many other chemical bonds. Sample preparation such as dissolving in a solvent, preparing a nujol mull, drawing film etc. is required in order to get sufficient light through a sample. Mid IR spectroscopy is generally used for non–water bearing materials such petrochemicals, plastics, polymers and chemicals. Mid IR spectroscopy is used mainly for qualitative analysis rather than quantitative analysis. For this reason, it is regarded as a highly valuable technique for both food-related research and quality control purposes in the food industry.

The Fourier-transform infrared spectroscopy (FTIR) is used the most often. The mid-infrared spectrum (4000–400 cm^{-1}) includes four regions: the X–H stretching region (4000–2500 cm^{-1}), the triple-bond region (2500–2000 cm^{-1}), the double-bond region (2000–1500 cm^{-1}) and the fingerprint region (1500–600 cm^{-1}). The fundamental vibrations of the X–H stretching region are generally due to O–H, C–H and N–H stretching. Vibrations of C≡C and C≡N bonds are mainly observed in the triple-bond stretching region. Whereas absorption bands corresponding to C=C, C=O and C=N occurs in the double-bond region. The "fingerprint" bands are connected mostly with bending and skeletal vibrations.

4.11.1. Application in Food Quality

The MIR spectroscopy plays a crucial role in research on the chemistry of fats and oils. An important result of early investigations of the IR spectra of fats and oils was the identification of an absorption band at 996 cm^{-1} which is characteristic for isolated Trans double bonds. Although the double bonds in naturally occurring fats and oils predominantly have the cis configuration, extensive cis–trans isomerization occurs during industrial catalytic hydrogenation processes, which are widely employed to convert oils to fats and to increase the oxidative stability of polyunsaturated oils.

The important mid-infrared bands associated with major food components (water, proteins, fats and carbohydrates) as shown in the Table 4.1.

Table 4.1: Mid-Infrared Bands of Major Food Components

Wavenumber (cm^{-1})	Assignment
	Water
3600–3200	O–H stretching
1650	H–O–H stretching
	Proteins
1700–1600	Amide I
1565–1520	Amide II

3000–2800 1745–1725 967	*Fats* C–H stretching C=O stretching C=C–H bending
3000–2800 1400–800	*Carbohydrates* C–H stretching Coupled stretching and bending

Source: Barbara Stuart(2005). Infrared Spectroscopy: Fundamentals and Applications

Likewise fluorescence, MIR spectroscopy is used to study adulteration of olive oil by other edible oils. The edible oils widely employed in virgin olive oil adulteration can be lower quality olive oil (refined or pomace olive oil) or other vegetable or seed oils such as corn, peanut, cottonseed, sunflower, soybean and poppy seed oils. Gurdeniz and Ozen wanted to demonstrate that MIR spectroscopy connected with chemometrics is a rapid method to detect and quantify adulteration of extra–virgin olive oil (EVOO) with vegetable oils (rapeseed, cottonseed, and corn-sunflower binary mixture). The adulteration of EVOO by different concentrations of palm oil was studied by using FTIR technique. Not only EVOO is adulterated by a low-quality edible oils but also virgin coconut oil (VCO), which possesses several biological activities such as antiviral and antimicrobial.

4.12. FAR INFRARED SPECTROSCOPY

The region below 400 cm^{-1} down to 10 cm^{-1} is defined as the far-infrared. The region below 200 cm^{-1} is not readily accessible and there are not many useful spectra-structure correlations in this region. However, compounds containing halogen atoms, organometallic compounds and inorganic compounds absorb in the far-infrared and torsional vibrations and hydrogen bond stretching modes are found in this region. In other words, the far–infrared spectral region is the range of wavenumbers where one finds many of the large amplitude anharmonic vibrations. These include both the symmetric and asymmetric internal torsional modes of many small organic and organometallic molecules, ring puckering vibrations of four- and five-membered rings, and the heavy–atom skeletal bending modes. Additionally, this is the spectral region where one finds lattice vibrations from which information is obtained on intermolecular forces that are of considerable interest to both chemists and the physicists (Figure 4.20).

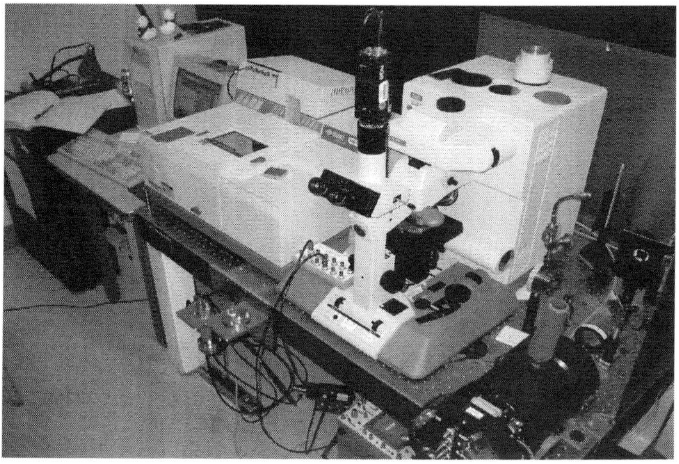

Figure 4.20: Far Infrared Spectrometer–Nicolet Magna 550 FTIR, (*Source: https://www.nist.gov/laboratories/tools-instruments/far-infrared-spectrometer*).

4.13. RAMAN SPECTROSCOPY

Raman spectroscopy is a vibrational spectroscopic technique that is complementary to IR measurements. When a photon of light collides with a molecule, the collision can result in the photon being scattered. The collision can either be elastic, where kinetic energy is conserved, or inelastic, where energy is lost. When the photon interacts with the molecule, the energy of the molecule is raised to an unstable virtual state. When most molecules return to their ground state, they return to the lowest vibrational state, and the scattered photon has the same energy as the incident light. This is known as Rayleigh scattering. However, a few molecules may return to a higher vibrational state. For those molecules, the scattered light will have less energy (lower frequency) than the incident light. The difference in frequency between the incident and scattered light is equal to the frequency of vibration of the molecule. This is known as Raman scattering. For Raman scattering to occur, a molecule must undergo a change in polarizability, but does not need to undergo a change in dipole moment.

Thus, symmetrical vibrations that cannot be observed by IR spectroscopy can be observed by Raman. As indicated earlier in this section, Raman is complementary to IR spectroscopy in that some vibrations are only Raman active, some are only IR active, and some are both. In Raman spectroscopy, the intensity of the scattered light depends on the intensity of the source; therefore, a laser is most commonly used as the source of radiation. To minimize interference from fluorescence, near-infrared lasers are frequently used. The intensity of the scattered light is measured at different frequencies from the incident light, and a plot of Δv in cm^{-1} vs. intensity is made to obtain the Raman spectrum.

Quantitative analysis can be performed with Raman spectroscopy, as the intensity of light scattered at a specific frequency is directly proportional to the concentration of scattering molecules present. Raman spectroscopy is applicable to both solid and liquid samples, often with minimal sample preparation for solids. Water is a very weak Raman scatterer, allowing low concentrations of organic molecules in aqueous solution to be measured. Conversely, water is a very strong IR absorber, which increases the complexity of applying IR techniques to aqueous samples. Finally, the Raman spectrum of a compound is often less complex than its mid or near–IR spectrum. These factors make Raman spectroscopy a useful alternative to IR spectroscopy.

4.13.1. Fourier-Transform Infrared Spectrometers

Fourier-transform infrared (FTIR) spectroscopy is based on the idea of the interference of radiation between two beams to yield an interferogram. The latter is a signal produced as a function of the change of pathlength between the two beams. The two domains of distance and frequency are interconvertible by the mathematical method of Fourier-transformation. The radiation emerging from the source is passed through an interferometer to the sample before reaching a detector. Upon amplification of the signal, in which high-frequency contributions have been eliminated by a filter, the data are converted to digital form by an analog-to-digital converter and transferred to the computer for Fourier-transformation.

There are three basic spectrometer components in an FT system: radiation source, interferometer, and detector. The same types of radiation sources are used for both dispersive and Fourier transform spectrometers. However, the source is more often water-cooled in FTIR instruments to provide better power and stability. In contrast, a completely different approach is taken in an FTIR spectrometer to differentiate and measure the absorption at component frequencies. The monochromator is replaced by an interferometer, which divides radiant beams, generates an optical path difference between the beams, and then recombines them in order to produce repetitive interference signals measured as a function of optical path difference by a detector. As its name implies, the interferometer produces interference signals, which contain infrared spectral information generated after passing through a sample.

The most commonly used interferometer is a Michelson interferometer. It consists of three active components: a moving mirror, a fixed mirror, and a beamsplitter. The two mirrors are perpendicular to each other. The beamsplitter is a semi reflecting device and is often made by depositing a thin film of germanium onto a flat KBr substrate. Radiation from the broadband IR source is collimated and directed into the interferometer, and impinges on the beamsplitter. At the beamsplitter, half the IR beam is transmitted to the fixed mirror and the remaining half is reflected to the moving mirror. After the divided beams are reflected from the two mirrors, they are recombined at the beamsplitter. Due to changes in the relative position of the moving mirror to the fixed mirror, an interference pattern is generated.

The resulting beam then passes through the sample and is eventually focused on the detector. For an easier explanation, the detector response for a single–frequency component from the IR source is first considered. This simulates an idealized situation where the source is monochromatic, such as a laser source. As previously described, differences in the optical paths between the two split beams are created by varying the relative position of moving mirror to the fixed mirror. If the two arms of the interferometer are of equal length, the two split beams travel through the exact same path length. The two beams are totally in phase with each other; thus, they interfere constructively and lead to a maximum in the detector response. This position of the moving mirror is called the point of zero path difference (ZPD). When the moving mirror travels in either direction by the distance l/4, the optical path (beamsplitter–mirror–beamsplitter) is changed by 2 (l/4), or l/2.

The two beams are 180° out of phase with each other, and thus interfere destructively. As the moving mirror travels another l/4, the optical path difference is now 2 (l/2), or l. The two beams are again in phase with each other and result in another constructive interference. When the mirror is moved at a constant velocity, the intensity of radiation reaching the detector varies in a sinusoidal manner to produce the interferogram output. The interferogram is the record of the interference signal. It is actually a time domain spectrum and records the detector response changes versus time within the mirror scan. If the sample happens to absorb at this frequency, the amplitude of the sinusoidal wave is reduced by an amount proportional to the amount of sample in the beam.

Extension of the same process to three component frequencies results in a more complex interferogram, which is the summation of three individual modulated waves. In contrast to this simple, symmetric interferogram, the interferogram produced with a broadband IR source displays extensive interference patterns. It is a complex summation of superimposed sinusoidal waves, each wave corresponding to a single frequency. When this IR beam is directed through the sample, the amplitudes of a set of waves are reduced by absorption if the frequency of this set of waves is the same as one of the characteristic frequencies of the sample.

The interferogram contains information over the entire IR region to which the detector is responsive. A mathematical operation known as Fourier transformation converts the interferogram (a time domain spectrum displaying intensity versus time within the mirror scan) to the final IR spectrum, which is the familiar frequency domain spectrum showing intensity versus frequency. This also explains how the term Fourier transform infrared spectrometry is created. The detector signal is sampled at small, precise intervals during the mirror scan. The sampling rate is controlled by an internal, independent reference, a modulated monochromatic beam from a helium neon (HeNe) laser focused on a separate detector. The two most popular detectors for a FTIR spectrometer are deuterated triglycine sulfate (DTGS) and mercury cadmium telluride (MCT).

The response times of many detectors (for example, thermocouple and thermistor) used in dispersive IR instruments are too slow for the rapid scan times (1 sec or less) of the interferometer. The DTGS detector is a pyroelectric detector that delivers rapid responses because it measures the changes in temperature rather than the value of temperature. The MCT detector is a photon (or quantum) detector that depends on the quantum nature of radiation and also exhibits very fast responses. Whereas DTGS detectors operate at room temperature, MCT detectors must be maintained at liquid nitrogen temperature (77 °K) to be effective. In general, the MCT detector is faster and more sensitive than the DTGS detector. A simplified optical layout of a typical FTIR spectrometer is illustrated below (Figure 4.21).

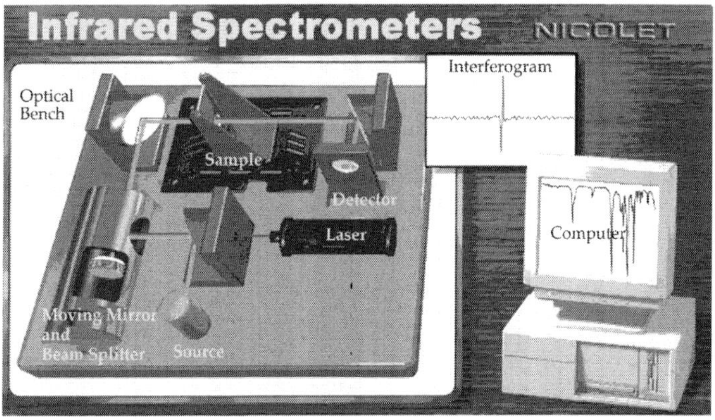

Figure 4.21: Optical chamber layout of FTIR Spectrometer. (*Source: http://lab-training.com/2015/05/29/useful-tips-to-ensure-top-performance-of-your-ft-ir-spectrometer/*).

4.13.2. Application

An area of application of FTIR spectroscopy that has been investigated is its use in the assessment of oil quality and stability. The reaction between unsaturated lipids and atmospheric oxygen under ambient conditions, termed lipid auto–oxidation, is a leading cause of deterioration of fats and oils, as well as of any lipid-containing food, as it gives rise to the off-flavors and unpalatable odors associated with oxidative rancidity. FTIR spectroscopy proved to be the most direct and accurate method of monitoring gross changes in the frying oil over time. A quantitative FTIR method was used for monitoring the oxidative state of frying oils, based on the determination of anisidine value (AV), a measure of aldehydes that are major secondary oxidation products in polyunsaturated oils, has also been reported. FTIR methods have also been developed to serve as alternatives to the peroxide value (PV) test, which is widely employed by the fats and oils industry to assess the oxidative status and stability of refined oils. This method entails measurement of the hydroperoxide OO–H stretching absorption, which is observed at 3444 cm^{-1} in the spectra of neat oils. FTIR spectroscopy was also used to understand in more detail the mechanisms of thermally induced oxidative processes (thermal oil degradation) in extra-virgin olive oils.

4.13.3. Dispersive Raman Spectroscopy

Dispersive Raman spectroscopy is the most common and basic technique, with intense monochromatic radiation, which is usually provided by continuous-wave gas lasers such as argon (488.0 and 514.5 nm), krypton (647.1 nm) ions, or diodes (780 or 830 nm) combined with a charge-coupled device (CCD) array detector with a single-grating spectrograph. The essential problem of the dispersive spectrometer lies with its monochromator. This contains narrow slits at the entrance and exit which limit the wavenumber range of the radiation reaching the detector to one resolution width. Samples for which a very quick measurement is needed, for example, in the eluant from a chromatography column, cannot be studied with instruments of low sensitivity because they cannot scan at speed. However, these limitations may be overcome through the use of a Fourier-transform infrared spectrometer.

Dispersive Raman spectroscopy has some instinct advantages:

(a) It is very suitable for aqueous-phase samples.

(b) It can be used for samples with elevated temperatures, even above 1000 °C.

(c) It allows black samples to be analyzed.

(d) With a 780- or 830 nm laser, it can help suppress the fluorescence of samples.

4.13.4. Spatially Offset Raman Spectroscopy

Spatially offset Raman spectroscopy (SORS) is based on earlier investigations into the Raman photon migration process. Unlike conventional Raman spectroscopy, Raman spectra for SORS are collected from spatial regions offset from the point of illumination on the sample surface. Due to wider lateral diffusion of photons emerging from greater depths, SORS contains different signals from sample layers at different depths. With a larger spatial offset, a greater suppression of surface layer signal will be achieved in SORS. Standard SORS does not automatically permit the simultaneous interrogation of diffusely scattering and transparent samples. Eliasson et al. demonstrated a variant of SORS to adapt SORS for a wide variety of packaging and samples analysis. In adapted SORS, the container acts not only as a container but also as a source of diffuse radiation illuminating a large area of liquid, which generates Raman photons emanating in all directions. This mechanism facilitates the effective illumination of the container wall and the liquid content directly under the Raman collection area.

4.13.5. Surface-Enhanced Raman Spectroscopy

When an electromagnetic wave interacts with a metal surface, the fields at the surface are different than those observed in the far field. If the surface is rough, the wave may excite localized surface plasmons on the surface, resulting in amplification of the electromagnetic fields near the surface. If one assumes that there is enhancement of the intensity of the incident and scattered fields (albeit at different wavelengths), then the possibility of a large enhancement of Raman scattering intensity arises. This notion is commonly has served as an important model in the understanding of Surface-Enhanced Raman Spectroscopy (SERS) since its discovery in the 1970s.

The SERS is among the most sensitive techniques available to study surface science. This technique can be coupled with a wide range of other surface techniques to study corrosion, catalysis, advanced materials, biology and sensors. In a recent development, a new approach has been introduced to enhance the Raman scattering intensity by combining Raman spectroscopy with scanning probe microscopy such as atomic force microscopy (AFM). Its central element consists of an illuminated AFM tip placed in the close vicinity of a substrate. The tip is operated in contact or tunneling mode. Upon illumination with a laser beam, localized surface plasmons are excited in the tip–substrate gap, producing a large, local enhancement of electromagnetic field compared to the incident radiation. In this way, Raman spectroscopy and scanning probe technique can be combined; this new technique is known as tip–enhanced Raman spectroscopy (TERS).

It is highly desirable to apply in situ SERS to obtain molecular level information about the breaking and formation of bonds in the reaction, observe the reaction intermediate on the surface and finally, to distinguish the reaction products. The microscopic structure and dynamics of interfacial water molecules is important in understanding the electrode/electrolyte interface and is still an ongoing issue in electrochemistry as well as surface science. However, the interactions between water molecules and electrolyte ions as well as between water and electrode surface are very complex.

4.13.6. Application

There is a great potential for applying SERS as a routine analytical technique in rapid detection of chemical or biological contaminants in foods due to the following reasons:

(a) Different chemical and microbiological samples yield unique Raman spectral patterns and the Raman intensity is related to the concentration of samples or the number of microbes. Therefore, it is possible to detect, quantify, and characterize different chemical and biological contaminants.

(b) SERS is very sensitive. The LOD of SERS could reach the ppm to ppb level depending on the properties of SERS substrates and tested analyte samples. Therefore, SERS is suitable for detection of trace amount of samples.

(c) SERS technique requires minimum sample preparation before measurement. Extraction of food contaminants from a food system could be simplified and large particles do not affect the final results significantly.

(d) The whole procedure of SERS detection is fast due to short sample preparation time and short measurement time (less than one min for measurement).

(e) The operation of SERS measurement is easy and straightforward.

(f) SERS equipment is cheaper and smaller than mass spectroscopy if LC/MS or GC/MS is used. With the development of SERS substrates and the use of a portable Raman, it is possible to have cost-efficient SERS methods for on-site investigation.

4.14. APPLICATIONS AND CASE STUDIES

The contextual analyses beneath identify with heated item and oat grain assessment, and are vital in outlining a wide assortment of picture preparing and machine vision methods, including a reasonable extent of those shrouded in Section 4.2 . Specifically, great utilize is made of edge location, Hough changes, surface investigation, morphology, thresholding, coordinated separating, middle sifting and theoretical example acknowledgment techniques, while the need for heartiness is underlined. The scope of uses will be expanded well past the limits of these contextual investigations by the audit in Section 4.4, however a significant number of similar standards will be kept in locate.

4.14.1. Case Study: Chocolate Cake let Inspection

Numerous cutting edge nourishments have high added esteem with respect to the first crude item–or even in respect to the fundamental cooked item. This applies especially to bread rolls and cakes, and at times even little cakes are wonders of creative make. Here we think about the assessment of a specific kind of cakelet, made with a round wipe base, to the focal point of which a spot of stick has been included, and the highest point of the item covered with chocolate. Chocolate is a costly ware, and consequently it is vital to control its measure that achieves the item: excessively, and the cost will be too high; too little, and the layer of chocolate will be broken and the customer won't be fulfilled. Moreover, the spot of stick must not spread too far from the focal point of the item, or else the trademark dull ring that shows up in the chocolate simply over the edge of the stick will be truant, deformed or misshaped, and the last item will lose its satisfying symmetry (Figure 4.12). Thus, while investigation is by and large concerned prevalently with imperfections or contaminants, here the presence of the item has additionally turned out to be critical. While it is hard to influence a PC to take on a similar mindset as a human, the accompanying PC implementable standards can be set down for agreeableness with this sort of cakelet:

1. The item ought to be round.
2. The breadth should exist in all around characterized limits or the pressing machine will be at risk to stick.
3. The spot of stick ought to be focal.
4. The spot of stick ought to be round and of the correct distance across.
5. The chocolate should cover the entire of the highest point of the cakelet.
6. The chocolate ought to have a reasonable surface (bestowed to it by the plunging lattice).
7. The cakelet ought to be the correct far up.
8. The cakelet ought not to be covered by, or notwithstanding touching different cakelets, as this implies it will more likely than not be adhered to them.

Checking every one of these standards for every item at the required rate of somewhere in the range of 20 items for every second is computationally requesting. It is additionally an exorbitant undertaking, since cameras, lighting frameworks and dismissal apparatus must be connected deliberately to the product offering. Moreover, it is faulty where to put the examination station.

Put toward the finish of the line, it can effectively screen the general nature of the item, yet this will be past the point where it is possible to find early blames, for example, absence of stick, so esteem included at a later stage (counting costly chocolate) will wind up being squandered. So there is a comment picked up by having an early investigation station and a last one, however this is rarely improved the situation sustenance items. (Be that as it may, the switch circumstance applies for bread, where the crude grains will probably be investigated than the last item).

The primary phase of examination is to find the item, and for this situation the Hough change is utilized and functions admirably. Specifically, it is hearty to items that are adhered together and to smears of chocolate on the transport. Once the item has been found, a round district of intrigue is characterized around it. At that point the level of chocolate cover can be surveyed exceedingly precisely and the general distance across can be estimated. Next, the surface can be estimated utilizing an approach known as an outspread force histogram, which contains data on the spiral dispersion of powers in the surface of the chocolate. This allows the outward presentation of the item to be evaluated; it additionally allows the dull ring around the edge of the stick district to be estimated for size and consistency. With these methodologies, every one of the tenets recorded above can be tried and the item hailed as 'adequate' or 'dismiss.' Note that it is the line supervisor's business to characterize the limits that must be determined to any one day: it is the machine's business to execute his desires utilizing the estimations made utilizing the vision framework.

At long last, a few comments about the lighting are all together. This must be set up utilizing four lights that were symmetrically put around the camera, to give sensibly uniform lighting over the item: awesome care over this was not important to the activity of the Hough change, which is very hearty. In any case, sensible care was as yet required in setting up the lighting, in order to limit the quantity of particular estimations to be made thus that straightforward averaging methods could be utilized to spare calculation.

REFERENCES

1. Batchelor, B. G., Davies E. R., & Graves M., (2004). Using X-rays to detect foreign bodies in Edwards M (ed.). Detecting Foreign Bodies in Food. Woodhead Publishing Ltd. Cambridge, UK.
2. Bennedsen, B. S., Peterson D. L., & Tabb, A., (2005). 'Identifying defects in images of rotating apples,' Computers and Electronics in Agriculture, *48*, 92 –102.
3. Blasco, J., Aleixos, N., Gómez, J., & Moltó, E., (2007). Citrus sorting by identification of the most common defects using multispectral computer vision. *Journal of Food Engineering*. doi.org/10.1016/j.jfoodeng.2007.03.027
4. Davies, E., (1997). Machine vision: Theory, algorithms, practicalities. Academic Press, London.
5. Caldwell G. Darwin, (2013). Robotics and Automation in the Food Industry: Current and Future Technologies. Woodhead Publishing.1st edition
6. Gerrard, D., Gao, X., & Tan, J., (1996). Beef Marbling and Color Score Determination by Image Processing. *Journal of Food Science, 61*(1), 145–148.
7. Gunasekaran, S., (1996). Computer vision technology for food quality assurance. Trends *Food Sci. Techn., 7*(8), 245256.
8. Gunasekaran, S., (2001). Non-Destructive Food Evaluation Techniques to Analyze Properties and Quality. Food Science and Technology Series, Marcel Dekker, New York, pp.105.
9. Kim, S., & Schatzki, T., (2000). Apple Watercore Sorting Using X-ray Imagery: I. Algorithm Development. *Transactions of the ASAE, 43*(6), 1695–1702.
10. Sapirstein, H., (1995). Quality Control in Commercial Baking: Machine Vision Inspection of Crumb Grain in Bread and Cake products. In Food Processing Automation IV Proceedings of the FPAC Conference. St. Joseph, Michigan, USA: ASAE.
11. Tan, F., Morgan, M., Ludas, L., Forrest, J., & Gerrard, D., (2000). Assessment of fresh pork color with color machine vision. *Journal of Animal Science, 78*(12), 3078–3085.
12. Wallin, P., & Haycock, P., (1998). Foreign body prevention, detection and control. Blackie Academic & Professional, London.Volume 4.

CHAPTER 5

OPTICAL SENSORS AND ONLINE SPECTROSCOPY FOR AUTOMATED QUALITY AND SAFETY INSPECTION OF FOOD PRODUCTS

"Food is simply sunlight in the cold storage"

—John Harvey Kellogg

CONTENTS

Because of expanding purchaser interest for sound, high caliber, and safe food, food preparing enterprises are investigating/receiving new non-dangerous, quick, and precise procedures for quality and wellbeing assessment of their items. Optical detecting and spectroscopic strategies have a high potential for mechanized continuous quality and wellbeing investigation of rural and food items. These systems have just been embraced effectively in a few food-preparing units for mechanized quality observing of their items. This chapter talks about different optical detecting strategies under three general classifications in particular: spectroscopic, fiber optic and picture detecting. Working standards, instrumentation, focal points, impediments, and confinements of these systems have been portrayed and different applications for quality checking in the horticulture and food handling enterprises have been looked into.

5.1. INTRODUCTION

Food material ought to be assessed crude and in addition prepared on-line, in-line, or at-line for its quality (shading, surface, enhance, shape, dietary esteem, piece, and usefulness) and wellbeing (substance and microbial sullying, outside material, pollution, and contaminated). Human investigation of food items is moderate and subjective, though concoction wet examination for food compositional examination is ruinous, tedious, and costly. The food handling industry is searching for elective in/online, quick, exact, non-ruinous, and more affordable methods for mechanized food quality observing. Optical detecting and spectroscopic methods meet these criteria and can be utilized as a part of computerized frameworks in the food preparing industry.

Optical sensors are gadgets used to gauge radiation motions as consumed, scattered, weakened, or bright light from the example with trademark optical properties for grouping/evaluation. The connection between test particles and emanated vitality is wavelength particular, along these lines conveys recognizable attributes. The locators in optical sensors change over the gathered radiometric light as electric signs, which are then recorded and utilized for assist investigation. The optical sensors can have single, numerous, line or cluster of identifiers in light of the system and end utilize. In optical-spectroscopic instruments, for the most part, single-point-based identifier components are utilized, while in optical imaging sensors, for the most part, a direct or two-dimensional (2D) exhibit of locators is utilized. In the event that the sensor utilizes a fiber optic procedure for conveying and accepting the transmitted light, it is called fiber optic detecting innovation. The decision of indicator is vigorously reliant on the particular utilization of a specific scope of electromagnetic radiation (e.g., bright (UV), unmistakable (VIS), close infrared (NIR), infrared (IR)).

A perfect identifier ought to have high affectability, extensive dynamic range, wide ghastly reaction, low float, and low warm commotion. Regardless of the kind of sensor, the greater part of the detecting systems comprises of a finder, a light source, and a wavelength channel/determination gadget. There are a few characterizations of optical sensors to be found in the distributed writing in light of greatness and nature of the amount to be estimated (mechanical, warm, electromagnetic, synthetic piece, radiation, and common biomedical estimations); spatial circulation (point, incorporated, dispersed, and semi appropriated); detecting innovation (optical fiber innovation (OFT), coordinated optical innovation (IOT), incorporated optoelectronic (IOE), and cross breed optic innovation); or adjustments (adequacy or force sensor, stage or interferometric sensor, polarimetric sensors, and spectroscopic sensors).

For effortlessness and essential comprehension of the subject, we have isolated optical sensors into three general classes in particular, spectroscopic, fiber optic and picture detecting strategies. Every one of these classes is examined alongside the standards, instrumentation, points of interest and weaknesses, and restrictions. Different uses of these methods for quality checking in the food handling industry have likewise been evaluated and displayed.

5.2. OPTICAL SENSING AND SPECTROSCOPIC TECHNIQUES

This section portrays the working standards and equipment segments of different spectroscopic, fiber optic, and picture detecting methods.

5.2.1. Spectroscopic Techniques

Point-based optical sensors are principally utilized as a part of spectroscopic instruments. They comprise of a solitary identifier component or a variety of finder, a radiation source, and wavelength channel gadgets. Wavelength channel or grinding gadgets scatter the light onto a material, and the identifier component records the measure of radiation assimilated (UV, IR, and IR) or scattered (Raman spectroscopy) after its collaboration with the material.

The indicator, otherwise called a photodetector or photodiode, is the central component of the detecting unit, changing over the deliberate information flag (radiation) into an electrical yield flag. A perfect photodetector ought to have a wide otherworldly perceptibility extend, high reaction, rapid (vast transfer speed), low clamor, high steadiness, high dependability, reduced size, and minimal effort.

The full electromagnetic range ranges from high vitality gamma beams and X-beams, UV, VIS, NIR, IR, and microwaves, to the low vitality radio waves (Figure 5.1). Distinctive kinds of spectroscopic methods are accessible for various districts of electromagnetic radiation. The ghastly attributes rely upon the material to be assessed/broke down, the area of electromagnetic radiation, and kind of instrument utilized for the investigation.

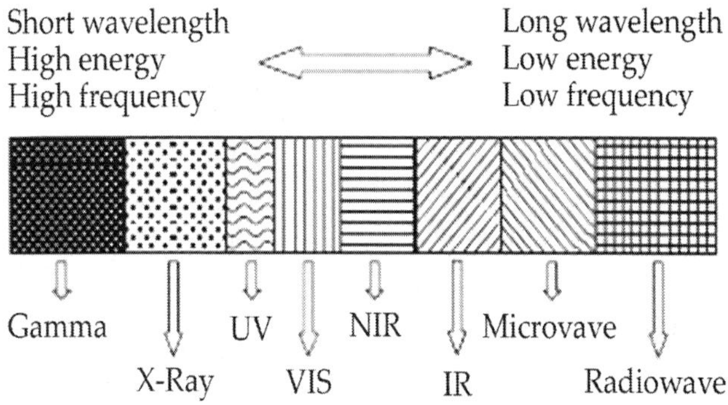

Figure 5.1: The electromagnetic spectrum, (*Source: Singh and Jayas, 2010*).

Particles in space can have vitality in a few structures, for example, vibrational, rotational, or electronic. Vibrational vitality causes occasional uprooting of molecules from their balance position, and rotational vitality causes pivot about a focal point of gravity. The vitality required for rotational change is low contrasted with the vitality required for vibrational progress (NIR– mid-IR). Rotational vitality is caught up in far IR and microwave districts, and the spectra in this locale are exceptionally expansive (with the exception of the gases). Vibrational movement is for the most part utilized as a part of the ghostly investigation of solids and fluids. Electronic vitality related with electron excitation in the UV-VIS area is considered in the UV spectroscopy. Infrared, Raman and NIR spectroscopy are a piece of vibrational spectroscopic strategies.

5.2.1.1. IR Spectroscopy

In IR spectroscopy, essential vibrations emerging from sub-atomic advances between neighboring vitality states (passable change in vibrational quantum number is ν = □} 1) are contemplated. Photons in the IR locale have bring down energies, comparing to covalent security extend and twist vibrations fit for delivering just crucial advances. The IR spectra are principally examined in the mid-IR area (650–4000 cm^{-1}), which has very much settled pinnacles. The IR spectra are exceptionally valuable in the auxiliary and subjective examination of food materials. IR spectroscopy has the capacity to break down strong, fluid, and gas tests; be that as it may; the example readiness necessity limits fast examination.

Previously, scattering IR spectroscopic instruments utilizing grinding gadgets constrained the application in light of moderate information accumulation rates. With the approach of Fourier change infrared (FTIR) methods the example can be examined and spectra can be recorded in just a couple of moments. FTIR spectroscopy has a few favorable circumstances, for example, high affectability, expanded splendor, high determination, and low foundation commotion since just an obstruction flag is utilized as a part of the Fourier change. A FTIR instrument, for the most part, comprises of an IR radiation source (alluded to as a light source and not as a warm source), a settled mirror, a moving mirror, a beam splitter, and a locator. The beam splitter parts the approaching radiation into two pillars and guides them individually to a moving and a settled mirror. These two bars are reflected back to the beam splitter, causing obstruction (productive impedance if the mirrors are equidistant, and dangerous obstruction in the event that one mirror moves away), which is detected by the identifier subsequent to going through the example.

The locator yield is a fluctuating cosine wave otherwise called an interferogram, which is intricate and hard to dissect, and subsequently, Fourier change is utilized to change this information into a wavelength and power. The PbS locators are utilized for the low IR phantom range. The mercury cadmium telluride (MCT) is a generally utilized photodetector in mid-IR that has high affectability, low clamor, and high otherworldly determination. The PbSe and InSb indicators are likewise utilized as a part of the mid-IR ghastly range. Quantum all around infrared photodetectors (QWIP) with wide IR otherworldly reaction are additionally accessible these days and are practical, however, have higher dim current and lower quantum effectiveness contrasted with the MCT locators. Deuterated triglycine sulfate (DTGS) is utilized as a part of IR warm locators.

5.2.1.2. Raman Spectroscopy

Raman spectra are likewise created by sub-atomic vibration, however, the vibrations in Raman spectroscopy are symmetric and the primary criteria is the polarizability of the mixes. Raman spectra are the aftereffect of scrambling of episode radiation on an example. On the off chance that the scattered light has an indistinguishable recurrence from the episode light, because of flexible impact between striking photons and vibrating particles, the diffusing is called Rayleigh dissipating. Be that as it may, on account of inelastic crashes, the scattered light has an alternate recurrence from the striking radiation, because of vitality trade amongst photons and atoms, and the disseminating wonder is then portrayed as Raman scrambling. Most of the vibrating particles are available in the ground vitality state, yet some are available in energized vitality states. At the point when an approaching photon strikes a particle of moderately low vitality display in the ground vitality express, the atom picks up vitality and the scattered photon loses its vitality, bringing about Stokes Raman diffusing.

At the point when a photon hits a particle with moderately high vitality, the photon picks up vitality from the atom, and the scattered photon has higher vitality, offering ascend to the counter Stokes diffusing wonders. Stirs dissipating lines are generally more grounded than hostile to Stokes lines because of the high vitality of the scattered photons. Stirs lines are utilized as a part of Raman ghastly investigation. In the Raman spectrometer, different excitation sources, (for example, noticeable excitation source in VIS excitation Raman spectroscopy, UV lasers in reverberation Raman spectrometers, and diode-bar-pumped Nd-YAG lasers in Raman NIR spectrometers) are utilized for excitations in different wavelength ranges. For VIS and UV excitation, charge-coupled gadget (CCD) indicators are utilized, while in NIR excitation germanium identifiers are the favored detecting component.

IR and Raman spectroscopic systems give reciprocal data for breaking down an example. Crests in IR spectra are delivered because of sub-atomic vibrations caused by an adjustment in dipole snapshot of the particle, and tops in Raman spectra are created because of vibrations caused by an adjustment in the electronic polarizability, that is, polar bonds (C– H, O– H, C=O, C==O, C– O, C– F, Si– O, and N-H) give solid IR ingestion tops however homonuclear apolar symmetric bonds (C==C, C=C, C– C, S– S) are poor safeguards of IR radiation yet give solid Raman Stokes lines. Raman spectroscopy isn't extremely prevalent for a few reasons, including little example volume, obstruction by fluorescence, low proficiency, low power of wanted Stoke's disseminating, and high force of the undesired Rayleigh dissipating.

5.2.1.3. NIR Spectroscopy

NIR spectroscopy is the most widely recognized among the vibrational methods as of now being utilized as a part of the agri-food industry. Suggestions (progress vitality for change from $\nu = 0$ to higher progress levels for instance, $\nu = 2, 3$) and mix groups (mixes of hints and basic vibration) that happen in the range 700– 2500 nm (14285.71– 4000 cm −1) are contemplated in the NIR spectroscopic examination (Osborne et al., 1993). Ingestions in the NIR locale are caused by vibrations of –CH,–OH, and–NH bond gatherings. Hints and blend groups are weaker than crucial retention groups in the IR ghastly locale, which is leverage as tests of thickness of a few mm –, for example, strong food, pressed food, and in place, entire seeds– can be investigated in a NIR spectrometer with no example arrangement.

A NIR instrument comprises of a light source (Tungsten incandescent light, quartz incandescent light, light producing diodes (LED)), a wavelength sifting gadget (crystal, grinding, impedance channels, and progressed electronically tunable channels (ETF)), and a diode cluster identifier (Silicon (Si), lead sulfide (PbS), lead selenide (PbSe), indium gallium arsenide (InGaAs)), and an information securing and preparing framework and programming. Lab and research-based instruments require full unearthly examination, consequently, crystal, grinding, or ETF are required. Two propelled ETFs, in particular, acousto-optical tunable channels (AOTF) and fluid gem tunable channels (LCTF), are usually utilized as a part of NIR examining instruments for full otherworldly examination in bordering wavebands. In a modern investigation setting, test examining at just a couple of discrete wavelengths (recognized from full checking research instruments) is required, which can be expert utilizing more affordable obstruction channels. Interferogram based FT-NIR instruments are additionally accessible with enhanced clamor level.

The unearthly information procured from a spectroscopic instrument can be dissected both quantitatively and subjectively, and adjustment (forecast) and arrangement models can be created utilizing different progressed chemometric apparatuses for multivariate examination. Before display improvement, the otherworldly information must be pretreated to evacuate/ beaten the clamor and non-linearity between unearthly flag (absorbance) and the centralization of sustenance material. This non-linearity may emerge because of a few physical and concoction marvels, for example, dissipating because of particles, interferents, sub-atomic connections, changes in refractive file at high fixations, stray light, and changes in test estimate (ground or entire) and hence way length, and the impacts of developing area and developing season on the product/food. A few pre-preparing techniques, for example, multiplicative disseminate remedy (MSC), de-inclining, standard ordinary variate (SNV), standardization, gauge rectification, and determination (limited contrast and Savitzky-Golay smoothing) are utilized as a part of preprocessing ghostly information.

5.2.1.4. UV Spectroscopy

UV spectra emerge because of electron excitation into high vitality levels. A few natural particles containing unshared electrons in sulfur, bromine, and iodine have ingestion tops in the UV locale. The UV locale is isolated into two subgroups, to be specific vacuum UV (100– 190 nm) and close UV (190– 400 nm) (Cooper, 1980). The spectroscopic instrument for the vacuum UV district requires outright vacuum conditions to take out/ diminish UV retention in this area and consequently isn't appropriate for in-line-preparing. The close UV area is regularly utilized as a part of UV spectroscopic examination. Optical material in UV cells for holding the example is made of quartz (for use over 210 nm) or silica (for utilize around 165 nm). An UV spectrophotometer comprises of a light source (deuterium or xenon light, LEDs, diffraction grinding, and an identifier (diode cluster or CCD finder). Unadulterated solvents (CCl 4, CHCl 3, CH 2 Cl 2, ethanol, methanol, water, and dioxane), with no ingestion in the UV ghastly region of intrigue, are utilized. The UV assimilation spectra are by and large expansive and subsequently not basic in follow ID, but rather they have potential for the quantitative examination of food items.

5.2.2. Fiber Optic Sensors

Optical fiber sensors offer a few favorable circumstances, for example, being aloof, dielectric, light weighted, conservative, negligible or no impedance, reasonable, safe, and simple to introduce. In fiber optic procedure, fiber materials (as thin wire) are utilized as a part of transmitting light through optical segments (channels, beamsplitters) that guide the light to the example and additionally detecting component or indicator. The fiber line comprise of three layers, specifically an inward center layer made of glass, silica or other fiber material, and a center layer (cladding) encompassed by an external defensive layer. The light goes through the inward center by methods for inner reflection. The fiber optic lines can convey top notch light to an identifier a few meters away regardless of whether the links are bowed, similar to some other wire, and are not influenced by the encompassing condition. A basic fiber optic detecting instrument comprises of an optical source, indicator and optical fiber line (single or multimode). LEDs or laser diodes are fundamentally utilized as light sources. Semiconductor gadgets, such photodiodes, are utilized as identifiers. Appropriated optical fiber detecting (DOFS) is utilized to outline dimensional information with spatial naming like other imaging strategies. Now and then fiber material itself is a piece of the detecting component or finder (inherent sensor), be that as it may, in the present setting, fiber material is alluded to and implied just for transmission of light to the example and identifier component. Definite data about the fiber optic innovation is past the extent of this chapter and perusers can discover the data somewhere else.

5.2.3. Imaging Sensors

In imaging sensors, a line or 2D cluster of sensors is utilized, rather than simply the single finder utilized as a part of point-based instruments. The imaging detecting framework comprises of a light source and channel gadgets, as utilized as a part of point-based instruments. The preferred standpoint with the picture sensor is that it gives the spatial naming of every pixel in conjunction with absorbance/reflectance esteem. The spatial determination of the imaging gadget relies upon the span of the locator component, that is, a little finder will give higher determination in any case, in the meantime, create more commotion as less extreme light achieves the indicator zone. 2D information obtaining at a few wavebands (consequently shaping a 3D hypercube) is conceivable, utilizing either grinding gadgets or tunable channels. In a shading imaging framework, information are gathered at red, green, and blue groups. In the event that 2D picture information are gathered at up to ten wavebands, this strategy is named multispectral imaging system. 2D picture information can likewise be gathered at many restricted touching wavebands, utilizing hyperspectral imaging. In this way, tests can be examined as spectra for every pixel or as a picture at every wavelength. The 3D hyperspectral information should be redressed for an indistinguishable sort of commotion from in spectroscopic information, by applying comparable preprocessing systems.

In any case, hyperspectral information likewise require revising geometric mutilations in pictures caused either by moving grinding instruments or optical blunders in ETFs. The subtle elements of adjusting hyperspectral imaging frameworks have been accounted for somewhere else. Progressively imaging applications, rapid information securing (filtering speed) by the sensor (camera) is required. The checked information from the locator (yield flag) ought to likewise be exchanged to the host PC for capacity and investigation. CAMERA Connection is a standard correspondence interface that is being utilized for fast, high determination, and great quality information transmission applications. Other correspondence interfaces, for example, firewire (as alluded as IEEE 1394) and GiGE VISION are additionally utilized. The host PC ought to have the capacity to deal with the measure of information exchanged by the correspondence interface.

5.3. APPLICATIONS IN THE FOOD INDUSTRY

Different uses of spectroscopic, fiber optic and picture detecting systems have been assessed and are introduced in Tables 5.1, 5.2, and 5.3, separately.

5.3.1. Applications of Spectroscopic Techniques

Infrared spectroscopy has been connected to the examination (organization, quality, and validness) of an assortment of sustenance items, for example, meat, cheddar, and apples, and this innovation can give fast portrayal of food items (Fagan and O'Donnell, 2008). Sivakesava et al. (2001) utilized mid-FTIR spectroscopy to identify sugar adulterants in squeezed apple. A direct discriminant investigation (LDA) classifier separated juice debased with beet and pure sweetener modify. Fractional minimum squares (PLS) and vital segment relapse (PCR) models were likewise created to decide the amount of beet and stick alter sugar in squeezed apple. Results showed the pertinence of IR methods for fast location and evaluation of contaminated in squeezed apple and different drinks. Inon et al. (2004) utilized FTIR spectroscopy to anticipate five wholesome parameters of drain, specifically add up to fat, add up to protein, add up to sugars (CH), calories, and calcium. The FTIR spectra were gathered altogether constricted reflection mode (ATR) from business drain tests covering wide parameter ranges and diverse drain writes, in particular, entire, semi-skimmed and skimmed, enhanced with calcium and vitamins, and changed by modifying lipid or sugar content. Phantom information were preprocessed and the adjustment set was chosen utilizing various leveled group examination (HCA), and forecast models were created utilizing PLS. Unearthly information accumulation utilizing ATR mode rather than transmission mode gave great flag to-clamor proportion as ATR was less impacted by the nearness of water in drain tests.

Utilizations of Raman spectroscopy have been accounted for validness location of matured drain, separation amongst glutinous and non-glutinous rice, confirmation of additional virgin olive oil and nectar. Afseth thought about the execution of Raman spectroscopy and NIR spectroscopy to decide the unsaturated fat organization and substance of principal constituents in a perplexing model food test made out of 70 unique blends of protein, water, and oil mixes and speaking to fish and meat. The model examples, and in addition unadulterated oil blends, were checked in Raman and NIR spectroscopy, and PLS models were created for forecast of unsaturated fat as far as iodine esteem and as substance of soaked, monounsaturated, and polyunsaturated unsaturated fats. Raman spectroscopy gave the best iodine esteem forecast precision; be that as it may, NIR gave the best exactness in deciding substance of primary constituents in the model examples.

Batsoulis et al. utilized FT-Raman spectroscopy to all the while decide fructose and glucose content in nectar. Reference information were gathered by an elite fluid chromatography (HPLC) strategy, and the PLS expectation show was produced utilizing Raman ghastly and HPLC reference information. The Raman spectrometer and standard HPLC strategy were observed to be measurably identical as far as exactness and reproducibility, however, HPLC requires more investigation time. Muik et al. (2005) utilized FT-Raman spectroscopy to specifically screen lipid oxidation in eatable oils. From the unadulterated part spectra, development of aldehydes was recognized and both immersed and unsaturated aldehydes were distinguished. The conjugated twofold bond frameworks development and cis to Trans isomerization of twofold bonds was seen in the C=C extending district with an extraordinary example for each oil. These distinctions were identified with unsaturated fat creation. This examination shows the ability of Raman spectroscopy to screen the oxidative corruption of lipids in consumable oils with no example arrangement necessity.

Several applications of NIR spectroscopy for on/in-line quality monitoring of meats, fruit and vegetables, grains and grain products, dairy and dairy products, fish and fish products, oils, and beverages have been reported (Huang *et al* ., 2008). NIR spectroscopy can potentially be used for multicomponent analysis (moisture, fat, protein, oil, etc.) for a variety of food products, such as biscuits, chocolates, cocoa powders, cheese, flour, cereals and cereal based products, milk and milk powder, oilseeds and oils, minced meat (beef, pork, turkey, chicken, lamb), snack foods, and animal feed (Benson, 2003). Nicolai *et al.* (2007) presented an overview of NIR instrumentation, chemometrics, calibration transfer, and various applications of NIR spectroscopy for soluble solid content (SSC) determination in a variety of horticulture products (apple, apricot, cherry, citrus fruit, grape, guava, kiwifruit, mandarin, mango, melon, nectarine, papaya, peach, pear, pineapple, plum) and quality evaluation (firmness, pH, dry matter, insoluble solids, stiffness, sugar, color, starch, acidity, sensory attributes, bruise, optimum harvest date, essential oils and their composition, carotenoids, defects, maturity, volatiles, pectins, hardness, density, citric acid, oil, moisture, oleic acid, linoleic acid, ripeness, chlorophyll, fresh weight, and contaminants) of fruits and vegetables (apple, apricot, cherry, citrus fruit, grape, guava, kiwifruit, macadamia nuts, mandarin, mango, melon, mushroom, nectarine, olive, onion, papaya, peach, pear, pepper, pineapple, plum, tangerine, and tomato).

Marquez *et al.* (2005) used NIR transmittance spectroscopy for on-line quality control and characterization of virgin olive oil. The NIR equipment was installed in a production line to collect spectra from real-time samples. Calibration models were developed for acidity value (AV), bitter taste (k225) and fatty-acid composition (FAME), using PLS regression and reference data. The results of PLS models demonstrated the potential of NIR spectroscopic technique for on-line and real-time quality monitoring and characterization of the oil during the oil-extraction process. Kawamura *et al.* (2007) developed an online NIR sensing system to monitor milk quality during milking. The NIR system was able to obtain spectra of unhomogenized milk during milking of individual cow and predict major milk constituents (fat, protein, and lactose), somatic cell count (SCC), and milk urea nitrogen (MUN) using a developed calibration model.

The precision and accuracy of the validated models showed the capability of the developed NIR systems for real-time quality monitoring of milk during milking. Chen *et al.* (2006) used NIR spectroscopy for rapid determination of organic acid composition in Japanese apricot juice. Spectral data were collected from juice samples of different organic acid composition, and calibration models were developed using PLS and validated using an independent validation set. The predicted values of organic acid composition (citric and malic acid contents) were in good agreement with reference values. The results demonstrated the capability of NIR techniques for quantitative compositional analysis of fruit juices.

A few uses of NIR spectroscopy for on/in-line quality observing of meats, leafy foods, grains and grain items, dairy and dairy items, fish and fish items, oils, and refreshments have been accounted for. NIR spectroscopy can possibly be utilized for multicomponent examination (dampness, fat, protein, oil, and so on.) for an assortment of food items, for example, bread rolls, chocolates, cocoa powders, cheddar, flour, grains and oat based items, drain and drain powder, oilseeds and oils, minced meat (hamburger, pork, turkey, chicken, sheep), nibble food, and creature encourage. Nicolai et al. (2007) exhibited a review of NIR instrumentation, chemometrics, adjustment exchange, and different utilizations of NIR spectroscopy for dissolvable strong substance (SSC) assurance in an assortment of cultivation items (apple, apricot, cherry, citrus natural product, grape, guava, kiwifruit, mandarin, mango, melon, nectarine, papaya, peach, pear, pineapple, plum) and quality assessment (solidness, pH, dry issue, insoluble solids, firmness, sugar, shading, starch, sharpness, tactile properties, wound, ideal gather date, basic oils and their piece, carotenoids, abandons, development, volatiles, pectins, hardness, thickness, citrus extract, oil, dampness, oleic corrosive, linoleic corrosive, readiness, chlorophyll, crisp weight, and contaminants) of foods grown from the ground (apple, apricot, cherry, citrus organic product, grape, guava, kiwifruit, macadamia nuts, mandarin, mango, melon, mushroom, nectarine, olive, onion, papaya, peach, pear, pepper, pineapple, plum, tangerine, and tomato).

Marquez et al. (2005) utilized NIR transmittance spectroscopy for on-line quality control and portrayal of virgin olive oil. The NIR gear was introduced in a creative line to gather spectra from constant examples. Adjustment models were produced for acridity esteem (AV), severe taste (k225) and unsaturated fat synthesis (Popularity), utilizing PLS relapse and reference information. The aftereffects of PLS models exhibited the capability of NIR spectroscopic method for on-line and continuous quality observing and portrayal of the oil amid the oil-extraction process. Kawamura et al. (2007) built up an online NIR detecting framework to screen drain quality amid draining. The NIR framework could acquire spectra of unhomogenized drain amid draining of individual cow and anticipate significant drain constituents (fat, protein, and lactose), substantial cell tally (SCC), and drain urea nitrogen (MUN) utilizing a created adjustment show.

The exactness and precision of the approved models demonstrated the capacity of the created NIR frameworks for ongoing quality checking of drain amid draining. Chen et al. (2006) utilized NIR spectroscopy for fast assurance of natural corrosive structure in Japanese apricot juice. Ghostly information were gathered from juice tests of various natural corrosive structure, and adjustment models were produced utilizing PLS and approved utilizing a free approval set. The anticipated estimations of natural corrosive synthesis (citrus and malic corrosive substance) were in great concurrence with reference esteems. The outcomes showed the capacity of NIR methods for quantitative compositional investigation of natural product juices.

UV spectroscopy has been used for detection of fruit juice and vegetable oil adulteration. Fruit juice and wine contain polyphenols whose concentration can be measured using the UV spectrum. This technique has been used to detect adulteration in Florida orange juice and authenticity of virgin olive oil (Meurens, 2003) (Table 5.1).

Table 5.1: Applications of Spectroscopic Techniques in the Food Industry

Technique	Product	Analysis	Reference
IR	Food products	Quality	Fagan and O'Donnell
			(2008)
IR	Apple	Adulteration	Sivakesava ***et al.*** (2001)
IR	Milk	Nutritional parameters	Iron ***et al.*** (2004)
Raman	Milk, rice, olive oil, honey	Authenticity	Meurens (2003
Raman and MR	Complex food	Fat composition	Afseth (2005)
Raman	Honey	Fructose and glucose	Batsoulis ***et al.*** (2005)
Raman	Oil	Lipid oxidation	Muik ***et al.*** (2005)
MR	Food products	Quality	Huang ***et al.*** (2008)
MR	Virgin olive oil	AV, bitter taste, fatty-acid composition	Marquez ***et al.*** (2005)
MR	Fruit and vegetables	SSC, quality	Nicolai ***et al.*** (2007)
MR	Food/feed	Moisture, protein, fat, oil	Benson (2003)
MR	Milk	Quality and composition	Kawamura ***et al.*** (2007)

MR	Apricot juice	Organic acid composition	Chen *et al.* (2006)
UV	Fruits and oil	Adulteration	Meurens (2003)

Source: G. Caldwell Darwin (2013), Robotics and Automation in the Food Industry: Current and Future Technologies, Woodhead Publishing

Table 5.2: Applications of the Fiber Optic Techniques in the Food Industry

Technique	Product	Analysis	Reference
NIR spectrometer	Cheese	Whey-fat concentration	Castillo *et al.* (2005)
Color sensor	Cheese	Curd syneresis	Everard *et al.* (2009)
NIR spectroscopy	Cheese	Curd moisture content and solids in whey	Mateo *et al.* (2010)
VIS reflection spectroscopy	Foal products	Quality	Lewis *el al.* (2008)
Fiber optic spectrometer	Meat	Emulsification process monitoring	Alvarez *el al.* (2009)

Source: G. Caldwell Darwin (2013), Robotics and Automation in the Food Industry: Current and Future Technologies, Woodhead Publishing

Table 5.3: Applications of Image Sensors in the Agricultural and Food Industries

Technique	Product	Analysis	Reference
Computer vision	Food	Quality	Brosnan and Sun
			(2004)
Computer vision	Food	Learning techniques	Du and Sun (2006)
Color imaging	Pizza	Classification	Du and Sun (2008)
Color imaging	Boiled food grains	Identification and classification	Anami and Burkpalli (2009)
Color imaging	Food products	Classification	De-Silva *et al.* (2005)
Color imaging	Grains, fruits, and flowers	Classification	Savakar and Anami (2009)
Color imaging	Food products	Quality	Zheng *et al.* (2006)

Hyperspectral imaging	Poultry	Quality	Lawrence ***et al.***
			(2003b)
Hyperspectral imaging	Chicken	Wholesomeness	Chao ***et at*** (2008)
Hyperspectral imaging	Apple	Fecal contamination	Kim ***et al.*** (2002)
Hyperspectral imaging	Apple	Bruise detection	Lu (2003)
Hyperspectral imaging	Cucumber	Chilling injury	Liu ***et at*** (2006)
Hyperspectral imaging	Strawberry	Moisture, total soluble solids, pH	ElMasry ***et** al.* (2007)
Hyperspectral imaging	Tomato	Ripeness	Polder ***et al.*** (2003)
Hyperspectral and color imaging	Wheat	Damage detection	Singh ***et al.*** (20I0a, 20 Ob)

Source: G. Caldwell Darwin (2013), Robotics and Automation in the Food Industry: Current and Future Technologies, Woodhead Publishing

5.3.2. Applications of Optic Fiber Sensing Techniques

Castillo et al. (2005) utilized optic fiber detecting innovation to decide whey-fat fixation in cheddar making. The light side scatter and transmittance spectra from the examples were gathered in the 300– 1100 nm wavelength run. Standardized ghastly information were corresponded with reference whey-fat fixation utilizing a power work. The model utilizing side scatter waveband proportion gave preferable expectation exactness over the transmission waveband proportion, exhibiting the capability of light scattered method in mechanized process control in the sustenance business. Everard et al. (2009) utilized an on-line fiber optic sensor and determined shading parameters to screen curd syneresis amid cheddar making. An on-line fiber optic shading sensor was introduced in a cheddar vat to screen curd syneresis. Shading parameters got from the sensor anticipated curd dampness and whey-fat substance amid syneresis. The outcomes demonstrated that shading optic fiber sensor can screen the syneresis procedure and enhance cheddar quality. Lewis et al. (2008) utilized optical fiber detecting innovation to screen sustenance quality amid cooking in mechanical expansive scale cooking stoves. Amid cooking food material changes its shading because of different compound responses. A shading sensor was utilized to screen the cooking procedure and unmistakable spectra were recorded at each stage. Different sustenance items, to be specific steamed skinless chicken filets, broil entire chickens, marinated chicken pieces, wieners, baked good, breadcrumb covering, and roast grilled chicken filets, were utilized as a part of the examination and obvious spectra were recorded.

A test, encased in stainless steel, comprising of a brightening optic fiber (appended to a tungsten incandescent light), and a fiber to manage diffuse reflected light to the spectrometer, was intended to cut the food material at rapid for examining without bringing about any harm. Dimensionality of the unearthly information was diminished by vital segment investigation (PCA) and characterization models were produced by manufactured neural systems (ANN) utilizing PCA lessened otherworldly information. Food items were effectively ordered into different cooked stages as crude, light, right, and dull. Alvarez et al. (2009) utilized an optical fiber sensor to screen solidness, amid the meat emulsification process, by noticeable light dispersing. Standardized backscatter spectra were determined and related with slashing circumstances and fat/lean proportions, which impacted meat emulsion strength. Expanding the separation amongst emanating and identifying optical strands diminished the scattered light force logarithmically.

Light spread additionally diminished because of increments in hacking term and fat/lean proportion. The fattier and over chopped emulsions were more insecure and caused all the more cooking misfortune. The ideal sensor parameters from this examination showed that different estimations gatherings (identifying filaments encompassing a transmitting fiber), divided over a moderately vast zone with under 2.50 mm outspread separation between the discharging and the distinguishing strands inside a gathering, would be more viable in emulsion observing. Mateo et al. (2010) utilized an on-line NIR optical sensor to decide curd dampness substance and solids in whey in cheddar making. An optic fiber line was utilized to produce and gather the backscattered light. The reflected backscattered light was transmitted through another fiber and collimating focal point, which centered the reflected light onto another fiber associated with an identifier. Spectra were recorded by the identifier and fitted to a cross-approval (forget one) relapse demonstrate for choice of valuable wavelengths utilizing difference as criteria.

The PLS models were created to foresee the curd dampness substance and whey solids utilizing single wavelength, single wavelength with other compositional and system parameters (time after gel cutting, gel cutting time, and drain fat level), and wide spectra. Single wavelength gave sensible outcomes, while single wavelength with different parameters gave the best forecast exactness. Whey strong could be anticipated just utilizing the full phantom range.

5.3.3. Applications of Image Sensing Technology

Imaging sensors (cameras) have been widely utilized as a part of a few applications in agribusiness and food ventures for quality and security investigation, harm location, arranging, evaluating, and observing of ongoing food preparing frameworks. Most basic imaging frameworks utilize sensors delicate to the VIS area of the electromagnetic range. Shading cameras utilized as a part of shading imaging frameworks for modern process assessment are the slightest costly cameras utilized as a part of machine vision innovation. Machine vision innovation has been utilized as a part of the appraisal of leafy foods (apple, oranges, strawberries, nuts, tomatoes, peaches, and pears), vegetable review (mushrooms, potatoes, and others), grain characterization and quality assessment (wheat, corn, rice, grain, oats, rye), and applications in sustenance items (pizza, bread shop items, cheddar, meat, and meat items) (Brosnan and Sun, 2004).

Three-chip CCD finders are utilized to get pictures at red, green, and blue groups of the VIS range and consolidated to mimic a close genuine nature picture of the protest as observed by human eyes. Shading pictures contain a considerable measure of data fit as a fiddle, size, shading, and surface of a question, which are quantitatively portrayed by separating shading, textural, and morphological (shape and size) highlights. These highlights are utilized as a part of example acknowledgment and in getting the hang of/preparing the classifiers/forecast models for future order and expectation of constituents in food items. The learning strategies utilized as a part of machine vision innovation are ANN, factual learning (SL), fluffy rationale, hereditary calculations, and choice trees; nonetheless, ANN and SL are the most broadly received learning techniques.

Shading, size, shape, and textural highlights have been separated and connected in the quality investigation of a few items in the agri-sustenance industry in particular apple, meat, chicken, pork, cocoa spread, angle, lettuce, maize noodle, nut, pizza, peach, tomato, cucumber, corn, cherry, orange, grape, rice, grains, and poultry (Zheng et al., 2006).

Anami and Burkpalli (2009) created shading based models to distinguish diverse bubbled grain writes and afterward arrange bubbled grain into full, medium, and half bubbled grains. Five sorts of grain, in particular, Bengal gram (Cicer arietinum), tur dal (Fragaria), green gram (Vigna emanate), dark gram (Vigna mungo) and red gram (Carolus Linnaeus) were utilized as a part of the examination. Shading highlights were separated to create distinguishing proof and grouping models. The outcomes from this examination showed the potential utilization of the shading imaging framework for mechanized observing of the cooking procedure in the food business. De-Silva et al. (2005) proposed a technique in view of a watershed division calculation to subgroup shading pictures of food tests (pizza, butterfly cakes, and scones) into gatherings of comparative shading attributes.

Each portion in a picture was independently broke down and its shading attributes were contrasted and an edge an incentive for every sustenance item. Tests falling underneath the edge esteem were rejected. The created calculation could identify the shading changes in the food tests with high exactness. Savakar and Anami investigated the likelihood of utilizing the shading imaging framework for order of grain into ten grain writes; arrangement of grapes into ten assortments; characterization of mangoes into five assortments; grouping of jasmine blossoms into ten assortments. Shading and textural highlights (morphological highlights were likewise included for grapes) were given as contribution to the back proliferation neural systems (BPNN). The grains were grouped with the most noteworthy precision. Mangoes gave the most reduced exactness because of the comparable shading and surface of various assortments.

Du and Sun (2008) utilized a shading machine vision framework to group the pizza base, sauce spread, and fixing. Picture highlights (shape highlights of the pizza base and shading highlights of pizza sauce spread and garnish) were removed and given as contribution to a help vector machine (SVM) for order (dimensionality of the shading highlights was lessened before being given a contribution to the SVM). The outcomes demonstrated the capacity of shading imaging frameworks in the programmed multi-grouping of pizza base, sauce spread, and garnishes. Choudhary et al. (2008) utilized a shading imaging framework to group Canada Western Red Spring wheat, Canada Western Golden Durum wheat, grain, oats, and rye.

Wavelet, shading, morphological, and textural highlights were removed from the shading pictures of non-touching single bits and order calculations were produced by LDA and QDA utilizing distinctive blends of separated highlights. Direct discriminant classifiers utilizing joined wavelet, shading, morphological, and textural picture includes as info gave the most elevated grouping precision (89.4–99.4%).

Shading pictures don't give data about the concoction creation and its appropriation in the question and depend for the most part on outer surface highlights of the protest under scrutiny. Hyperspectral imaging gives the phantom data in a spatially settled way, by filtering an example at many limited wavelengths. Hyperspectral imaging has been effectively connected in pharmaceutical definitions to dissect compound conveyance and measure dynamic pharmaceutical fixings (Programming interface). This method has shown high potential for constant mechanized quality observing in the farming and sustenance preparing ventures. Hyperspectral imaging has been utilized to distinguish fecal and ingesta defilement on poultry remains (Lawrence et al., 2003b). The poultry remains were checked utilizing a NIR hyperspectral imaging framework and furthermore a NIR spectrophotometer. Otherworldly information were investigated utilizing PCA scores and loadings to recognize critical wavelengths.

Picture proportions at huge wavelengths (434, 517, 565, and 628 nm) in different wavelength mixes were utilized to get wavelength proportion pictures. Histogram extending and fecal thresholding were connected in the mapped proportion pictures for subjective and quantitative examinations, and 100% exactness in a subjective model and over 96% in quantitative model were accomplished. Utilization of just four wavelengths diminished the dimensionality of hyperspectral information, making it multispectral investigation for continuous applications. Dimensionality of hyperspectral information can be lessened by applying the multivariate picture examination (MVI) procedure, which utilizes PCA to distinguish noteworthy wavelengths for multispectral investigation. Multivariate picture investigation will evacuate the requirement for extra NIR spectroscopic instruments and will conquer framework variety impacts. Chao et al. (2008) built up a high throughput on-line otherworldly imaging framework for healthiness review of chicken in a fast business chicken preparing line. The line-examine framework comprised of an electron-duplicating charge-coupled gadget (EMCCD) indicator and a line-check spectrograph.

Hyperspectral pictures of healthy and unwholesome chicken bodies were gained in reflectance mode and changed over into relative reflectance utilizing a standard. The relative reflectance pictures were then broke down to advance the area of intrigue (return for money invested), the district giving the most noteworthy ghastly contrasts amongst healthy and unwholesome chicken corpses over all wavebands, and huge wavebands were distinguished for multispectral investigation. In the multispectral investigation, noteworthy wavelengths and two-waveband proportions were utilized and arrangement calculations were produced by applying a fluffy rationale participation works that recognized 99% healthy and 94% unwholesome chickens at the very least arranging rate of 140 winged creatures for every moment (BPM).

The framework was equipped for achieving food security standard and rapid creation necessities, achieving in this way expanding generation productivity. Kim et al. (2002) utilized multispectral imaging, in light of noteworthy wavelengths recognized from hyperspectral imaging information, to distinguish fiscally debased apples. A basic edge was connected to a NIR band (noteworthy wavelength) picture to make a cover for expelling the foundation. At that point, PCA was connected to foundation expelled pictures to segregate the uncontaminated apple surface territory from defiled surface zone with the assistance of essential part (PC) score pictures (pseudo shading maps of PC scores). Lu (2003) utilized NIR hyperspectral imaging to distinguish the wounds on apple surfaces. Crisp apples and wounded apples were checked and a base commotion division (MNF) change and PCA were connected to the chosen return on initial capital investment. A changed picture was produced by duplicating first and third PC pictures.

Mean pixel force esteems in the changed picture relating to the dim and light zones in MNF picture were figured for the correlation. The territories on the wounded surfaces had higher mean pixel power. Separation precision of the framework was influenced by apple assortment, wound advancement organize (from first to 47th day of wound improvement), and ghastly determination. Comparable parameters ought to be thought about in future model advancements. Chilling damage in cucumbers was likewise identified by hyperspectral imaging by applying both ghostly and picture investigations (Liu et al., 2006). A double band picture proportion (811 nm/756 nm) gave preferred segregation over PC score pictures and accurately ordered over 90% of the cucumbers.

El Masry et al. (2007) utilized hyperspectral imaging to foresee dampness content, add up to solvent solids, and pH in strawberries. The return on initial capital investment was chosen in the wake of evacuating the foundation by making a parallel cover, and after that arrived at the midpoint of spectra from return for capital invested in each hypercube were acquired. Arrived at the midpoint of spectra were utilized as a part of creating PLS adjustment models. Multilinear relapse (MLR) forecast models were additionally created utilizing β coefficients from PLS investigation. Surface investigation was done to characterize strawberries into unripe, ready and overripe stages. The variety in the surface was related with light power or dim levels. A shading picture for each example was framed by removing red (650 nm), green (500 nm), and blue (450 nm) band pictures from a hypercube.

Shading pictures were then examined by gray level co-event network (GLCM) to decide textural includes and utilized for grouping by discriminant examination. Hyperspectral imaging has been additionally been researched for arranging tomatoes (Polder et al ., 2003). The dimensionality of hyperspectral information was decreased by autonomous segment examination (ICA), otherwise called a visually impaired source detachment strategy. Unsupervised arrangement models were produced by pseudo-shading mapping of autonomous part scores like PCA scores. The ICA needs to indicate the autonomous segments previously beginning calculation, this requires a more drawn out time to merge, and may not be reasonable for quick online reviews. Hyperspectral imaging has been examined for its potential applications in the grain taking care of and handling ventures for grain quality assessment.

Singh et al. (2010 a, 2010b) utilized hyperspectral imaging to identify midge-harmed and bug harmed wheat pieces. Multivariate picture examination was utilized to distinguish critical wavelengths, and factual and histogram highlights from the huge wavelength NIR picture were extricated and given as contribution to measurable classifiers and BPNN classifiers. Execution of the classifiers was contrasted and shading imaging frameworks. Shading, textural, and morphological highlights from the shading pictures of the sound harmed wheat pieces procured utilizing the shading imaging framework were separated and utilized as a part of arrangement. The most noteworthy arrangement exactness was accomplished by consolidating the NIR and shading imaged highlights.

Shading pictures don't give data about the concoction creation and its appropriation in the question and depend for the most part on outer surface highlights of the protest under scrutiny. Hyperspectral imaging gives the phantom data in a spatially settled way, by filtering an example at many limited wavelengths. Hyperspectral imaging has been effectively connected in pharmaceutical definitions to dissect compound conveyance and measure dynamic pharmaceutical fixings (Programming interface). This method has shown high potential for constant mechanized quality observing in the farming and sustenance preparing ventures. Hyperspectral imaging has been utilized to distinguish fecal and ingesta defilement on poultry remains. The poultry remains were checked utilizing a NIR hyperspectral imaging framework and furthermore a NIR spectrophotometer.

Otherworldly information were investigated utilizing PCA scores and loadings to recognize critical wavelengths. Picture proportions at huge wavelengths (434, 517, 565, and 628 nm) in different wavelength mixes were utilized to get wavelength proportion pictures. Histogram extending and fecal thresholding were connected in the mapped proportion pictures for subjective and quantitative examinations, and 100% exactness in a subjective model and over 96% in quantitative model were accomplished. Utilization of just four wavelengths diminished the dimensionality of hyperspectral information, making it multispectral investigation for continuous applications. Dimensionality of hyperspectral information can be lessened by applying the multivariate picture examination (MVI) procedure, which utilizes PCA to distinguish noteworthy wavelengths for multispectral investigation. Multivariate picture investigation will evacuate the requirement for extra NIR spectroscopic instruments and will conquer framework variety impacts.

Chao et al. (2008) built up a high throughput on-line otherworldly imaging framework for healthiness review of chicken in a fast business chicken preparing line. The line-examine framework comprised of an electron-duplicating charge-coupled gadget (EMCCD) indicator and a line-check spectrograph. Hyperspectral pictures of healthy and unwholesome chicken bodies were gained in reflectance mode and changed over into relative reflectance utilizing a standard. The relative reflectance pictures were then broke down to advance the area of intrigue (return for money invested), the district giving the most noteworthy ghastly contrasts amongst healthy and unwholesome chicken corpses over all wavebands, and huge wavebands were distinguished for multispectral investigation. In the multispectral investigation, noteworthy wavelengths and two-waveband proportions were utilized and arrangement calculations were produced by applying a fluffy rationale participation works that recognized 99% healthy and 94% unwholesome chickens at the very least arranging rate of 140 winged creatures for every moment (BPM). The framework was equipped for achieving food security standard and rapid creation necessities, achieving in this way expanding generation productivity.

Kim et al. (2002) utilized multispectral imaging, in light of noteworthy wavelengths recognized from hyperspectral imaging information, to distinguish fiscally debased apples. A basic edge was connected to a NIR band (noteworthy wavelength) picture to make a cover for expelling the foundation. At that point, PCA was connected to foundation expelled pictures to segregate the uncontaminated apple surface territory from defiled surface zone with the assistance of essential part (PC) score pictures (pseudo shading maps of PC scores). Lu (2003) utilized NIR hyperspectral imaging to distinguish the wounds on apple surfaces. Crisp apples and wounded apples were checked and a base commotion division (MNF) change and PCA were connected to the chosen return on initial capital investment.

A changed picture was produced by duplicating first and third PC pictures. Mean pixel force esteems in the changed picture relating to the dim and light zones in MNF picture were figured for the correlation. The territories on the wounded surfaces had higher mean pixel power. Separation precision of the framework was influenced by apple assortment, wound advancement organize (from first to 47th day of wound improvement), and ghastly determination. Comparable parameters ought to be thought about in future model advancements. Chilling damage in cucumbers was likewise identified by hyperspectral imaging by applying both ghostly and picture investigations. A double band picture proportion (811 nm/756 nm) gave preferred segregation over PC score pictures and accurately ordered over 90% of the cucumbers. El Masry et al. (2007) utilized hyperspectral imaging to foresee dampness content, add up to solvent solids, and pH in strawberries.

The return on initial capital investment was chosen in the wake of evacuating the foundation by making a parallel cover, and after that arrived at the midpoint of spectra from return for capital invested in each hypercube were acquired. Arrived at the midpoint of spectra were utilized as a part of creating PLS adjustment models. Multilinear relapse (MLR) forecast models were additionally created utilizing β coefficients from PLS investigation. Surface investigation was done to characterize strawberries into unripe, ready and overripe stages. The variety in the surface was related with light power or dim levels. A shading picture for each example was framed by removing red (650 nm), green (500 nm), and blue (450 nm) band pictures from a hypercube. Shading pictures were then examined by gray level co-event network (GLCM) to decide textural includes and utilized for grouping by discriminant examination. Hyperspectral imaging has been additionally been researched for arranging tomatoes (Polder et al., 2003).

The dimensionality of hyperspectral information was decreased by autonomous segment examination (ICA), otherwise called a visually impaired source detachment strategy. Unsupervised arrangement models were produced by pseudo-shading mapping of autonomous part scores like PCA scores. The ICA needs to indicate the autonomous segments previously beginning calculation, this requires a more drawn out time to merge, and may not be reasonable for quick online reviews. Hyperspectral imaging has been examined for its potential applications in the grain taking care of and handling ventures for grain quality assessment. Singh et al. (2010 a, 2010b) utilized hyperspectral imaging to identify midge-harmed and bug harmed wheat pieces. Multivariate picture examination was utilized to distinguish critical wavelengths, and factual and histogram highlights from the huge wavelength NIR picture were extricated and given as contribution to measurable classifiers and BPNN classifiers.

Execution of the classifiers was contrasted and shading imaging frameworks. Shading, textural, and morphological highlights from the shading pictures of the sound harmed wheat pieces procured utilizing the shading imaging framework were separated and utilized as a part of arrangement. The most noteworthy arrangement exactness was accomplished by consolidating the NIR and shading imaged highlights.

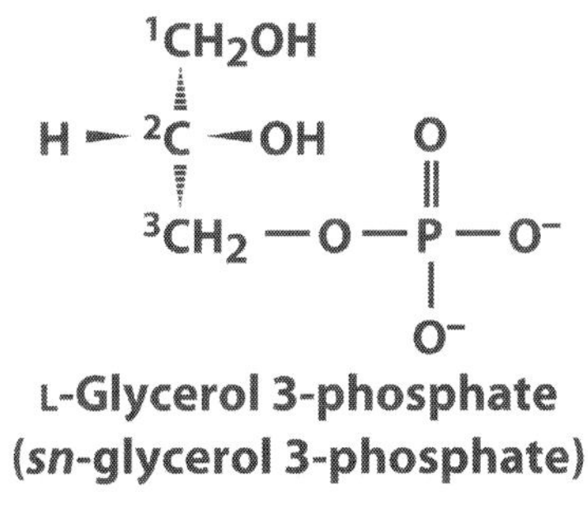

Figure 5.2: sn-glycero-3-phosphate, (*Source: http://slideplayer.com/slide/9309297/*).

5.4. OCCURRENCE OF PHOSPHOLIPIDS

All living organisms, plants, animals or microorganisms are made up of minute functional units, called cells. These cells are surrounded by an external membrane, but numerous membrane structures also exist within the cell. The principal structure-forming substances in these highly functional units (i.e. cells) are the phospholipids. Phospholipids are therefore present in all living organisms and can be obtained from a wide variety of raw materials in the natural world. Essentially, however, the only raw materials suitable for commercial use are oilseeds and egg yolk. Plant raw materials have a low phospholipid content, which (as a percentage of dry matter) does not exceed 2.5%. Raw materials of animal origin, by contrast, have much higher phospholipid content. In dried whole milk, for example, it is around 2% and in dried egg yolk around 17%. The phospholipid composition and fatty acid profile generally gives us a clear indication of the origin of the product.

The phospholipids are the functional ingredients of all lecithins both from a technological and physiological point of view. The broader range of processes and technologies available offer the possibility to produce tailor-made functional ingredients which are used in various special food applications especially in food supplements, functional food, cosmetics and pharmaceutical application products. Every phospholipid has its own specific application profile. Extensive scientific studies have demonstrated the effects on the human body that can result from the nutritional action of phospholipids such as phosphatidylserine as a brain cell nutrient, phosphatidylcholine for liver cell regeneration, soya phospholipids for lipid reducing effects and hydrogenated phospholipids as a basis for production of stable liposomes.

5.5. LECITHIN

Lecithin is a naturally occurring fatty substance referred to a phospholipid. Although the term lecithin is sometimes used generically to refer to a wide variety of lipids (including choline and glycine), it is most commonly associated with phosphatidylcholine, a precursor for choline. It is naturally found in several foods, including egg yolks, soybeans, nuts, and whole grains, as well as in organ meats. Because lecithin is not considered an essential nutrient, no recommended daily allowance (RDA) has been established. Generally, dietary intake of lecithin ranges from 1–5 g/d, but some have proposed that current trends toward low-fat diets may be producing lecithin deficits (Potter et al., 2009).

Lecithins are among the most widely used emulsifiers in the food industry. However, although used in a vast variety of applications, it is difficult to find systematic data about the emulsification and application properties of lecithin. For the user of lecithin, this information should be of specific interest, as lecithin is not a uniform, standard material but a natural mixture of a series of surface active components that are contributing to the overall emulsifying performance. In addition, there is a wide variety of materials that, according to legislation, are called 'lecithin' but vary significantly in their composition and thereby functionality. And the term 'lecithin' might have a substantially different meaning within commercial and food legislation areas. Lecithin from soy, sunflower, and canola is produced to comply with food-grade regulatory specifications. The chemical analyses include acetone-insoluble matter, hexane–insoluble matter, acid value, moisture, peroxide value, and viscosity. The principal groups of diacyl-phospholipids that can be distinguished by the nature of 'Z.'

As further sub-groups of phospholipids, besides glycerophospholipids, sphingosine-phospholipids and glycolipids also need to be mentioned. One of the best-known representatives of sphingosine-phospholipids is sphingomyelin that occurs exclusively in animal organisms. It consists of one molecule each of ceramide, phosphoric acid and choline (ceramide phosphocholine) (see figure 5.3). Glycolipids occur mainly in plants and microorganisms, although small quantities are also found in animals, particularly in the brain. 1,2-Diacyl-sn-glycerol is linked to the carbohydrate unit via 3-hydroxy group by means of α- or β-glycosidic bond. The most common representatives of this group are monogalactosyl and digalactosyl-diglycerides (MDGD, DGDG).

Based on this, we shall discuss the extraction of lecithin from these three different crops.

5.5.1. Soy Lecithin

Soybeans are the primary commercial source of vegetable lecithin. Solvent extraction of soybeans is carried out in large crushing plants with capacities of 1000–6000 metric tons per day, yielding 18% oil. The extracted oil is filtered to optimize the oil degumming efficiency as well as the production of lecithins with low residual dirt (earth, seed protein residue), expressed as hexane insoluble The water degumming process may give the best quality of standard lecithin. The gums removed from the oil in the enzymatic degumming step may contain high amounts of hydrolyzed lysophospholipids with new functional properties. The gums with 20–50% moisture are removed from the oil by continuously operating decanters. Lecithin is dried in rotating film evaporators within 1–2 minutes to a content of <1% moisture, achieving long shelf life and good liquidity. The product is rapidly cooled, preventing post-darkening. Standard lecithin is adjusted for oil content and viscosity;

and it is used for processing modified PL, such as hydrolyzed, acetylated, hydroxylated, fractionated, or deoiled lecithins.

5.5.2. Sunflower Lecithin

Oilseed-type sunflower seeds have mostly black hulls that tend to stick to the meat. Sunflower crushing requires dedicated crushing plants since the seeds are smaller than soybeans. Dehulling efficiency may influence the natural wax content of 0–1% in lecithin. Hence, additional expelling (pressing) is often used to squeeze out oil and reduce the original oil content of the seed (~40%) down to 15% in the press cake; the press cake is then hexane–extracted to reduce the oil to under 0.5%. Both press oil and extraction oil streams are combined for degumming. Sunflower lecithin yields are about 0.3% on seed basis. Intensive filtration of the non-degummed oil is important for obtaining lecithin with low-impurity HI contents of less than 0.3%. Sunflower lecithin is a preferred alternative to soy lecithin in Europe. The FA composition of sunflower lecithin is attractive. Its taste-flavor profile is pleasant (nutty-like), and its emulsifying properties are good.

Liquid sunflower lecithin is a perfect replacer for Soy fluid lecithin and thus it carries the same performance. In addition, fluid lecithin is irreplaceable as a wetting agent, antioxidant, instantizing agent, lubricant, and a release agent. The main sunflower growing regions are Argentina, France, Hungary, Ukraine, and Russia. Sunflower lecithin was approved by the FDA in 2009 and is the only raw form of lecithin available (Figure 5.3).

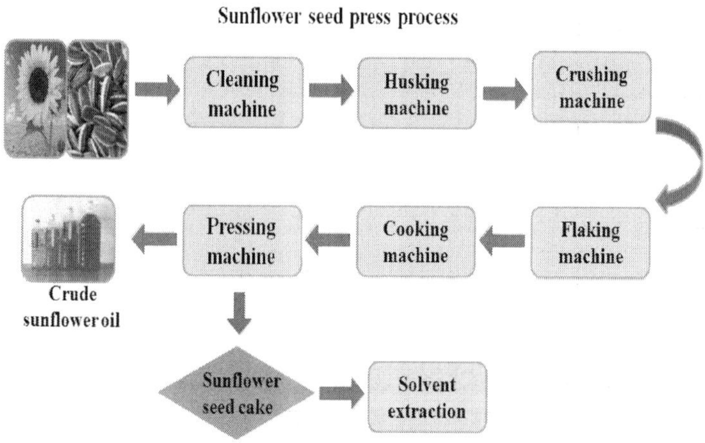

Figure 5.3: Sunflower oil press process, (*Source: http://www.edibleoilrefinery-machine.com/SOLUTION/sunflower_oil_press_machine_242.html*).

5.5.3. Canola (Rapeseed) Lecithin

Rapeseed is a bright yellow flowering member of the family Brassicaceae. It has been cultivated for centuries by ancient civilizations in Asia and more recently in Europe. Subsequently, during the Industrial Revolution, when the superior lubricant qualities of the oil extracted from rapeseed were appreciated, its cultivation became increasingly popular, but its use as an edible oil is only very recent (Shahidi, 1999). Compared to other oilseeds, a peculiarity of rapeseed is that it includes different species, though all belong to the same genus Brassica. In fact, different botanical species, namely, *B. napus, B. campestris*, *B. rapa*, *B. carinata,* and *B. juncea*, which are closely related and very similar in appearance, are all cultivated in different parts of the world and commonly referred to as "rapeseed." A second peculiarity of rapeseed is that it is one of the few oilseed crops that can be cultivated in cool temperate climates. Rapeseed is known by a variety of common names, including, rape, oilseed rape colza, mustard, rapa, and turnip rape, but lately, it is increasingly referred to by the name "canola." Canola seed crush and lecithin production are quite similar to sunflower processing; most often an expelling step is used to squeeze out oil from an original 40% to below 20%, followed by solvent extraction. Since canola contains chlorophyll components, lecithin from canola may have a slight greenish tone. This is masked, however, in food recipes because the usual dosage content is less than 0.5%. Canola varieties of rapeseed produce little or no erucic acid (C22:1) and thioglycosides, in contrast to lecithin produced from high-erucic acid rapeseed. The lecithin is used in food, feed, and technical industries. Canola lecithin is produced in Canadian, European, and Chinese crushing plant.

5.5.4. Animal Lecithins

Unlike the production of vegetable lecithins, there is no comparable industrial process for the production of animal oils. Various methods are therefore used for producing lecithins of animal origin. Egg lecithin, the most important product in this category, is usually produced by combined extraction with ethanol and acetone. Due to these different processes, egg lecithins are more expensive in their production and are therefore not competitive with plant lecithins in technological driven applications of the food industry. They, however, play a key role, for example, in infant nutrition under physiological aspects (they can deliver high amounts of arachidonic and docosahexaenoic acid) and is highly purified and isolated versions as emulsifiers in pharmaceutical emulsion systems.

5.6. MONO AND DIACYLGLYCEROLS

Mono- and diacylglycerols are the most widely used synthetic emulsifiers in the food industry. They are present in small quantities in natural fats and oils as a result of hydrolysis, which also releases fatty acids. They consist of esters synthesized via catalytic transesterification of glycerol with triglycerides, with the usual triglyceride source as hydrogenated soybean oil. Mono- and diglycerides are also synthesized directly from glycerol and fatty acids under alkaline conditions. Monoacylglycerols, which contain two free hydroxyl groups, exhibit stronger surface activity than diacylglycerols. In the laboratory, monoacylglycerols may be prepared by reaction of a fatty acyl chloride with glycerol in the presence of pyridine, which acts both as a solvent and an organic base. However, the corrosivity of acyl chlorides and the toxicity of pyridine are problematic for commercial application of this approach. For example, the isopropylidene (acetonide) protective group can block the 1 and 2 positions of glycerol while esterification can be performed on the 3–position. Glycidol, an epoxide derived from glycerol, may also be used as a basic material to produce pure monoacylglycerols. Diacylglycerols may be used as intermediates in the synthesis of regioselective and chiral triacylglycerols and Phospholipids. The two most prevalent commercial preparations of mono- and diacylglycerols are:

(1) Direct esterification of glycerol with a fatty acid.

(2) Glycerolysis of natural or hydrogenated fats and or oils.

The glycerolysis procedure is more economical because fats are cheaper than fatty acids and less glycerol is required. Fats and fatty acids are insoluble in glycerol and, in the absence of solvent; elevated temperatures are required to force the reaction to proceed. Direct esterification may be catalyzed either by acids or bases. The ratio of glycerol to fatty acid determines the concentrations of mono-, di- and triacylglycerols in the final product. Higher levels of glycerol produce higher concentrations of monoacylglycerols. In a typical batch procedure, fatty acid, glycerol, and catalyst are stirred at 210–230 °C. Water is continuously removed by distillation, causing the equilibrium to shift toward products. Progress of the reaction is monitored by periodic measurement of the acid value. Neutralization is more critical when batch distillation is used than for rapid short path/ short time processes. In the batch process, the reaction time varies from 1 to 4 hours and in the continuous process, the time may be less than ½ hour. The enzymatic preparation of mono- and distearin by glycerolysis of ethyl stearate and direct esterification of glycerol in the presence of a lipase from *Candida Antarctica* has also been reported.

5.6.1. Molecular Distillation

Monoglycerides may be separated from the di- and triglycerides and glycerol via high–molecular thin film molecular distillation. The concentration by molecular or short-path distillation takes place under high vacuum condition (i.e. down to 10−8 mmHg), which allows lower distilling temperatures (ca. 140–170 °C) and shorter residence time of the feed liquid. In combination with optimal efficiency in heat and mass transfer, minimal product decomposition and maximum product quality is reached. A typical composition of the distillate contains 95% monoglycerides, 3–4% diglycerides, 0.5–1% free glycerol and 0.5–1% free fatty acids.

Monoglycerides produced in this way contain equilibrium of 1-monoglycerides and 2-monoglycerides. The ratio between the two isomers is dependent on temperature. The rate constant of the equilibrium is low at room temperature and depends on fatty acid composition, crystal form, and traces of basic catalyst present. The content of 1-monoglyceride in commercially distilled monoglycerides is usually 90–95% (Figure 5.4).

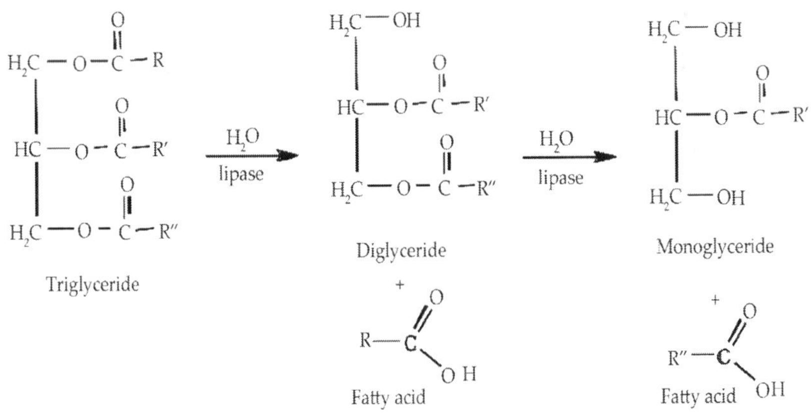

Figure 5.4: Hydrolysis of a triacylglycerol to form mono-and diacylglycerols, (*Source: http://www.namrata.co/digestion-of-lipids-a-quick-revision/*).

5.7. DIACETYL TARTARIC ESTERS OF MONOGLYCERIDES (DATEM)

Diacetyl tartaric esters of monoglycerides (DATEM) are created when monoglycerides are combined with diacetyl tartaric acid anhydride. The DATEM are glycerol derivates esterified with edible fatty acids and mono and diacetyl tartaric acids. In principle, two different processes exist for the manufacturing of these species. The first one is the esterification of mono- and diglycerides with tartaric- and acetic acid in the presence of acetic acid anhydride. The other process is the reaction of diacetyl tartaric acid anhydride in the presence of acetic acid with mono- and diglycerides. First, the diacetyl tartaric acid anhydride is manufactured via the reaction of diacetyl tartaric acid with acetic acid anhydride. The acetic acid liberated in this reaction has to be separated by distillation.

Then the reaction with monoglycerides is possible. The DATEM is an anionic emulsifier, meaning that it carries a negative charge. This property allows DATEM to interact with proteins readily. Consequently, DATEM has been found to be an excellent dough strengthener when used in breads and other baked products. The DATEM include glycerol derivatives esterified with edible fatty acids and mono- and diacetyl tartaric acid, generally permitted for the use in foodstuffs and as dough conditioners for all baked products, particularly yeast-leavened products, such as white bread. The hydrophilic/ lipophilic balance (HLB) value is 8–10. The optimal dosage is between 0.25 and 0.50% (flour basis). It also comes as a sticky viscous liquid, or with a consistency like fats, or yellow waxes, or in lakes or powder form. In comparison with the mono- and diglycerides, and its starting materials, DATEM is more hydrophilic in nature.

5.8. SUCROSE ESTERS

Sugar esters synthesized from renewable sources such as fatty acids and sucrose constitute an interesting group of nonionic surfactants. The food industry, as detergents, oral-care products, and medical supplies, and also in agriculture as insecticides. Sucrose esters have very wide HLB, excellent physical properties, and surface activity. Sucroesters were investigated to possessed emulsion stabilization activity, as sucrose laurate and sucrose palmitate have the capacity to stabilized O/W emulsions.

Fully esterified sucrose fatty acid esters have been widely investigated as synthetic fat replacements and their synthesis has been reviewed (Swanson and Swanson, 1999). Partially esterified sucrose esters are versatile emulsifiers for food products. The distribution of mono- di- and triesters, and therefore the HLB, may be controlled by the ratio of fatty acid and sucrose in the reaction mixture. The degree of saturation and chain length of the fatty acid also influence the functional properties of the product. As with other polyol starting materials, sucrose fatty acid esters are prepared by interesterification. However, sucrose undergoes caramelization reactions above 140 °C. High temperatures cannot be used to force homogeneity of the two-phase reaction mixture. One approach is to carry out a base-catalyzed interesterification with fatty acid methyl esters in a solvent, such as Dimethylformamide (DMF) or dimethyl sulfoxide (DMSO). The major disadvantage of this method is the difficulty of completely removing the high-boiling, toxic solvent.

Another simulated approach is the use of high levels of soap or other surfactants to promote miscibility of the phases. Excess soap may be removed by neutralization to the fatty acid, followed by short path distillation. We can utilize solvent extraction, such as in an ethyl acetate/water mixture, as an alternative. Sucrose octaacetate, an oil-soluble derivative of sucrose may be used as a starting material to promote a homogeneous reaction. Reaction of sucrose with methyl esters can be performed with a high-shear, mixer to improve contact between the insoluble phases. A continuous process, where the reaction mixture is passed through an immobilized solid catalyst, has been described. A two-component emulsifier system of sucrose esters and monoacylglycerols may be obtained by interesterification of sucrose and triacylglycerols (Nakamura et al., 1986). Enzyme-catalyzed interesterification may be used to produce regioselective isomers of sucrose esters (see figure below). R_1, R_2, and R_3 are the acylation positions in lactulose, where each R is a lauryl residue. When the synthesis of monoesters occurs, the acylation positions are 1-O-, 6-O-, or 6′-O-, diester: (1-O-, 6-O-), (1-O-, 6′-O-), or (6-O-, 6′-O-), and triester: (1-O-, 6-O-, 6′-O (Figure 5.5).

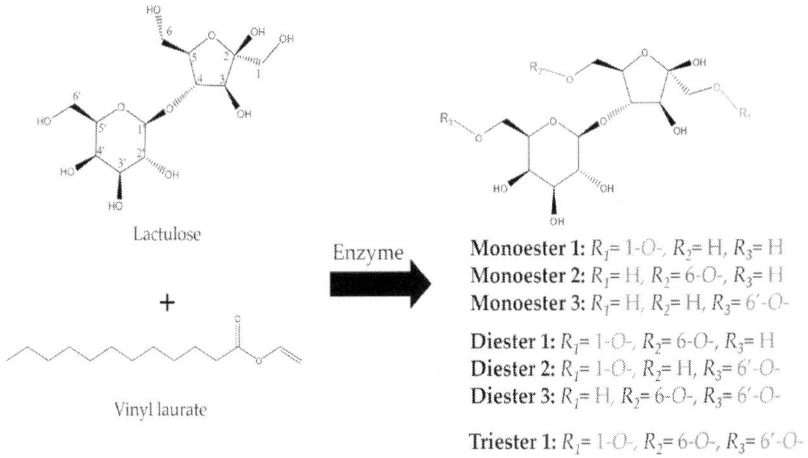

Figure 5.5: Studied enzymatic reactions, (*Source: http://www.mdpi.com/2073-4344/7/9/263/htm*).

5.9. SODIUM AND CALCIUM STEAROYL LACTYLATE

A surfactant with a carboxylic acid functional group may be nonionic, or if reacted with sodium or calcium hydroxide, converted into an anionic molecule. Lactic acid is a bifunctional molecule, which can self-condense to form an oligomer or react with a fatty acid to form stearoyl lactylic acid. Reaction with sodium or calcium hydroxide forms sodium or calcium stearoyl lactylate.

In a typical preparation, lactic acid is neutralized with sodium or calcium hydroxide and excess water is removed by distillation. Iron is highly detrimental to the quality of the product. Consequently, raw materials should have minimum iron content and the reactor should contribute no leachable iron. Stearic acid is added and esterification is carried out at 160–180 °C. Higher temperatures lead to side reactions, which produce dark colors and disagreeable odors and flavors. Water of reaction is removed by distillation and acid value is monitored until a minimum value is obtained. Color of the final product may be improved by bleaching with 30% hydrogen peroxide followed by heating to destroy excess peroxide. The final product is characterized by acid value, saponification number and total lactic acid.

5.10. SORBITAN ESTERS

Sorbitan esters of fatty acids are derived from a reaction of sorbitol and a commercial grade fatty acid. Sorbitol, the starting material for sorbitan esters along with mannitol, was originally produced from the hydrogenation of sucrose. However, currently, a more economic and pure form of sorbitol is produced commercially from d-glucose which itself is derived from corn syrups (maize) or tapioca. D-glucose is reacted with hydrogen gas under pressure in the presence of a catalyst such as nickel phosphate or ruthenium dichlorobiphenyl phosphine. The product of this hydrogenation process is d-glucitol, or what is commonly known as Sorbitol. Commercial grade sorbitol typically contains minor amounts of other polyols such as mannitol. Sorbitol belongs to a group of compounds called sugar alcohols or polyols, which have all the properties similar to that of glycerol. In the formula below, each Z independently denotes a hydrogen atom or C(O)R-, and each R mutually independently denotes a hydrocarbyl group of 9 to 99 carbon atoms, more preferably 11 to 23 carbon atoms. Examples of sorbitan esters include sorbitan stearates and sorbitan oleates (Figure 5.6).

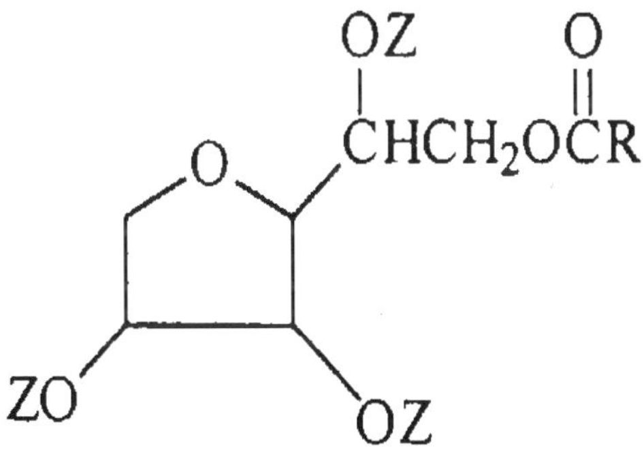

Figure 5.6: Structure of Sorbitan Ester, (*Source: https://encrypted.google.com/patents/EP0967255A1*).

5.11. FUTURE TRENDS

For online estimations a few components, for example, close uncontrolled example introduction, variety out yonder amongst test and sensor, variable example sum passing the sensor, non-uniform example (appearance, molecule measure, thickness, dampness), and variable surrounding conditions (temperature, mugginess, light power), may influence the execution of the alignment/grouping model (Benson, 2003). Thusly, on-line frameworks ought to be intended to withstand these variables, and strong alignment arrangement models ought to be created utilizing a delegate test consolidating vast example/item variety. Creation and visual attributes of rural items are affected by developing area, developing season, and developing year. Accordingly, these parameters ought to be joined into adjustment display advancement, consequently requiring an extensive agent test. The reference information utilized as a part of preparing the adjustment model ought to likewise be exceptionally exact generally the forecast model will give mistaken outcomes.

5.12. CONCLUSION

Determination of a strategy for mechanized process observing in the sustenance business relies upon the particular kind of use–food material, sort of investigation, ease in inspecting, rate of process, encompassing condition, condition of material (strong, fluid, hot, frosty, and so on.). Infrared spectroscopy isn't exceptionally appropriate for in/on-line quality and compositional examination of sustenance items for a few reasons, including low reflectivity of solids over 2500 nm (henceforth poor flag to-clamor proportion), extreme and fine ingestion groups that are troublesome for quantitative investigation, and test readiness necessity.

NIR is viewed as the most reasonable ghostly locale for in/on-line investigation in the sustenance business permitting fast and multi-constituent examination (e.g., dampness, fat, oil, and protein, fiber, starch, sugar salt, and strong substance) all the while with almost zero example arrangement. Fiber optic sensors can possibly filter (in back diffuse mode) the strong sustenance materials in mechanical food preparing setups utilizing encompassing (e.g., warm) safe fiber tests that can infiltrate the example without making any harm the items. Machine vision frameworks utilizing shading cameras have been fruitful in arranging/examining food items utilizing their surface attributes (shading, shape, and size).

Multispectral imaging frameworks in light of wavelengths chose from hyperspectral imaging can review and examine food items at fast and give reflectance/absorbance data with spatial naming. With propels in optical instrumentation, PC equipment, and computational strategies, optical detecting and spectroscopic system are ready to end up a target device for mechanized food security and quality examination in the agri-food industry.

REFERENCES

1. Afseth, N. K., (2005). 'Raman and near-infrared spectroscopy for quantification of fat composition in a complex food model system.' *Appl Spectrosc. DOI: 10.1366/000370205774783304*
2. Alvarez, D., Castillo, M., Payne, F. A., & Xiong, Y. L., (2009). 'A novel fiber optic sensor to monitor beef meat emulsion stability using visible light scattering,' *Meat Sci, 81*, 456–466.
3. Anami, B. S., & Burkpalli, V. C., (2009). 'Color based identification and classification of boiled food grain images.' *Int J Food Eng, 5* (7), Article 7
4. Birnbaum, H., (1981). Bakers Digest, pp.6–18.
5. Caldwell G. Darwin, (2013). Robotics and Automation in the Food Industry: Current and Future Technologies. Woodhead Publishing, 1st edition
6. Griffin, W., Griffin, M. C., (1949). Soc, *J. Cosmetic Chemicals, 1*, 311.
7. Hasenhuettl, R., & Hartel, R., (eds)., (1997). Food Emulsifiers and their Applications, [PDF] International Thompson Publishing, New York, 1997. Page 19-21
8. Stauffer, C., (1999). Emulsifiers, 1st ed. Eagen Press, St. Paul.
9. Stauffer, C., (2005). Emulsifiers for the food industry. Chpt. 8 in Bailey's Industrial Oil and Fat Products, ed. F. Shahidi, pp. 229-267. John Wiley & Sons, Inc., Hoboken.
10. Tamstorf, S., (1983). *Grindsted Technical Paper. 9*, 1-27
11. Walstra, P., & Vliet, T., (2008). Dispersed systems: basic considerations. Chapter 13 in Fennema's Food Chemistry, ed. S. Damodaran, K.L. Parkin, O.R. Fennema, pp. 783-848. CRC Press, Boca Raton.

CHAPTER 6

SUPERVISORY CONTROL AND DATA ACQUISITION (SCADA) AND RELATED SYSTEMS FOR AUTOMATED PROCESS CONTROL IN THE FOOD INDUSTRY: AN INTRODUCTION

"Civilization depends on our expanding ability to produce food efficiently, which has markedly accelerated thanks to science and technology."

—Nina Fedoroff

CONTENTS

Supervisory Control and Data Acquisition (SCADA) is turning into the standard for modern control frameworks. In spite of the fundamentals of SCADA going back to the 1960s, this control method has just turned out to be clearer in food handling conditions in the course of the most recent decade. The acceptance of SCADA is because of increments in the cost of work, decreases in the cost of control frameworks, and the general advances in food handling frameworks requiring further developed control frameworks. This chapter gives an outline of SCADA, some abnormal state cases, and what ought not too out of the ordinary of those with plans to execute a SCADA framework into a food processing condition.

6.1. INTRODUCTION

Supervisory Control and Data Acquisition, otherwise called SCADA, is a term used to depict an arrangement of mechanical control measures that have been advancing more than a very long while to enhance the control of complex procedures. At first an expansion of programmable rationale controllers (PLC), SCADA has developed to end up exceptionally coordinated inside frameworks from assembling offices to huge utilities, for example, water handling offices or atomic power plants. Albeit executed in a few food processing offices, numerous have been ease back to receive such new advancements, on account of the many-sided quality of such mixes and defense of the cost-to-profit proportion.

6.1.1. Need for SCADA and High-Level Overview

The requirement for more mind-boggling control frameworks is driven by the expanding development of control advancements inside modern frameworks and the interest for frameworks to be progressively more proficient. Because of this abnormal state of intricacy, it is inconceivable for people alone to control the activities in a way that can fulfill the requests of the end client. For example, a high-volume activity where a formula is to be controlled painstakingly for quality purposes may expect choices to be made each second as different stream rates and temperatures change inside the procedure. SCADA tends to this by making an ace terminal unit (MTU) that screens a few lower-level activities, alluded to as RTUs (remote terminal units). These RTUs ordinarily work at high information rates inside the nearby subsystem, and report process data to the MTU so parameters can be acclimated to streamline the general procedure. It is at the MTU where the abnormal state control calculations can be actualized to advance the execution of the framework.

For instance, consider an average high-volume preparing activity where bread is being delivered. For this situation, there are a few stages that must happen all through the office including: (1) the exchange of flour and water into a blender, (2) batter blending and massaging, (3) aging, (4) framing, (5) preparing, (6) cutting, and (7) bundling. One can envision the quantity of machines required to encourage this activity, which will incorporate transports, blenders, slicers, broilers, and bundling machines, among numerous others. Additionally, one can envision the numerous parameters that influence this task, for example, water temperature, mixture heating temperature, preparing time, shade of cooked merchandise, temperature of the office, cost of energy, storage room, bundling materials, and cutting activity speed, to give some examples. The unpredictability engaged with dealing with this activity is gigantic, as any parameter may considerably affect the nature of the completed item. The convenience of the choices is likewise basic to quality. For instance, if the bread shading turns out to be excessively dull leaving the stove, all items in the broiler may wind up unusable, bringing about the loss of thousands of items. It is evident that an abnormal state controller could increase the value of this procedure. Figure 6.1 outlines an abnormal state piece chart of how SCADA could be executed in a run of the mill heating activity.

Notice in Figure 6.1 that the term human-machine interface (HMI) is presented as an expansion of the primary terminal unit. At some level, the requirement for a human administrator is required to screen the general procedure. Notwithstanding, a few HMIs can be available all through the procedure with fluctuating levels of authorizations and control that empower people to either enter process information, screen the execution of subsystems, or to roll out improvements as required.

6.1.2. General Benefits and Drawbacks of SCADA

The past case makes a system in which a few procedure enhancements are instantly conceivable. In any case, these advantages can't be considered without additionally considering the negative perspectives a SCADA framework can present.

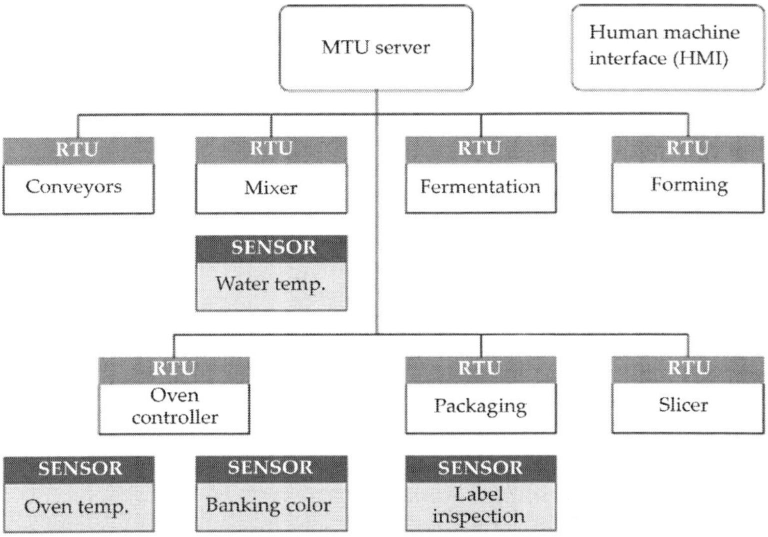

Figure 6.1: National SCADA network for baking process, (*Source: G. Caldwell Darwin (2013), Robotics and Automation in the Food Industry: Current and Future Technologies, Woodhead Publishing*).

The list beneath covers the significant focuses to consider when choosing whether to actualize a SCADA framework or not:

Benefits

- storage of lots of information.
- customization of information showed.
- ability to coordinate some great sensors.
- ability to propelled control calculations.
- ability to screen few kinds of information.
- ability to see information remotely.
- better consistency of items.
- automatic information logging to free up representatives for different undertakings.
- automated cautioning frameworks.

Drawbacks

- system many-sided quality increments in programming and additional equipment.

- possible requirement for new aptitudes inside an association.
- new security dangers made that can bargain the wellbeing of the procedure or restrictive process information.
- overall cost of execution.

As noted in past segments, unmistakably a coordinated SCADA framework can enable a procedure to work at an unheard of level of effectiveness yet the downsides, for example, security dangers, are not to be neglected.

6.2. HISTORY OF SCADA

The thought for SCADA left the requirement for an innovation that could screen various diverse climate and barometrical stations and hand-off that data to a concentrated station keeping in mind the end goal to better foresee climate conditions. Before long the potential advantages to industry were seen and service organizations began to utilize SCADA frameworks to control basic parts of foundation in 1960s. By the 1980s, both water and power organizations were sending SCADA on a huge scale, controlling everything from Extra High Voltage (EHV) control substations to the stream in oil pipelines. Amid this time, every application needed to have custom-composed conventions to control the correspondence between the numerous RTUs and the MTU. This soon prompted various diverse conventions being actualized that were not perfect with each other. This condition of tangled correspondence conventions stayed until the mid-1990s when an unmistakable requirement for an open SCADA correspondence convention was figured it out. Amid this era, two contending conventions were produced: IEC 870 and DNP3. The first was presented by the International Electrotechnical Commission (IEC), and afterward the second by Harris Controls Division, Distributed Automation Products. The correspondence standard created by the IEC was in the end named IEC 60870.5 and was intended to be exceptionally broad in order to be relevant in a wide range of SCADA applications. The standard created by Harris Controls Division is known as the Distributed Network convention Version 3.0 (DNP3). DNP3 was centered more around particular businesses, for example, power, and oil, than IEC 60870.5. Since the conventions origin, DNP3 has turned out to be all the more broadly acknowledged and upheld by industry around the globe, barring Europe where IEC 60870.5 is as yet a noteworthy contender. With the expansion of the Internet amid the late 1990s and mid-2000s, security of the correspondence framework turned into the need. The requirement for an attention on security in SCADA frameworks was regularly pushed forward by the administration with various records being composed regarding the matter. The present province of SCADA security is talked about in more detail in a later area.

6.3. SCADA STANDARDS AND APPLICATIONS

Following SCADA back to its underlying foundations uncovers the size of advances made in the previous era. The essential move has been from various restrictive bundles and correspondence conventions to far-reaching support for a few gadgets and process writes. The outcome is countless offering an expansive scope of help for a few gadgets. The other union point is the conventions that SCADA systems are based upon. Three conventions have turned into the business guidelines because of their open nature, unwavering quality, and seller bolster–IEC 60870-5-101/104, IEC 61850, and DNP3. Lower level control conventions, for example, Modbus and Allen Bradley DF1, still exist among numerous RTUs, however the abnormal state control executed by SCADA programming bundles commonly bolster gadgets utilizing these lower level conventions.

While thinking about SCADA programming, it is normal that the accompanying highlights be available:

- Support for different gadget conventions (Modbus Organization 2006), Profibus (International 1998 etc.).
- PC-based usage.
- remote terminal help.
- easy expansion of more gadgets to the system.
- fully adaptable graphical UI (GUI) for HMI.
- alarm definition.
- open access to chronicled information.
- Maintenance.
- security measures display.

6.3.1. Examples of Installations in Industry

The unwavering quality, adaptability, and versatility of SCADA have settled on it an appealing decision while mechanizing complex frameworks. SCADA has just been utilized effectively in a wide scope of enterprises, going from vitality generation to industrial facility computerization to horticulture. For each situation, the particular application required that different kinds of equipment be associated and impart proficiently so as to play out their task. The accompanying contextual analyses will jump into the particular difficulties looked in every one of the applications and exhibit the execution picks up by applying SCADA.

6.3.2. Thermal Power Plant

Thermal power plants are unpredictable frameworks including various boilers, turbines, and pumps, which should be observed and controlled with a specific end goal to accomplish crest execution. Security is likewise of foremost worry in basic foundation offices like power plants. This level of security and many-sided quality joined with the necessity that every segment capacities productively freely of the others and additionally in conjunction with alternate segments, makes SCADA an engaging arrangement.

The execution of SCADA in a power plant anticipates data overburden for control laborers and permits the solidification and translation of information, to show signs of improvement worldwide to get a point of view on the activities and conjecture conceivable issues. One not so distant progress in control is to utilize SCADA to incorporate the power plant with a shrewd lattice to take into account ongoing tweak of energy creation construct straightforwardly upon current request.

6.3.3. Water Treatment Implementation

SCADA keeps on being an exquisite answer for control of utilities, similar to the case with the case of a water treatment office in Hawaii. Outlined in 1975, the Schofield Wastewater Treatment Plant was controlled physically with substantial utilization of labor supported by cautions on different controls inside the task. A nitty-gritty contextual investigation was produced to delineate how a SCADA-based framework could be coordinated inside calendar and budgetary evaluations. In this specific case, Allen Bradley PLCs were utilized with a Rockwell Automation SCADA checking framework. The length of the exertion was two years and the cost of the establishment was $575 000 US, and all the more imperatively, the establishment did not bring about any downtime of the office.

6.3.4 Computer Production

Worldwide PC makers have utilized SCADA to take care of a portion of the issues related with creating adjustable PCs rapidly and productively in extensive volumes. Frequently, similar to the case in numerous different ventures, custom applications made for particular equipment are created with next to zero idea concerning how they will interface with the other hardware in a manufacturing plant. This makes superfluous many-sided quality and makes it hard to enhance the assembling procedure. This is a brilliant case for SCADA to be utilized to enhance working proficiency and consider more point by point control over the whole generation process.

In a specific PC generation office, the organization expected to fabricate modified PCs on a quick mechanical production system. The procedure should have been controlled remotely, with the goal that a client could put in a request for a PC on the web and the PC would start being amassed rapidly in a manufacturing plant the world over. The sequential construction system was completely mechanized from introductory establishment through dispersion utilizing SCADA. The organization likewise needed to have the capacity to screen the get-together process remotely from both PC and Internet empowered phones. The blend of checking the procedure and adding new things to the mechanical production system remotely made the security of the framework be put at the bleeding edge. Security was worked in through a layered approach, with each segment that connected to the Internet first going through the organization's firewall. Expanded security safeguards were taken with respect to the cell phones in light of the fact that the hazard for misfortune or burglary was impressively higher than for the other PC-based checking frameworks.

When this new SCADA framework was introduced, the organization discovered expanded profitability and decreased downtime because of the expanded deceivability of every one of the frameworks inside the plant. Support laborers would regularly perform settles before specialists on the sequential construction system knew about an issue, enhancing dependability as well as framework wellbeing. One other preferred standpoint that was found because of the establishment of the SCADA framework was the capacity to screen the work in advance continuously, furnishing both administration and clients with a superior comprehension of the creation procedure. After the SCADA framework was introduced, the organization thought about it to have a huge degree of profitability, with over $3.7 million being spared every year per plant by means of the expanded activities deceivability alone with the aggregate yearly profit for a speculation being in overabundance of

$40 million for the organization.

6.4. SCADA IN FOOD PROCESSING

The general idea of SCADA has been shrouded in detail, however, there are particular worries to the food processing industry that have influenced the selection and usage of SCADA. An essential concern is the commonness of low net revenues inside the food business. On account of poultry processors, the current obtaining of Gold Kist, Inc.–a noteworthy maker of poultry items–by Pilgrim's Pride, Inc. can to a great extent be credited to the increasing expense of sustain in North American markets. Since net revenues are much lower in food preparing than in numerous different businesses, awesome care must be taken before the expensive execution of full-scale SCADA reconciliation.

Furthermore, because of the idea of food preparing tasks, an expansive data innovation (IT) staff may not be available to help SCADA frameworks, which means the speculation required could increment considerably. Finally, worries with security are vital because of sustenance wellbeing oversight and the potential pessimism that can come about because of a vast review. In spite of these difficulties, numerous food processors are starting to execute SCADA control frameworks in their activities with a specific end goal to exploit the numerous advantages that they can give.

The biggest advantages that SCADA can convey to the food handling industry are diminishments in labor and increments in quality. Unmistakably work diminishment builds the upper hand of a task, yet this can't be accomplished at lost quality. The view of the item by the buyer remains a target and quality additionally reaches out into the region food well-being. Associations, for example, the Food Standards Agency, or the Food and Drug Administration cover significant reviews that can influence a large number of items and shoppers. By fusing better controls and checking of a food handling task, it is doable that quality and along these lines security could be definitely enhanced while at the same time cutting working expenses.

6.4.1. Cost versus Benefit

There are numerous establishments of SCADA frameworks, yet the genuine cost to introduce and keep up an intricate framework should be considered preceding making a dedication. The past case, identified with a SCADA establishment at a waste-water preparing office in Hawaii, reported a cost of $575K USD, however huge scale frameworks, for example, the move up to SCADA interfaces for Florida's District's Water Management System can be as high as $292M USD. It merits saying that the last case included upkeep more than 10 years. There are situations where basic SCADA establishments, for example, those utilized as a part of water system frameworks, can be as meager as $3– 22K (2001 US). The accompanying table gives a thought of the costs identified with a little framework, for example, a solitary plant versus the costs identified with a huge scale task including SCADA combination over a substantial number of locales. The way to detract from the introduction of information in Table 6.1 is that the cost of equipment is fairly little in contrast with foundation refreshes, IT support, and upkeep. Likewise, the table is based upon cases outsourcing huge numbers of the combination undertakings though utilizing existing inward hotspots for establishment and foundation things could lessen general expenses.

Remember that Table 6.1 alludes to the general assessed cost of a full establishment on a total and exceptionally mechanized SCADA framework. A more viable approach is to gradually incorporate new hardware into a SCADA procedure utilizing a straightforward yet versatile interface at first. As the framework turns out to be more perplexing, the MTU Server and HMI can be moved up to help a bigger number of RTUs and new advances. Backpedalling to the heating illustration talked about in Section 6.1.1 and depicted in Figure 6.1, an underlying establishment could comprise of an essential MTU server, HMI interface, and two noteworthy RTU gadgets, for example, a broiler controller and review gadget to investigate the items leaving the stove. The greater part of the advantages of SCADA could be acknowledged, for example, ongoing information logging, and remote checking, quality alerts, and so on, for the preparing procedure. This should be possible at a cost of under $50K USD thinking about equipment, programming, and establishment and joining costs.

Table 6.1: Estimates of Installation and Maintenance Costs of SCADA

	Single facility (USD)	Multiple facilities (USD)
Non-recurring		
Installation/labor	$150K	524M
Hardware	S70K	$3M
Software	S40K	56M
Professional integration of system	SI6OK	$13M
Recurring (annually)		
Software	SIOK	5300K
Hardware	510K	51.2M
IT	S8OK	55.3M
Infrastructure	510K	512M
Totals		
Non-recurring	5420K	5112M
Recurring (annually)	S 110K	17M
Cost of implementation over 5 years	5970K	5198M

Source: Source: G. Caldwell Darwin (2013), Robotics and Automation in the Food Industry: Current and Future Technologies, Woodhead Publishing

Various advantages of SCADA have been talked about, yet there are a few advantages that are one of a kind to the food processing industry. For example, oversight by food facilities, for example, the United States Department of Agriculture or the World Health Organization frequently requires various measurements to be accounted for to controllers. The movement of catching and revealing these measures can be tedious and troublesome in complex frameworks. Fusing history-catch capacities into a SCADA-based system could reduce a portion of the exertion required around there and give incredible enhancements to manual procedures. Advantages identified with nourishment wellbeing could give extra esteem.

With food wellbeing concerns continually rising, any extra observing of parameters identified with food security could give huge incentive to the administrator. For instance, having stringent controls over temperatures in the process can give greater quality to the customer, good mechanisms of pathogens, and improve vitality use. Advantages identified with vitality reserve funds can be of most extreme significance in food preparing because of the idea of working huge stoves, coolers, and temperature-controlled plants. The potential cost investment funds identified with warming venting and aerating and cooling (HVAC) alone could legitimize the establishment of a SCADA framework.

6.4.2. Security and Operation Concerns

A noteworthy downside to SCADA is the capability of helplessness made by a security breach or other disturbance in the task of the MTU. These issues can emerge from the below:

- mischievous assaults
  - Internal (disappointed worker, etc.).
  - external (psychological oppression).
- natural causes
  - power surges/blackouts.
  - operator blunder.
  - hardware breakdown.
  - software configuration defects.

SCADA designers are sensitive to these issues and have joined a few highlights to battle a large number of these dangers, for example inserting firewalls and different interruption location frameworks. In any case, there have been various SCADA-related assaults or disappointments reported. The list below is just an example:

- Intentional
  – January 2000: Mariachi Shire Sewage Spill: flooding of 264 000 gallons of crude sewage.
  – March 1997: Worcester Air Traffic Communications Attack: an individual from people, in general, turned off the phone arrange in the zone which traded off airport regulation.
  – 2000–Gas Pipelines in Russia: SCADA network traded off, however, risk killed before harm.
- Unintentional
  – August 2005: Automobile Plants in Zotob Worm: 13 automobile plants closed down for 1 h.
  – August 2003: CSX Train Signaling System and the Sobig Virus: shutdown of basic flag and dispatching frameworks

These attacks are well documented in critical infrastructure areas, but it is clear that food processing attacks or failures could end with potentially disastrous results. Concerns are the basic encryption techniques used for password-protection services, which are becoming easier to undermine as was demonstrated with using low-cost graphics processors to crack seven character passwords in a matter of hours making them ‘hopelessly inadequate’ (BBC, 2010). Additionally, many organizations are offering wireless solutions for industrial control hardware due to the obvious convenience of implementation. Although there are proven encryption techniques concerning wireless protocols, wireless transmission of data is inherently less secure than wired transmission of data. Overlooking proper security protocol at a single access point could create a more substantial vulnerability if using a wireless protocol. A breach in security can result in the loss of personal data, corporate data, or affect the throughput of an operation and safety.

These assaults are very much archived in basic framework regions, yet plainly food processing assaults or disappointments could end with possibly awful outcomes. Concerns are the essential encryption methods utilized for secret word security administrations, which are getting to be less demanding to undermine as was exhibited with utilizing minimal effort designs processors to break seven character passwords in an issue of hours making them ‘pitifully lacking.’ Furthermore, numerous associations are putting forth remote answers for modern control equipment because of the undeniable comfort of usage. Despite the fact that there are demonstrated encryption systems concerning remote conventions, remote transmission of information is naturally less secure than wired transmission of information. Disregarding legitimate security convention at a solitary access point could make a more significant helplessness if utilizing a remote convention. A break in security can bring about the loss of individual information, corporate information, or influence the throughput of a task and wellbeing.

In spite of the fact that these cases exhibit real disappointments in SCADA frameworks, these frameworks can be ensured if legitimate safety efforts are set up. Assets, for example, those offered by The Center for SCADA Security from Sandia National Labs or the US President’s Critical Infrastructure Protection Board, can empower more educated basic leadership with respect to SCADA. These security assets regularly stretch that the most vital bit of security is to keep up approaches and an arrangement for SCADA security. This incorporates the correct preparing of workforce, duty of assets, staff security controls, assigning parts and duties, and making punishments for those that don’t take after conventions. To develop the dedication of assets, it is foremost that assets are devoted not just for the establishment of SCADA equipment/programming yet in addition for future innovation acquirements. As dangers develop, so should the SCADA framework advance to decrease the adequacy of those dangers.

6.5. LABORATORY STUDY: IMPLEMENTATION OF SCADA

Analysts at the GTRI Food Processing Technology Division created a little scale SCADA condition to evaluate numerous SCADA programming bundles. This was a basic framework including a supervisory level controller with a few control level elements. Through the usage, an assessment of three SCADA programming bundles was created to feature particular difficulties and advantages of utilizing these projects. The three projects utilized were Cimplicity, Citect, and VTSCADA. Cimplicity is an item from GE Intelligent Platforms, Citect was as of late gained by Schneider Electric, and VTSCADA is a Trihedral item.

Scientists found that the little scale framework could without much of a stretch be reached out to incorporate more gadgets and show a more elevated amount of control between different bits of gear with any of the three programming bundles. Posting of u three programming bundles with a concise review taken from the producers are given beneath:

CIMPLICITY–CIMPLICITY is a customer/server based representation and control arrangement that causes you imagine your activities, perform supervisory computerization and convey dependable data to more elevated amount systematic applications. With the most recent highlights of CIMPLICITY 8.1, including a capable illustrations motor, dynamic time taking care of and the extra choice computerized graphical replay (DGR) 2.1, administrators and designers have the power and security to unequivocally screen and control each part of their condition, hardware, and assets. The outcomes: quicker responsiveness, decreased expenses and expanded gainfulness.

Citect (Electric, 2011b)–'The CitectSCADA scope of computerization programming has been conveying world-class arrangements and administrations to clients around the globe for more than 35 years. We work in the conveyance of arrangements that are creative, solid, and financially savvy. Our offerings associate individuals in business and industry to constant data in ways that enhance business methodologies and choices.'

VTScada (Trihedral 2011)–'Intended for the water and wastewater industry, VTScada is a product apparatuses layer that includes effective remote resource administration highlights to a visual label framework (VTS) for telemetry applications. Industry particular reports, auto-produced realistic presentations and programmed chronicled information logging give a stage to speedy application advancement without bargaining on highlights. Applications extend from remain solitary workstation applications to bound together, region-wide frameworks tying plants and SCADA centrals together.'

These different SCADA programming arrangements were looked into amid the research center usage, and the key focuses are partaken in Table 6.2. The three programming bundles go from a less difficult interface with less adaptability when utilizing VTSCADA, to more unpredictable and configurable frameworks, for example, Cimplicity and Citect. Contingent upon the span of the SCADA arrange, level of capability of IT workforce, and other general prerequisites, an association has numerous alternatives accessible for SCADA programming instruments. Moreover, there are a few integrators accessible to aid all levels of SCADA usage, with a few suppliers concentrated on the sustenance and refreshment industry.

6.6. FUTURE TRENDS IN SCADA

SCADA frameworks will keep on offering similar preferences in coming a very long time with an emphasis on facilitating the weight of upkeep and task for the client while offering more valuable highlights. It is additionally basic that security be at the front line of SCADA advancements since dangers continually develop. Particularly for sustenance preparing applications, the eventual fate of SCADA as an innovation isn't as imperative as the selection of SCADA control. Organizations and plant directors must be mindful of the effect of expanded utilization of SCADA for food processing activities–particularly to security as it applies to sustenance wellbeing. Food preparing tasks will start to work all the more effectively by utilizing further developed modern controls, and the business will see a move in rivalry moving towards all the more innovatively propelled handling plants. Besides, the requirement for all the more mechanically propelled plants will be more common in business sectors, for example, Europe, and the United States, where compensation keep on rising consequently driving weights to consolidate more mechanization into food handling offices. Increments in mechanization will prompt expanding weights to execute progressed modern controls, and SCADA will wind up ordinary in effective food preparing tasks without bounds (Table 6.2).

Table 6.2: Various Attributes of VTSCADA, CIMPLICITY, and CITECT

	VTSCADA (v9.1.03)	CIMPLICITY (v8.00)	CITECT (v7.10r1)
Install	Routine installation.	Detailed installation involving SQL database configuration.	Routine installation.
Overview	• Aspects of the program are accessed from application pages • The programming and run-time environments are blank user screens with a menu button at the bottom left that allows navigation between screens (Overview. Alarms, etc.) • The programmer and the user can access parameters on application pages • A configuration toolbox provides tools for defining program parameters • Uses VTS to define all parameters	• Workbench is the front end of the program, and provides a logically organized way to access all programming information • All parameters are listed in a tree structure on the left of the screen, similar to Windows Explorer • Screens are created in a separate screen editor, called CimEdit, that can access data points as defined in the Workbench - Uses a tree structure to allow the programmer to define variables	• Opens three screens for the programmer to work with: — Project Explorer — Project Editor — Graphics Builder window • Project Explorer allows the programmer to navigate within and between projects • Project Editor allows the user to define and change program parameters • Graphics Builder allows the programmer to design screens for the end user
Control level Remote level	All permit communication using standard protocol (i.e. drivers can be created to allow for integration.	Modbus over TCRIP). If product does not support	standard protocol, custom Multiple pans of the software support remote monitoring.
	YTS Internet Server states that **IE 6 or** 7 is required.	Uses Wcbview as webserver when only monitoring. Uses terminal services when actual setpoints need to be changed.	
General usability	Simplest program of the three to use. Minimal setup to generate a Modbus network and read data.	Complex software, yet flexible for integrating custom scripts using the	CITECT is highly configurable, necessitating a certain degree of
com-ments	Once the programmer understands the tag system, commands are relatively clear and logical. Good help file and good technical support.	Imegrator's Toolkit. The Workbench is logically put together, but requires significant learning from the users. Some inconsistencies in menu file structure among programming modules.	complexity. The GUI is attractive and consistent between all three programming windows.

Source: G. Caldwell Darwin (2013), Robotics and Automation in the Food Industry: Current and Future Technologies, Woodhead Publishing

REFERENCES

1. Bailey, D., & Wright, E., (2003). Practical SCADA for Industry. Burlington, MA, Newnes.Page Number: 1-6
2. Barr, D., (2004). Supervisory Control and Data Acquisition (SCADA) Systems. National Communications System. Chantilly, VA 20151, Communication Technologies, Inc. 76.
3. Clarke, G. R., Reynders, D., & Wright, E., (2004). Practical Modern SCADA Protocols Burlington, MA, Newnes.
4. Caldwell G. Darwin., (2013). Robotics and Automation in the Food Industry: Current and Future Technologies. Woodhead Publishing.1st edition

CHAPTER 7

GRIPPER TECHNOLOGIES FOR FOOD INDUSTRY ROBOTS AND METHODS OF ULTRASOUND

"There's a big difference between industrializing production of tractors and industrializing production of food. We like technology, but we really like technology that allows us to do better what nature does itself."

—Joel Salatin.

CONTENTS

This chapter gives a review of the difficulties of grasping food questions, the primary responses to these difficulties and a few signs of conceivable outcomes offered by new holding innovations. Regular squeeze holding advancements are examined and recently created techniques in light of infiltrating, suction and stop strategies are depicted. The essential administering conditions for squeezing and suction grasping are displayed. The sterile nature of the diverse techniques is examined. At long last, a subjective assessment of the appropriateness of the diverse techniques in sustenance taking care of is introduced.

7.1. INTRODUCTION

Since the innovation of the primary mechanical robot by George DeVol in 1959, the gripper has been an indispensable part. The entire thought of robots depends on the idea of some non-human gadget that can perform errands regularly done by people. It can be in assembling or whatever other human action where a machine is utilized to supplant the human activity. In a large number of these cases, the undertaking includes grasping some protest and accomplishing something with it. It may be a basic transportation starting with one place then onto the next, or a more mind-boggling assignment including the utilization or something to that effect of hardware and preparing.

The cutting-edge thoughts and executions of robots have extended from the first mechanical applications for the principal robots in light of DeVol's patent. The first mechanical robots were genuine material taking care of gadgets and nothing more. Hence the gripper was a fundamental piece of the robot.

However, not at all like the robot arm and control framework, the grippers have not been the subject of the fundamental research exercises in mechanical autonomy. There has been explore on mechanical hands with human-like fingers, yet at the same time, the greater part of research appears to have been focused on robot control hypothesis, arm structures, kinematics and flow, and on robot tactile frameworks.

Grippers can be significantly basic. A gadget with two mechanically activated fingers that can squeeze a question is all that is required in numerous basic taking care of cases. Since there are such a large number of uses where this gripper is adequate, the examination exertion has been spent on the all the more fascinating and numerically difficult parts of mechanical autonomy.

The well-working, straightforward two-finger gripper is excessively basic, making it impossible to challenge the brains of the specialists.

Be that as it may, circumstances are different. All the basic cases are so very much considered. Rather the automated world is confronting challenges in dealing with delicate, limp, 'unstructured' articles. These are the articles that are found in extraordinary numbers in food industry. Dealing with foods grown from the ground, and meat from ocean or land creatures, postures issues that are not quickly feasible with the straightforward grasping innovation of the original mechanical technology.

Gradually, in the period from around the new thousand years, knowledge has been produced into more complex holding strategies and advancements for these objects of delicate, limp or unstructured character. As these new holding strategies and advances turn out to be surely knew and created to mechanical gauges, new open doors for robotization in countless applications will show up.

In this chapter, the term 'hard grasping' will allude to any holding errand where the question has an all-around characterized shape and is sufficiently solid to withstand the power of a squeezing or comparative grasping technique. Ordinarily, grasping a metal screw or plate can be thought to be a hard holding case.

'Delicate grasping' will be utilized to portray any holding undertaking where the protest is of delicate materials that can be effortlessly affected by the power of the gripper, and by any quickening, and likewise the question might be of sporadic shape that will shift from protest question of comparative kind. Holding a bit of meat is a decent case of delicate grasping.

Food articles are frequently delicate hold objects, regardless of whether numerous food articles could be taken care of with hard grasp innovation. This chapter will cover the two kinds of grasping advances, yet it will put most accentuation on delicate holding.

7.2. GRIPPER CHALLENGES IN FOOD PROCESS AUTOMATION

All in all food materials show challenges that are not regularly observed with non-eatable items. Some of these difficulties are additionally found in items other than food, outstandingly delicate quality and limp conduct of materials and certain froth plastics and in elastic items. In any case, a large number of the difficulties in food dealing with are interesting when the mix of difficulties is considered.

Certain all around created gripper arrangements can be exchanged from the non-food division into food taking care of. Be that as it may, every uncommon necessity in the food area must be considered. This thought has prompt some intriguing and one of a kind gripper answers for programmed treatment of food materials.

7.2.1. Soft Materials

Most of the food material is delicate. This originates from the idea of food. Food material is either vegetable or meat. Meat is constantly delicate while vegetable food can be delicate or hard. The hard food material causes less issues from the holding side. Be that as it may, all delicate vegetable food materials are potential issues in taking care of.

There are two noteworthy issues with delicate materials:

1. The material transforms into a form with a goal that a settled width squeezing gadget essentially loses its grasp on the object.
2. The grasping power connected can harm the surface or inside the object with the goal, that quality is diminished. This, for the most part, prompts bring down offering cost or shorter existence of the object.

Consequently, delicate materials require a grasping technique that wipes out these issues.

For meat from land creatures, the issues of sort two are irrelevant. Meat is for most part sufficiently strong to endure ordinary powers from squeeze holding. The non-abrasiveness does prompt the issue of the main sort for meat. This won't not make any difference on the off chance that it were not for the accumulation of issues because of uneven surfaces and non-uniform shapes.

Natural products are in an alternate classification. Delicate natural products require exceptionally delicate dealing with. Indeed, even semi-hard organic products like apples and tomatoes are extremely delicate to gouging and neighborhood weight. When all is said in done the squeezing grasp isn't reasonable for any of these.

Different vegetables are less demanding to deal with–nuts and foundations of various types are for the most part more tolerant to holding powers.

7.2.2. Uneven Surfaces

All normally developed items have uneven surfaces. For a few examples, the unevenness is just slight, while others have unpleasant surfaces. The surface quality can impact the scope of holding alternatives.

Hard objects with unpleasant surfaces are typically the most effortless to grasp. Such objects are tolerant to high grasping powers. Delicate items with unpleasant surfaces, then again, exhibit challenges. Kiwi organic products are extremely touchy to marking and nearby weight, and the skin is harsh, making it hard to utilize vacuum grasping procedures.

Meat from both fish and land creatures likewise exhibits issues because of uneven surfaces. Vacuum systems are hard to apply as a result of the unusual surface structure much of the time.

7.2.3. Non-Uniform Shapes

Every natural objects have blemished geometrical shapes. It is just manufactured items that can display consummate external geometry. Some food materials come in this classification; most outstanding are chocolate items that are thrown in kicks the bucket that deliver impeccably normal shapes. Be that as it may, all normally developed items demonstrate a few inconsistencies.

Numerous processed object are correspondingly non-uniform. Distinctive cuts of meat from creatures will dependably have some inconsistency regardless of whether the slicing procedure is set up to deliver exceptionally customary shapes; however, the inclination of numerous clients is that these items ought to be unpredictable to give a 'home-made' appearance.

Non-uniform shape is the greatest test for gripper plan and techniques. In any case, lately the improvement of innovation that backings the grasping errand has given numerous new alternatives to dependable holding arrangements.

7.2.4. Hygienic Requirements

Since food is to be eaten there is an outright, no bargain, necessity for sterile treatment of food in collecting and preparing. Sullying of any sort must be evaded. Essentially there are three noteworthy classes of sullying to be worried about:

1. *Poisonous tainting*: in the collecting and preparing no harmful substance should interact with the food. Hints of poisonous substances can hold fast to the item and in this way be transmitted to the buyer. This suggests every single grasping gadget and techniques ought to be of non-harmful materials. This is essential to consider since some building materials have harmful impacts.
2. *Bacteriological and parasitic sullying*: microorganisms of various sorts have the likelihood of developing where there is natural material, and positive stickiness and temperature. Organisms can likewise discover good living conditions in sticky regions. These conditions can without much of a stretch be available in any food handling facility. Every grasping technique that depend on physical contact with the protest do likewise create the danger of leaving little measures of the dealt with material on the holding gadget. Temperature and moistness will frequently be great for microscopic organisms and parasites on the premises if there are people in the preparing zone, because of the necessity for adequate working conditions for people. Consequently, all holding gadgets ought to be planned with the goal that the amassing of natural material is decreased to irrefutably the base, and that proficient cleaning techniques exist for the gear.
3. *Staining*: some dealing with and holding strategies may prompt staining, by leaving little hints of material that won't not be destructive but rather will be obvious. Neighborhood weight from grasping activities may prompt staining due to little changes in material structure. Regardless of whether such staining has no impact on the supplement estimation of the food, it will frequently be considered as a quality deformity by the buyer.

The first and second classes of sterile necessities are outright. The second rate class is a 'decent to-have' prerequisite however it doesn't speak to a genuine wellbeing risk.

7.3. GRIPPING PHYSICS

All grippers ought to play out the essential capacity of exchanging the fundamental power from the robot arm to the object with a specific end goal to move the object. Any development includes speeding up, through gravity and the nearby development way. The increasing speed in the nearby way is a component of any adjustment in extent and bearing of the speed vector.

Quickening because of gravity is constantly guided opposite to the neighborhood even plane. For vertical developments, this quickening adds to the nearby way increasing speed for a development upwards. This will ordinarily give the biggest power applied amid a taking care of task.

Some gripper strategies require at any rate fractional encasing of the object. This is the situation with squeezing and encasing grippers. Others follow up on a solitary surface of the object. This is the situation with sticking, under pressure and surface impact grippers. For food items the accompanying choices for constrain exchange through the gripper exists:

Squeezing: the gripper has at least two fingers that apply power to the dealt with object. Straight fingers depend on grinding power between the object and fingers to exchange the essential power to keep the object in a protected grasp amid handling. Notched or roundabout formed fingers can likewise exchange drive through neighborhood opposite power vectors. Since a vast power is constantly applied on the object this is a run of the mill hard hold. Discharge is gotten by moving the fingers from the object.

Encasing: the gripper depends on fingers with expansive surfaces that encase the object mostly or totally. The fenced in area must be finished as in all conceivable escape headings from the gripper are shut by reasonable surfaces. The required power transmission to the object is gotten from the vector entirety of all typical power vectors of the surfaces in contact with the object. For this situation, there is no power bigger than the power of gravity and the increasing speed powers from the gripper dividers that follow up on the object so this is a delicate grasp.

Discharge is acquired by moving the encasing surfaces adequately away to make an opening sufficiently substantial for the object escape

Sticking, entrance: the gripper has a few strongly pointed pins that are squeezed into the protest in a way that makes a locking design. Typically this implies an arrangement of pins indicating inwards the focal zone of the object

There are two types of pins:

1. Short ones for surface entrance of level articles.
2. Long ones for profound infiltration of unpredictable items.

The articles must be of sorts that are not hurt by stick infiltration. In food processing, this can be the situation for certain meat items. Discharge is gotten by maneuvering the pins again into the gripper body.

Under pressure, suction: these grippers are regularly called vacuum grippers. They don't, however, depend on add up to vacuum, yet rather on a specific measure of under pressure in the grasping zone. Under pressure implies that the neighborhood weight in the gripper region is lower than the encompassing weight. This makes a power because of the weight distinction, which powers the object against the holding territory. Discharge is gotten by evacuating the under pressure; at times a slight overpressure is required for predictable discharge.

Surface impacts, surface stage changes: holding can be acquired by chilling off a contact territory to a temperature well underneath the point of solidification of water. At the point when this territory is gotten contact with a wet protest, a thin ice film is made quickly on the question surface in contact with the frosty gripper surface. This ice makes an adequate holding power for some articles. Meat does ordinarily have enough water content at first glance to empower this grasping activity. Discharge is gotten by warming the contact territory, or by cutting endlessly the question by methods for a thin blade.

7.4. PINCHING AND ENCLOSING GRIPPERS

The squeezing gripper is the first gripper idea for robots. It is an exceptionally straightforward and vigorous holding strategy, and it is extremely effective for countless grasping undertakings. There are a few variations of this holding rule, from two solid finger pneumatically worked grippers to different jointed finger variations. For useful robot applications the two and three hardened finger variations command. Figure 7.1 shows cases of these gripper types.

7.4.1. Two-Finger Solutions

Ordinary two-finger grippers resemble a kaleidoscopic body with two fingers reaching out from one side. The fingers are either moving in parallel or pivoting around a pole at their base. Regularly they are impelled by pneumatic power, however electric and water driven incitation does likewise exist. The fingers are either level confronted or have round or v-grooves.

Two-finger level confronted finger grippers depend on grating to produce the fundamental power for holding the object. The power required relies upon the course of the speeding up in the development. Figure 7.2 demonstrates the four distinctive main power expresses that can happen contingent upon the course of development. This is known as a power fit grasp.

In all cases the gripper fingers transmit a power *F* opposing sliding along the gripper surface:

$F_f = nF\mu$ (1)

where:

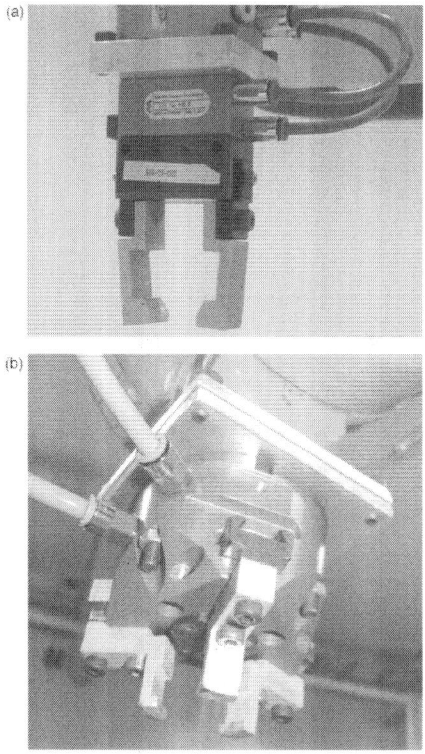

Figure 7.1: (a) A typical two-finger gripper (b) a typical three-finger gripper, (*Source: G. Caldwell Darwin (2013), Robotics and Automation in the Food Industry: Current and Future Technologies, Woodhead Publishing*).

F_f: the grinding power restricting sliding, F: the ordinary power on the finger cofront, μ: the erosion coefficient between the object and the fingers, n: the quantity of fingers.

To abstain from sliding, F_f for a two-finger gripper must fulfill the accompanying conditions relying upon the heading of development:

Vertical up:

$$F_f > m(g + a) \qquad (2)$$

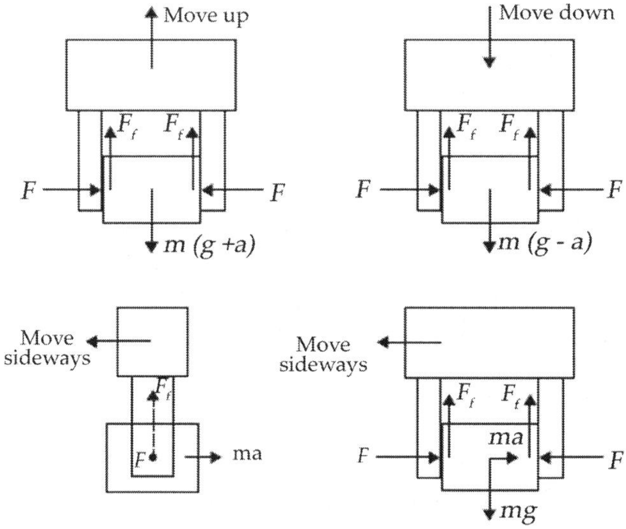

Figure 7.2: The powers following on an object in various speed cases: *m* is the mass of the dealt object, *a* is the quickening of the object because of the robot movement and *g* is speeding up of because of attractive energy, (*Source: G. Caldwell Darwin (2013), Robotics and Automation in the Food Industry: Current and Future Technologies, Woodhead Publishing*).

Vertical down:

$$F_f > m(g \quad a) \qquad (3)$$

Horizontal parallel to finger face:

$$F_f > m\sqrt{g^2 + a^2} \qquad [7.4]$$

Horizontal normal to finger face:

$$F_f > mg \qquad (5)$$

For the level development ordinary to the finger faces, the speeding up constrain from the development that expand the aggregate power on one face and diminish it with a similar sum on the other face. Be that as it may, the whole of powers will stay steady rendering the condition substantial for all increasing velocities.

The furrows will give holding power through shape encasing activity. This is called shape fit holding. The holding power is the resultant of the individual ordinary power vectors from the contact surfaces. The figuring for this sort of grasp is displayed in the area for multi-finger holding.

The hold finger drive *F* for pneumatically worked grippers is ordinarily reliant on the supply weight to the gripper. As a rule, the power will surpass the level for least F_f by an extensive edge, regularly a few requests of extent. These grippers will constantly near a position where the response drive from the squeezed question levels with the finger constrain *F*. On account of dealing with hard metallic, wooden or plastic segments this is of no result. For delicate food materials, the case is extraordinary. The holding power should frequently be acclimated to indisputably the sheltered least to stay away from harm because of neighborhood weight on the items.

7.4.2. Two-Finger Servo Gripper

One answer for the issue of extreme holding power is the utilization of an electric servo gripper. This sort of gripper is controlled to give a predefined gripper hole or to grasp with a predetermined hold finger drive *F*.

Remotely these grippers vary little from pneumatically worked grippers with the exception of that the actuator lodging is to some degree greater, see Figure 7.3. They are additionally more costly, yet the additional cost can without much of a stretch be defended by the higher flexibility of this gripper write.

The servo gripper requires a computerized order indicating the needed finger hole and hold constrain if drive control is executed. This order is regularly transmitted by means of serial connection from the control framework.

Figure 7.3: Two-finger servo gripper with v-groove finger for object centering, (*Source: G. Caldwell Darwin (2013), Robotics and Automation in the Food Industry: Current and Future Technologies, Woodhead Publishing*).

7.4.3. Multi-Finger Grippers

Multi-finger grippers are of three fundamental classes:

1. Basic hardened finger grippers with three or four fingers.
2. Numerous firm fingers that halfway encase the object.
3. Human-like finger grippers with jointed fingers.

Equations (1), (2) and (3) are legitimate for grippers with any number of level face hardened fingers when the gripper is utilized with its centerline vertically situated. These grippers work in part as shape fit grippers when worked in different positionings. Frame fit grippers depend on halfway encasing the question with the fingers, additionally, some of the time utilizing notched or formed fingers.

Frame fit grasping does not require high power from the fingers to make the friction force to hold the protest. This grasping technique depends on the fingers' capacity to withstand the typical power at the contact focuses at any position and speeding up.

This will dependably prompt lowermost extreme finger grasp compel than constrain fit holding. The (human-like) grippers should preferably emulate human taking care of best. A significant number of the grippers in this class have not as much as the aggregate degrees of flexibility that we find in the human hand, just a couple of research center illustrations approach mechanically emulating the human hand. Hence the adroitness of the human hand can't be completely copied. Besides, the affectability of the human hand, through material and power input through the human nerve framework, has demonstrated exceptionally hard to imitate. Also, the minor many-sided quality of the human-like finger grippers has prompt to a great degree high cost, making them financially unsuited for modern use at the present phase of advancement.

7.4.4. Enclosing Grippers

Encasing grippers are composed with the goal that they constitute an almost shut box that grasps the entire question when shut. In this manner, they can deal with all varieties of an item as long as it fits inside this case. The state of this container is ordinarily not a rectangular crystal; a 'vessel shape' is at times a more proper portrayal. The gripper does not apply a squeezing power unless the protest totally fills the container volume. As a rule, the power to move the question is applied from the container divider supporting the protest toward the path inverse to the consolidated power vectors of attraction and speeding up in the neighborhood development.

The encasing gripper won't give correct situating and introduction unless the object fills the container totally. In many applications, it will along these lines be appropriate just if the conveyance position and introduction isn't basic, or if extra arrangement gadgets are utilized at the conveyance station. Much of the time of bundling this is no issue since either there is room that permits variety of conveyance into the bundle, or some basic guide dividers are utilized to lead the protest into its last position. Figure 7.4 demonstrates a case of an encasing gripper.

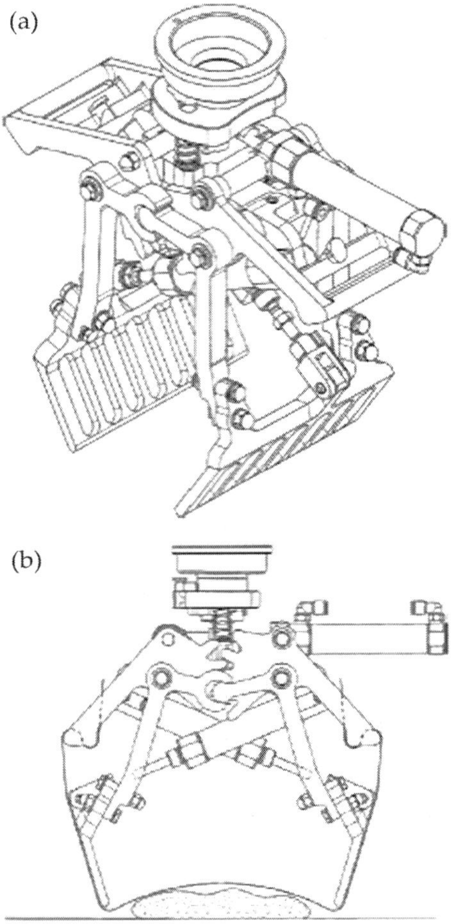

Figure 7.4: (a) Enclosing gripper, (b) enclosing gripper mechanism that minimizes volume swept in the gripper closing or opening action, (*Source: AEW Delford*).

7.4.5. Hygienic Performance of Pinching and Enclosing Grippers

All the grippers of the squeezing and encasing class have a type of moving mechanical components outside the gripper base lodging. Moving fingers require direction, aides, and shafts going through dividers. These components are dependably a worry as for sterile prerequisites. They ought to be composed so that there are no restricted openings, furrows or openings to the inside of the gripper where little particles of the sustenance material and dampness can gather. It is a test to configuration squeezing grippers to dodge such molecule gathering territories. Specifically, parallel finger grippers are hard to configuration as indicated by sterile prerequisites of this kind on account of the guideways fundamental for the finger development.

Another perspective is the geometry of the gripper fingers. They ought to have a smooth surface and no sunken regions that will gather particles of any sort. Some gripper fingers have a serrated surface on the genuine holding face. For some food applications, this is unacceptable since the serration grooves are molecule authorities that are hard to keep clean.

The system of a gripper needs grease. Finish fixing against long-haul spillage is hard to get. A superior arrangement is then to utilize bearing materials that are greased up by water, or polymer-against-metal heading that can work without ointments.

At long last, the materials that will be presented to the food took care of must be of cleanly satisfactory quality. By and by that infers stainless steel and polymers as materials for fingers and external lodging dividers of a gripper.

7.5. PENETRATING (NEEDLE) GRIPPERS

Entering grippers can speak to a practical option for dealing with food products much of the time. They are appropriate for foodstuffs that are delicate and can be infiltrated without destructive impact on their quality. Basically, they are usable for meat, yet it is conceivable that they could be utilized for delicate cakes also. These grippers depend on the utilization of a few needles that infiltrate the surface of the protest make a mechanical hold. This punctures the surface, and the fundamental necessity for the utilization of such grippers is that this surface entrance is permitted. Both clean and quality contemplations are critical here.

Some meat preparing depends on needles to infuse handling liquids like salt, flavors or taste enhancers. In these cases, as puncturing of the surface is fundamental, extra puncturing from a gripper is then of no outcome. In different cases, the puncturing leaves denotes that are not really detectable. This gripper compose is an intriguing option since it can give an exceptionally solid and exact grasp.

The upsides of entering gripper are:

- Single side hold that makes it simple to grasp and convey in limited space.
- Very solid grasp because of the physical infiltration of the gripper into the tissue of the object.
- Precise situating because of the physical interlock between the needles and the object.
- High increasing velocities and exchanges paces can be gotten because of the solid hold acquired.

The burdens are:

- Puncturing of the surface can cause bacterial and other defilement. Systems to guarantee cleaning of the needles and alternate surfaces that interact with the object are obligatory.
- Puncturing leaves openings in the object. These gaps speak to visual quality debasement. Now and again such quality debasement isn't acknowledged.
- High speeding up can now and again prompt nearby tearing of the tissue of the object

7.5.1. Short Needle, Skin Penetrating Grippers

Short needle grippers are produced for material taking care of. Figure 7.5 demonstrates a case of this gripper compose. The gripper has numerous needles, regularly more than ten, which are stretched out at a 45° to the gripper's object confront. The needles must work combine shrewd in inverse ways. The needles expand 1–2 mm while holding, and withdraw totally behind the gripper's face while discharging the object. All needles are worked at the same time by the mechanical activity of a solitary pneumatic barrel inside the gripper.

There are no reports of this kind of gripper being utilized for grasping meat or some other foodstuff. In any case, one could think about it as a contender for grasping dainty cuts of frankfurter, cured meat or comparative. Specifically, if these cuts are permeable it could be a reasonable option.

Figure 7.5: Short needle gripper, (*Source: G. Caldwell Darwin (2013), Robotics and Automation in the Food Industry: Current and Future Technologies, Woodhead Publishing*).

7.5.2. Deep Penetrating Needle Gripper

Profound entering needle grippers were created for the particular motivation behind taking care of meat. Meat from both land creatures and fish has been focused on. Specifically, treatment of fish filets has been examined.

The capacity of the gripper is to use no less than two sets of needles that are inclining towards each other, normally at a point of around 45° to the gripper confront. At the point when completely expanded these needles relatively meet at a point 3– 5 cm before the gripper surface. A regular distance across for the needles is 2 mm. Each needle is worked exclusively by a little pneumatic barrel. Figure 7.6 demonstrates a huge needle gripper; Figure 7.7 demonstrates a case of a measured needle gripper where every module has two sets of needles.

The individual activation of the needles makes the gripper less delicate to bones in the object. Any needles that meet bone inside the meat piece grasped will stop while the others keep on extending. Along these lines most extreme conceivable grasp compel is acquired.

Because of the profound infiltration, these grippers will give an extremely solid holding power. The grasp will be lost just if the neighborhood strain on the tissue surpasses its break quality. For land creature meat and most fish meat, this is no issue. However, as indicated by reports some fish species have a free meat surface in the spring season, making the utilization of this gripper standard less suitable for these cases.

Figure 7.6: Long needle gripper, (*Source: SINTEF*).

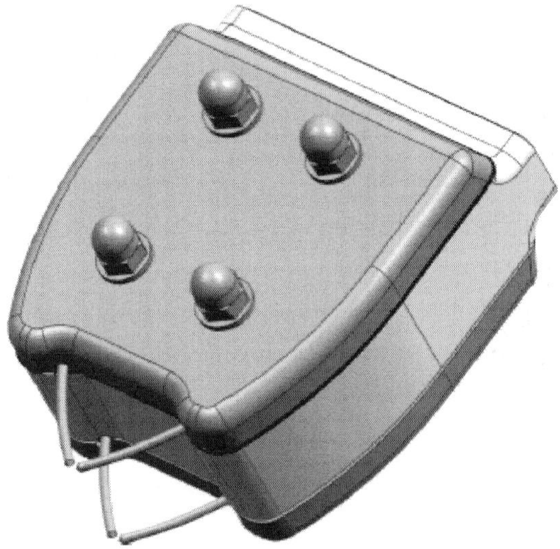

Figure 7.7: Modular long needle gripper, (*Source: Gjerstad et al., 2006*).

7.5.3. Lifting Performance of Needle Grippers

The fundamental execution parameter for an infiltrating gripper is its capacity to hold the object amid exchange. This is identified with the break quality of the material dealt with. It is exceptionally hard to make solid scientific models for the holding capacity since the crack system of meat when all is said in done isn't very much reported. Since meat is an inhomogeneous natural material, the outstanding designing models from inorganic materials can't be connected. Some streamlined models accepting homogeneous structure around the needles have not been effective in clarifying watched holding quality.

Be that as it may, some exploratory information from the treatment of fish filets demonstrate safe areas for taking care of increasing speed. Tests on treatment of cod and salmon filet indicate little contrast between these species, and dealing with most extreme increasing speed up to 2.5 m/s 2 appears to make no noticeable harm to the meat tissues.

7.5.4. Hygienic Performance of Needle Grippers

Clean necessities are a test for the short needle grippers. The openings for all needles are places where deposits from the dealt with materials will gather. A regular cleansing of all needle gaps with some antibacterial liquid is by all accounts important to get sheltered sterile task.

For long needle grippers, the sterile difficulties are met by building an extraordinary flushing framework into the gripper. This framework considers flushing of the needles as they are withdrawn after each holding cycle. Flushing can either be by purified water, or by some antibacterial liquid.

Sterile necessities make it obligatory to utilize just stainless steel needles on a wide range of needle grippers. The lodging ought to be either stainless steel or polymer. Sunken zones ought to be kept away from on every single outer surface to maintain a strategic distance from accumulation of buildups from the handled materials.

7.6. SUCTION GRIPPERS

Suction grippers have the upside of having the capacity to grasp on one surface as it were. Also, the physical size of the gripper at the robot's arm end can be made moderately little. Be that as it may, including vacuum producing gadgets and tubing to the gripper increases the volume and the multifaceted nature of these grippers. This is just a direct unpredictability increment in contrast with pneumatically worked squeezing grippers or needle grippers. The general evaluation is then that suction grippers can give the most reduced aggregate gripper arrangement.

Suction grippers are settled in the treatment of strong inorganic materials. They are dependable, non-nosy and function admirably in a kept space. There are likewise a few cases of suction grippers used to deal with chocolate, treats and other little food objects. Specifically bundled products are effectively taken care of.

The principal disadvantage of standard suction grippers is their affectability to air spillage into the suction glasses. On the off chance that the stream of air from this spillage is higher than the stream limit of the vacuum pump or ejector, the suction under pressure will vanish and the hold is lost. This is a notable wonder and restrains the utilization of conventional suction grippers to non-permeable materials.

Fresher advancement has in part tackled this issue. Grippers in view of direct under pressure yet high wind streams can make adequate grasping power for taking care of materials and comparable permeable materials. These grippers typically utilize the Coanda impact ejector as a vacuum generator since this ejector will deal with extensive wind streams.

Trial of gripper arrangements utilizing standard suction containers and vacuum generators for holding fish filet pieces have given baffling outcomes. This is because of the spillage as a result of the uneven surfaces of the fish filets. Tests with Coanda grippers then again are all the more encouraging. The spillage is by all accounts inside the range that these grippers will deal with.

Another variety of the suction guideline is to utilize the under pressure made by fast liquid stream. In this gripper, a wind stream guided towards the protest grasp is avoided by the object stream in a little hole between the object and the front face of the gripper. As indicated by the Bernoulli condition this fast stream makes a diminishing in the nearby weight in this hole. The weight distinction to the encompassing weight can be generous and will make a decent hold compel opposite to the gripper's front surface. Be that as it may, since there must be a ceaseless wind stream in the hole, the coefficient of erosion between the object and the gripper is basically zero here. The gripper can't apply any power parallel to the gripper's front surface. This is a downside since extra contact indicates have be set up to guarantee appropriate control of the held question's position and introduction.

7.6.1. Strong Under pressure Grippers

Figure 7.8a demonstrates the suction glass end of a traditional gripper framework. Figure 7.8b demonstrates a commonplace circuit graph of a similar framework. Much of the time a few suction glasses are associated with one vacuum gadget. Be that as it may, if there is high danger of air spillage into at least one of the suction mugs, isolate ejectors for each suction container can be connected. Another alternative is to utilize stream delicate cut off valves. These valves sense the stream in the line and close if the stream turns out to be too high. A valve in each vacuum line will then detach a spilling suction container from whatever remains of the framework.

The gripping force is calculated by Equation (6):

$$F_S = nA_s \Delta p \tag{6}$$

where F_S is the power from the suction gripper, n is the quantity of suction glasses, A_S is the region of one suction container, Δp is the weight distinction (under pressure) in the suction glass.

7.6.2. Moderate Under pressure, High-Flow Grippers

In the Coanda impact gripper, a Coanda ejector is utilized to make a direct under pressure. This ejector will pass on an extensive volume of air at this direct weight contrast. The ejector can be joined with the suction glass so that there is no requirement for in excess of a supply line and a control valve to control the wind current to the gripper.

The under pressure acquired with a Coanda impact gripper is one request of extent not as much as what can be gotten in standard suction grippers. In any case, since it acknowledges a spillage stream and can work on an expansive territory suction glass it will at present have the capacity to lift objects dealt with in the typical nourishment generation. Figure 7.9a demonstrates a planar Coanda impact gripper holding a bit of cowhide, Figure 7.9 demonstrates the same gripper holding a bit of material. Note that the unused suction glasses don't impact the lifting activity of the dynamic suction containers.

The lifting limit of a Coanda gripper can be evaluated from the suction bend of the particular gadget. The suction bend is reliant on the ejector geometry and estimate and must be resolved tentatively. Figure 7.10 demonstrates this bend for a planar gadget like the one appeared in Figure 7.9, yet with a thickness four times as extensive.

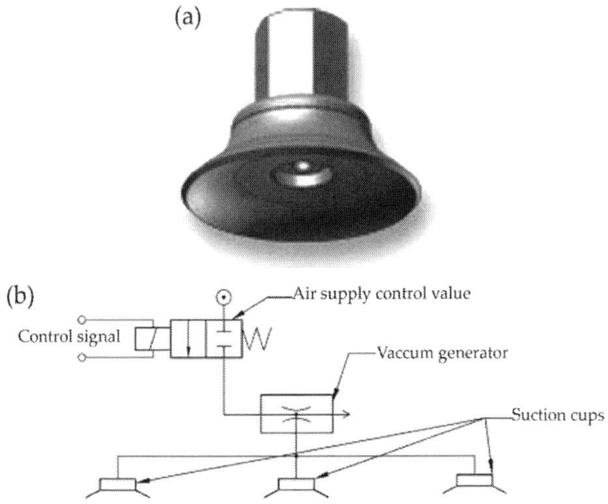

Figure 7.8: (a) A single suction cup, (b) diagram of high under pressure suction gripper control system, (*Source: G. Caldwell Darwin (2013), Robotics and Automation in the Food Industry: Current and Future Technologies, Woodhead Publishing*).

7.6.3. Bernoulli Grippers

The Bernoulli gripper has the primary shape appeared in Figure 7.11. This is a noncontact gripper which could have imperative cleanliness points of interest. The holding impact is reliant on the genuine nearby airspeed in the hole between the gripper surface and the held object. The grasping power is given by coordinating the differential spiral zone increased by the under pressure acquired from the disentangled Bernoulli Equation [7.7]. The improvement is to accept that there is no adjustment in air thickness and temperature as it courses through the hole.

$$F_g = \frac{1}{2}\rho\frac{Q^2}{2\pi h}\left[In\frac{r_{ext}}{r_{\text{int}}} - \frac{1}{2}\left(\frac{r_{ext}^2 - r_{\text{int}}^2}{r_{\text{int}}^2}\right)\right]$$

The air spilling out of the focal supply channel does likewise make an unpleasant power as it is diverted outspread way by the grasped question. Equation [7.8] gives that power.

$$F_r = r\frac{Q^2}{\pi r_{\text{int}}^2}$$

where F_g is holding power, F_r is loathsome power, ρ is surrounding air thickness, Q is wind stream, h is height of the air gap, r_{ext} is external radius of the Bernoulli gripper, and r_{int} is internal radius of the Bernoulli gripper. With an extensive air hole, the wind stream of the gripper is resolved principally by the stream protection of the supply channel. The wind current is then consistent with a steady supply weight, and the grasping power increments as the hole tallness is diminished. At a specific little hole stature, the stream protection of the air hole will command, and the wind current will begin to diminish, decreasing the holding power. At the point when the hole stature approaches zero, the unpleasant power will overwhelm. In this manner, the hole tallness never ends up zero. There is accordingly no immediate contact between the gripper and the protest. This holding guideline has been exhibited for dealing with delicate sustenance like cut cucumbers and tomatoes. Figure 7.12 demonstrates a chart of regular qualities the entirety of grasping and horrible powers.

(a)

(b)

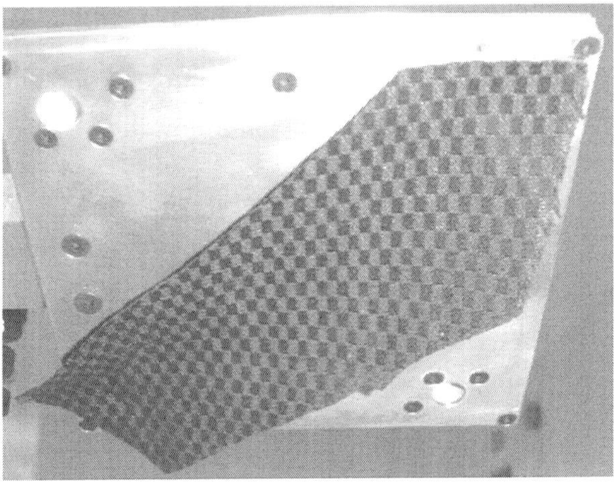

Figure 7.9: (a) A Coanda impact gripper holding a thin leather sheet (b) a Coanda impact gripper with four suction heads holding a bit of material, two of the suction containers are revealed, (*Source: G. Caldwell Darwin (2013), Robotics and Automation in the Food Industry: Current and Future Technologies, Woodhead Publishing*).

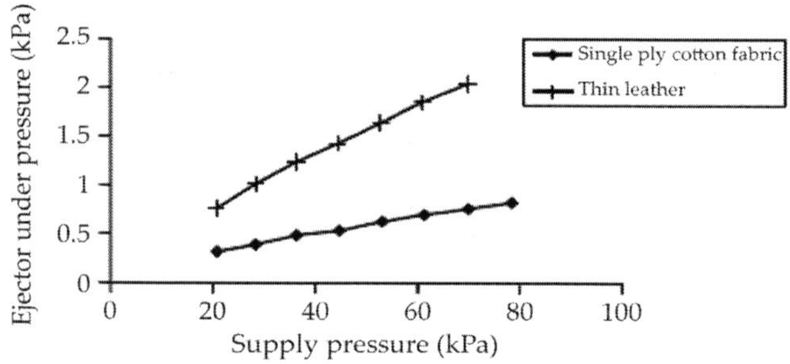

Figure 7.10: Obtainable under pressure for a Coanda gripper (9 x 20 mm throat size) for a non-porous and porous material, (*Source: Lien and Davis, 2008*).

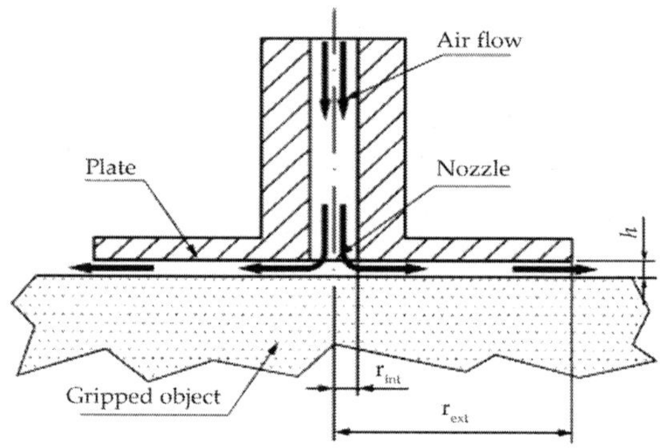

Figure 7.11: The working principle of a Bernoulli gripper, (*Source: G. Caldwell Darwin (2013), Robotics and Automation in the Food Industry: Current and Future Technologies, Woodhead Publishing*).

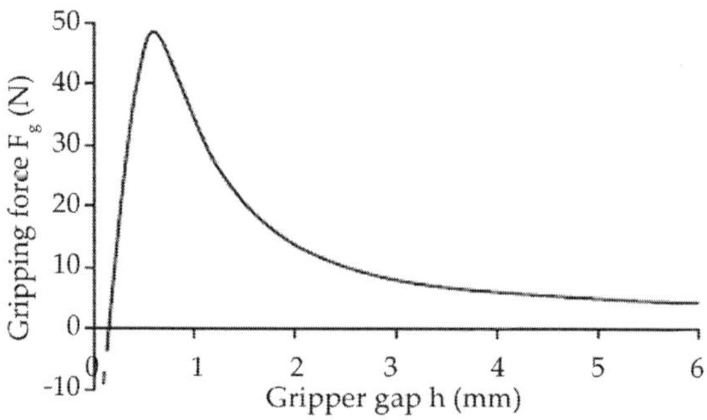

Figure 7.12: The gripping force of a Bernoulli gripper as a function of the gripper gap, (*Source: Dini et al., 2009*).

7.6.4. Lifting Performance of Suction Grippers

The lifting power relies upon the possible under pressure. There is a checked contrast between the standard suction container frameworks and the Coanda ejector grippers. The Coanda ejector may not give enough under pressure to lift overwhelming articles. The Bernoulli gripper comes nearer to the standard suction container as for lifting limit. The Coanda impact gripper does then again work with permeable and uneven surfaces for light questions because of its ability to keep up under pressure even with an extensive air spillage through the protest or along the outskirts of the suction glass. Run of the mill figures for realistic under pressure are given in Table 7.1.

7.6.5. Hygienic Performance of Suction Grippers

Clean issues are extremely important much of the time of taking care of uncovered foodstuff. The gripper ought not present any danger of pollution of the item. The two materials and the gripper configuration assume a critical part. The material utilized as a part of territories that interact with the item ought to be non-dangerous and cleanly endorsed. Stainless steel and numerous plastic materials are worthy.

Table 7.1: Under pressure Range and Lifting Capacity for Standard, Coanda and Bernoulli Suction Grippers

Gripper type	Typical underpressure range (N/m²)	Lifting force with e90 nun suction area (N)
High underpressure standard ejector	5000-80000	30-500
Low underpressure Coanda ejector	500-5000	3-30
Bernouli gripper	1000-8000	6-50

Source: G. Caldwell Darwin (2013), Robotics and Automation in the Food Industry: Current and Future Technologies, Woodhead Publishing

Table 7.2: Hygienic Characteristics of Suction Grippers

Gripper type	Collection of organic residues	Risk of bacterial growth	Cleaning
High underpressure standard ejector	Will collect organic residues in suction device and on the suction cup surface	Large	Difficult to clean the inside of the ejector or vacuum pump. Surface cleaning by washing
Low underpressure Coanda ejector	Will collect organic residues on suction cup surface but not in ejector	Moderate	Self cleaning ejector, surface cleaning by washing
Bernouli gripper	No collection of residues	None	Not required except for general equipment cleaning

Source: G. Caldwell Darwin (2013), Robotics and Automation in the Food Industry: Current and Future Technologies, Woodhead Publishing

The most widely recognized wellspring of pollution is buildups left on the gripper from the treatment of articles. These deposits will prompt the development of microscopic organisms or parasites in the event that they are not evacuated by customary cleaning. Specifically, treatment of exposed meat will leave deposits. Juices from the meat and little particles can without much of a stretch total in gaps, grooves and the inside of the gripper. Suction grippers are defenseless in this regard since they depend on air streaming into the gripper. Legitimate cleaning measures are in this manner required. Table 7.2 demonstrates a synopsis of the sterile qualities of the three kinds of suction grippers.

7.7. SURFACE EFFECT (FREEZE) GRIPPERS

One extremely fascinating thought for sterile holding is to utilize the impact of relatively prompt solidifying that happens when a chilly body is carried into contact with a wet protest. This solidifying frames a surface bond by the ice shaped on the interface territory between the cool body and the wet question. The icy body needs to have a temperature ordinarily around −10°C to cause this snappy solidifying activity. The rule is in modern use for taking care of little electronic segment were a drop of water is added to give dampness. A gripper for substantial articles was protected in 1988. In spite of the fact that grippers of this compose has been produced and shown effectively in inquire about research centers there are no reports of these being utilized yet in the food business. The fundamental preferred standpoint of these grippers is the way that they demonstration single sided and can be intended to high sterile norms. However, they likewise have a stamped downside in a to some degree eccentric discharging activity.

7.7.1. Design of Freeze Grippers

The general rule of freeze grippers originates from the way that warm vitality spill out of the surface of a object into a reaching colder surface will cool the question surface. On the off chance that the object surface is wet and the gripper surface has an adequately low temperature there will be a prompt arrangement of ice in the contact territory. This ice will tie the protest the gripper.

The seed of ice development relies upon the temperature of the gripper surface, the warm limit of both the gripper and the object and the wetting limit of the gripper surface. The measure of dampness on the surface of fish filets or filets of land creatures is generally enough. In any case, pieces that have been allowed to sit unbothered in dry zones for quite a while may not be sufficiently damp for a sheltered hold.

The solidifying procedure is extremely hard to demonstrate. A few vital parameters are normally not notable at the particular grasping moment, for example, the measure of water at first glance to hold, the warmth exchange coefficient and the wetting property of the gripper surface. Indeed, even a decent model won't give redress answers if these parameters are not known with high precision.

Rather these sorts of grippers have been examined tentatively to get some data about the useful capacity of this gripper family.

A standout amongst the most essential capacity parameters is an ideal opportunity to shape an adequately solid solidified film between the object and the gripper. This parameter is autonomous of the discharge techniques utilized, yet it is subject to the sort of protest be grasped. Examinations with pork meat, hamburger, codfish, and salmon have given the exact information in Table 7.3.

Table 7.3: Gripping Attachment Times and Temperatures for a Peltier Element Freeze Gripper with Stainless Steel Gripping Surface

Contact time (s)	Grip at temperatures below (°C)	No grip at temperatures above (°C)
0.2	—10.5	—8
0.4	—7	—5
1	—2	—0.5

Source: Lien and Gjerstad, 2008.

7.7.2. Freeze Grippers with Mechanical Release

Freeze grippers with mechanical discharge have an arrangement of littler metallic holding zones called solidify cushions. These freeze cushions are associated with a low-temperature supply that keeps it frosty as long as vital for a decent grasp. Liquid warmth exchange mediums are normally connected, brackish water or sorts of liquid that stay unfrozen at temperatures beneath −10°C are utilized as a warmth exchange medium between the gripper's freeze surfaces and a refrigeration unit.

Figure 7.13 demonstrates a freeze gripper with mechanical discharge created by SINTEF. The freeze cushions are round territories of stainless steel that are inserted in a plastic body. They are altogether associated with the cooling liquid in an internal assembly of the gripper. The discharge instrument comprises of a stainless steel outline with a blade segment that moves over the surface of the stop cushions when the edge is pivoted by its mounting shaft. The pole is pivoted by a little pneumatic actuator inside the gripper body.

To acquire a grasping activity the gripper's freeze cushions are squeezed against the protest. A film of ice is then framed promptly between the object and the freeze cushions. This ice makes the holding power of the gripper.

Discharge is acquired by moving the discharge cut over the stop cushions between the surface of the cushions and the protest. It will then rub the ice film off the stop cushions, hence discharging the question from the gripper.

7.7.3. Reversible Heat Flow Grippers

The reversible warmth stream gripper is worked around a Peltier component which creates low temperatures on the gripper surface. The Peltier component is a semiconductor gadget that will produce a warmth stream when an electric current goes through it. The bearing of warmth stream is controlled by the course of the electric current.

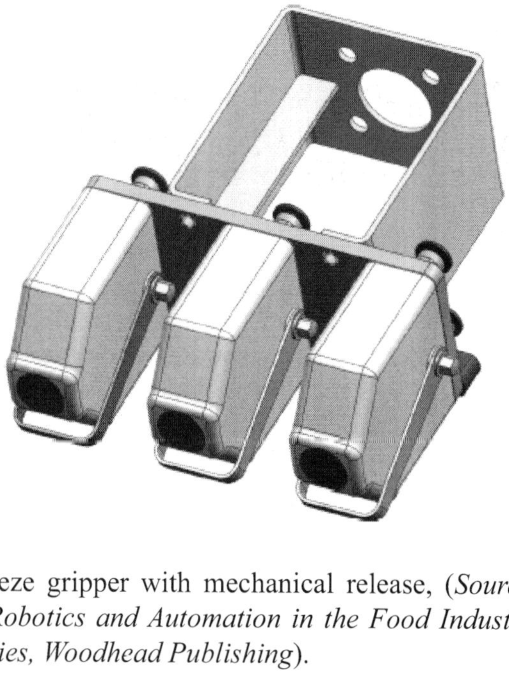

Figure 7.13: Freeze gripper with mechanical release, (*Source: G. Caldwell Darwin (2013), Robotics and Automation in the Food Industry: Current and Future Technologies, Woodhead Publishing*).

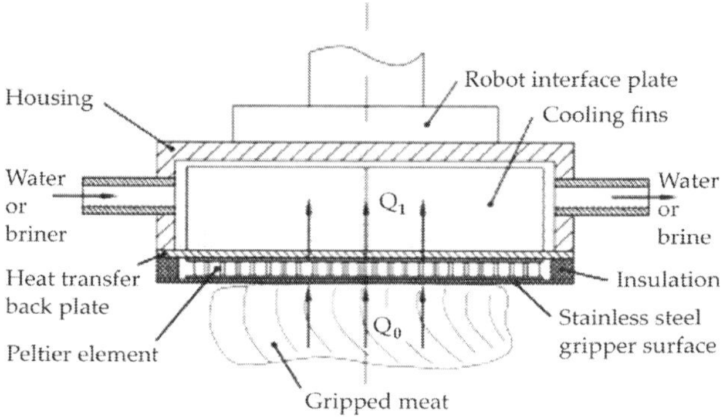

Figure 7.14: The principle of the Peltier element freeze gripper. Q_0 is the heat flow from the object; Q_1 is the heat flow into the cooling liquid, (*Source: G. Caldwell Darwin (2013), Robotics and Automation in the Food Industry: Current and Future Technologies, Woodhead Publishing*).

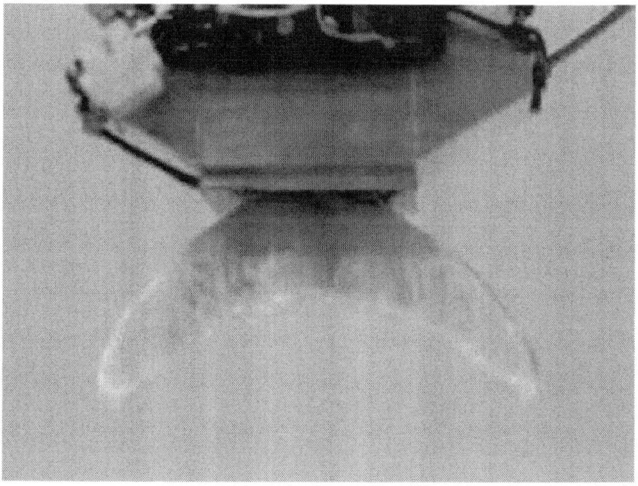

Figure 7.15: A Peltier element freeze gripper holding a piece of salmon fillet, (*Source: G. Caldwell Darwin (2013), Robotics and Automation in the Food Industry: Current and Future Technologies, Woodhead Publishing*).

Altering the electric current course will likewise alter the course of the warmth stream. This property is extremely helpful in a freeze gripper. Freeze grasping is gotten by sending current one path through the Peltier component. Turning around the present will discharge the object by dissolving the grasping ice.

Figure 7.14 demonstrates a cross segment of a reversible warmth stream gripper. It needs an inward chamber with a liquid that fills in as a steady temperature reference for the Peltier component. Icy water is a helpful liquid. It must be pumped through the gripper consistently to expel warm from the warm side of the Peltier component when the gripper is worked.

Figure 7.15 demonstrates a reversible warm stream gripper holding a bit of salmon filet. The gripper has the upside of a clean grasping surface and no mechanical moving parts.

An elective answer for discharge by turned around warm stream is to infuse a little measure of steam into the solidified zone. In this gripper arrangement, a focal opening is utilized to lead the steam into the solidified area. A slight overpressure of the steam will constrain it out in the hole between the gripper surface and the grasped question as the ice liquefies away. The stop activity of the Peltier component must be killed amid the discharge.

The holding power of a stop gripper is adequate to hold bits of meat of fish or land creatures. Tests have demonstrated that the holding power for angle filet is bigger than the burst power of the fish meat tissue. Analyses with grasping of materials demonstrated that extensive holding power could be gotten even with just incomplete ice arrangement on the gripper confront. In this analysis, ice secured 10–30% of the gripper– material contact zone. The deliberate crack power for the holding ice was in the range 9–15 N/cm^2.

The major problem with this type of gripper is the release time. While a typical release time for a pinching gripper is in the order of 0.1–0.2 s, the Peltier element gripper with current reversal has a release time in the order of 0.5–1.0 s. The time delay challenge can be overcome though by starting the release heat flow reversal a fraction of a second before arrival at the delivery place.

The real issue with this sort of gripper is the discharge time. While a run of the mill discharge time for a squeezing gripper is in the request of 0.1–0.2 s, the Peltier component gripper with current inversion has a discharge time in the request of 0.5–1.0 s. The time postpone test can be overcome however by beginning the discharge warm stream inversion a small amount of a moment before landing in the conveyance put.

7.7.4. Hygienic Performance of Freeze Grippers

The freeze grippers with mechanical discharge confront an indistinguishable clean difficulties from the squeezing grippers. There are control ways and inciting instruments that must be composed with the goal that buildup can't gather. Specifically, in the restricted hole between the blade and the stop surface, there will be an extraordinary danger of deposit gathering. This zone must be flushed with a cleaning liquid much of the time to dodge microorganisms and organism development.

The reversible warmth stream grippers then again have a greatly improved potential for to a great degree sterile activity. The grasping surface is a smooth stainless steel surface. There are no moving parts, and the lodging can be made totally level or curved to make outside cleaning simple.

The grippers with steam discharge will remove a little measure of steam through openings in the grasp surface. Since steam is a decent cleaning liquid, it is self-cleaning.

The Peltier component gripper has no openings in the lodging confronting the held protest, no moving part and no liquid stream in the gripper activity. From a clean perspective, it is the ideal gripper.

7.8. SELECTION OF THE APPROPRIATE GRIPPING TECHNOLOGY

In any computerization undertaking, the prerequisites for the treatment of the object must be mapped precisely to choose the ideal gripper arrangement. In delicate taking care of this turns out to be especially essential. Delicate treatment of food items poses challenges regarding mechanical taking care of and to sterile task. These difficulties prompt clashing prerequisite particulars. A few prerequisites will support one holding innovation while others will preclude a similar innovation.

A prerequisite rundown will then need to put weights on the distinctive necessities to have the capacity to fulfill the most critical ones.

A subjective synopsis of the reasonableness as for various components of the gripper advances said in this section is given in Table 7.4. The table has been created in view of sustenance taking care of difficulties. The order is unrefined A+ shows that the innovation functions admirably. The sign 0 demonstrates neither awful nor great, while – shows unsatisfactory quality. For a few elements, for example, uneven grasp and exact situating, the appropriateness is a go/no go issue. In any case, for grasping on uneven surfaces the appropriateness may not be so clear. There is unquestionably a variety in unevenness. On the off chance that the surface is exceptionally uneven, Coanda and stop grippers will end up inadmissible while encasing and long needle are appropriate regardless of what level of unevenness the errand presents.

Table 7.4: Qualitative Performance Parameters for Different Gripper Technologies

Gripper type	Hard surface	Uneven surface	Irregular shape	Soft structure	Porous	Fragile	Precise positioning	Single side grip	Dry surface	Moist surface	Heavy object	Grip marks	Hygienic performance
Two/three finger	+	+/0	0—	0—	0	—	+	—	+	+	+	0/	0/—
Enclosing	+	+	+	+	+	+	—	—	+	+	+	+	0/—
Short needles	—	—	—	+	+	—	+	+	+	+	—	—	—
Long needles	—	+	+	+	+	—	+	+	+	+	+	—	0/—
Standard suction	+	—	+	+	—	+	+	+	+	—	+	0	—
Coanda suction	+	+/0	+	+	+	+	+	+	+	+	—	+	+/0
Bernoulli suction	+	—	+	+	—	+	—	+	+	+	+	+	+
Freeze	+0	+0	+	+	+	+	+	+	—	+	+	+	+

+: Favorable according to this parameter

0: Neutral for this parameter

—: Unfavorable or unsuitable according to this parameter.

Source: G. Caldwell Darwin (2013), Robotics and Automation in the Food Industry: Current and Future Technologies, Woodhead Publishing.

7.9. METHODS OF ULTRASOUND

Ultrasound can be used for food preservation in combination with other treatments by improving its inactivation efficacy. There have been many studies combining ultrasound with either pressure, temperature, or pressure and temperature.

(a) Ultrasonication (US) is the application of ultrasound at low temperature. Therefore, it can be used for the heat sensible products. However, it requires long treatment time to inactivate stable enzymes and/or microorganisms which may cause high energy requirement. During ultrasound application, there may be rise in temperature depending on the ultrasonic power and time of application and needs control to optimize the process.

(b) Thermosonication (TS) is a combined method of ultrasound and heat. The product is subjected to ultrasound and moderate heat simultaneously. This method produces a greater effect on inactivation of microorganisms than heat alone. When thermosonication is used for pasteurization or sterilization purpose, lower process temperatures and processing times are required to achieve the same lethality values as with conventional Processes.

(c) Manosonication (MS) is a combined method in which ultrasound and pressure are applied together. Manosonication provides to inactivate enzymes and/or microorganisms by combining ultrasound with moderate pressures at low temperatures. Its inactivation efficiency is higher than ultrasound alone at the same temperature.

(d) Manothermosonication (MTS) is a combined method of heat, ultrasound, and pressure. MTS treatments inactivate several enzymes at lower temperatures and/or in a shorter time than thermal treatments at the same temperatures. Applied temperature and pressure maximizes the cavitation or bubble implosion in the media which increase the level of inactivation. Microorganisms that have high thermotolerance can be inactivated by manothermosonication. Also, some thermoresistant enzymes, such as lipoxygenase, peroxidase and polyphenol oxidase, and heat-labile lipases and proteases from Pseudomonas can be inactivated by manothermosonication.

7.10. APPLICATIONS

7.10.1. Meat Treatment

One of the most important quality attributes affecting consumer satisfaction and positive perception of beef is its tenderness. Inconsistency in beef tenderness has been rated as one of the major problems faced by the meat industry. Despite long-term research interest in achieving consistent eating quality of meat, it still remains as an elusive goal in meat science. One possibility is the use of ultrasound which can cause physical disruption of materials through cavitation related mechanisms such as high shear, pressure and temperature and formation of free radicals (Jayasooriya et al., 2007). Applications of ultrasound to provoke changes in physical and chemical properties of meat and meat products have attracted the interest of research workers for the past few decades because it is a purely physical technique, providing an alternative to chemical or thermal means of processing.

Ultrasound was tested for its ability to induce membrane cell disruption that could increase meat tenderness either directly, through the physical weakening of muscle structure, or indirectly, by the activation of proteolysis either by release of cathepsins from lysosomes and/or of Ca^{++} ions from intracellular stores so that it may activate the calpains.

The results of different studies may not be comparable because of differences in the type of muscles used, the animal age, the ultrasonic equipment and efficiency, intensities, frequencies, and the duration of ultrasound treatments. However, research in the last two decades has shown the potential benefits of ultrasound treatment as an alternative technology for modifying properties of meat and meat products. The meat industry will benefit from further research and development on the applications of ultrasound in modifying physical and/or chemical properties of meat.

Figure 7.16: Ultrasonic cutters in action, (*Source: http://www.newtech-ltd.co.uk/*).

7.10.2. Inactivation of Microorganisms

The most common techniques currently used to inactivate microorganisms in food products are conventional thermal pasteurization and sterilization. Thermal processing does kill vegetative microorganisms and some spores; however, its effectiveness is dependent on the treatment temperature and time. However, the magnitude of treatment, time and process temperature is also proportional to the amount of nutrient loss, development of undesirable flavors and deterioration of functional properties of food products. High power ultrasound is known to damage or disrupt biological cell walls which will result in the destruction of living cells. Unfortunately, very high intensities are needed if ultrasound alone is to be used for permanent sterilization. However, the use of ultrasound coupled with other decontamination techniques, such as pressure, heat or extremes of pH is promising.

The most frequently studied microorganisms, not only in the field of power ultrasound but also among other methods of food preservation are *Saccharomyces cerevisiae* and *Escherichia coli*. The former has been found to be less resistant to ultrasound than other vegetative cells, which is mostly attributed to its larger size. The inactivation of this microorganism has been proven in such food models as water, phosphate buffers, and sabouraud broth. The inactivation of *Staphylococcus aureus*, *Pseudomonas fluorescens*, *Listeria monocytogenes* and *E. coli* has been proven in water and phosphate buffers, as well as in foods such as UHT milk. Ultrasonication in combination with heat was performed to study the inactivation of *Listeria innocua* and mesophilic bacteria in raw whole milk.

7.10.3. Sonocrystallization

Crystallization is a considerably important process in many food industries such as chocolate, butter, margarine, whipped cream and ice cream. To obtain food products with specific sensory attributes (e.g., texture, hardness, smoothness, mouthfeel), fat crystallization must be controlled by temperature, cooling rate and application of shear or ultrasound. Power ultrasound in the range of 20 kHz and up to the MHz range has contributed as an effective tool for influencing the crystallization of liquids and melts (i.e., sonocrystallization), which is used in fat fractionation such as separating stearin (high melting) and olein (low melting) from a triglyceride oil. Ueno et al. investigated the effect of power ultrasound (at 20 kHz and 100 W for 2 s) on the crystallization behavior of model triglycerides (tripalmitin and trilaurin). Ultrasonication decreased the crystallization induction times of both triglycerides, increased nucleation rate due to which yielded small crystals and increased hardness. This effect could also be tuned by controlling sonication time, power, duration of the acoustic pulse and crystallization temperature.

7.10.4. Sterilization/Pasteurization

Conventional thermal pasteurization and sterilization are the most common techniques currently used to inactivate microorganisms and enzymes in food products (see chapter 7, food preservation methods). Unfortunately, the intensities of treatment, time and process temperature are also proportional to the amount of nutrient loss, development of undesirable flavors and deterioration of functional properties of food products. Ultrasound provides a method of improving such processes by virtue of the effects of cavitation. The use of ultrasound in pasteurization continues to be of great interest to the dairy industry. It has proved effective for the destruction of *E. coli*, *Pseudomonas fluorescens* and *Listeria monocytogenes* with no detrimental effect on the total protein or casein content of pasteurized milk. The mechanism of microbial killing is mainly due to the thinning of cell membranes, localized heating, and production of free radicals. Investigation on ultrasound effectiveness have also shown the inactivation of enzymes such as pectinmethylesterase, polyphenoloxidases, and peroxidases responsible for deterioration of fruit and vegetable juice and various enzymes pertinent to milk quality. In combination with heat, ultrasound can accelerate the rate of sterilization of foods, thus lessening both the duration and intensity of thermal treatment and the resultant damage. The advantages of ultrasound over heat pasteurization include minimization of flavor loss, greater homogeneity of treatment and significant energy savings.

7.11. HIGH HYDROSTATIC PRESSURE (HPP)

The HPP is the application of uniform and even ultra-high pressures to the product from all sides within a few minutes. This method causes no damage or distorts to the foods as long as the treated product is not hollow or having an empty space inside. During the pressurizing time, a major reduction of microorganisms and a denaturation of proteins could occur without influencing molecular bonds. This process can inactivate the inappropriate bacteria, extend the shelf life of the product and provide an improvement to the criteria of the product. In addition, chemical reactions that cause the destruction of vitamins or produce off-flavors can be reduced under High-pressure conditions. There are three major types of industrial high-pressure operating systems, a batch, continuous, and semi-continuous system. However, high-pressure systems either batch or continuous are suitable for high-pressure pasteurization. The Batch mode has the advantage that can process both liquid and solid products, and in this case, products are usually pre-packed first whereas continuous and semi-continuous systems can be applied only to liquid or pumpable products.

The general principles that explain the behavior of foods under effects of high pressure include the Le Chatelier, Isostatic pressing, and microscopic ordering principles.

(a) Le Chatelier's Principle states that a chemical system under equilibrium condition would experience a reaction change, accompanied by a decrease in volume when enhanced by pressure and vice versa.

(b) Isostatic Pressing (Pascal's Principle) the pressure is transmitted in a uniform manner in all directions. Following the decompression, the material returns to its initial shape.

(c) Microscopic Ordering Principle states that, at a constant temperature, increasing the pressure mutually increases the degree of ordering of the molecules of a substance. As a result, pressure, as well as temperature, exerts antagonistic forces on molecular structure.

The main objective of any non–thermal technology is to maximize the freshness and flavor qualities of the foodstuffs while achieving the required level of food safety. High hydrostatic pressure (HHP) meets with these requirements and today it being incorporated in many companies as an alternative to conventional heat treatment procedures.

Figure 7.17: High Pressure Processing Equipment, (*Source: http://blog.hiperbaric.com/en/salmonella-spoilage-how-hpp-could-have-helped*).

7.11.1. Effects of HHP on Proteins

HHP technology has been used fundamentally to reduce the microbial load and increase the safety and shelf life of treated foods with superior nutritional and sensory properties to those thermally treated. Nevertheless, the effect of HHP on proteins has raised interest and studies have been carried out to elucidate it. High Hydrostatic Pressure treatments affect the non-covalent links (ionic, hydrophobic, and hydrogen links) of proteins, which means that the secondary, tertiary, and quaternary structures can be unfolded and dissociated while the primary structure remains stable.

Messens et al. reported that it is necessary to apply a pressure of around 150 MPa to observe changes in the quaternary structure, and it is necessary to apply more than 200 MPa to significantly modify the secondary and tertiary structures. Owing to these changes, Liu et al. and Tabilo-Munizaga et al. studied the application of this technology to develop industrial applications to confer unique characteristics to foods (gel formation, emulsions, foams, new flavors, and textures) or to seek a fat replacement. The possibility of using these new products as fat replacements has encouraged in-depth studies of stabilizing and gelling agents, agents that are usually incorporated in foodstuffs to give stability, texture, and palatability. It is important to note that the changes depend directly on the type of protein used (disulfide bridges, linked by hydrophobic interactions, isoelectric points) and the HHP treatment (pressure, time, and temperature). All these studies make HHP a promising technology for the revalorization of waste and agro-industrial by-products.

7.11.2. Spore Inactivation

Spores are cellular forms that some microorganisms have developed as a response to adverse environmental situations in order to survive. Spores are characterized by their high resistance to different environmental stresses and preservation treatments. The most important spore-producing genera are *Clostridium*, *Bacillus*, and *Alicyclobacillus*. The initial spore load present in foods can be significantly reduced by HHP in combination with mild temperatures. In various published studies, 3.5 decimal log reductions have been reported for *Clostridium* sporogenes and 5.7 decimal log reductions for *Bacillus coagulans* by HHP at temperatures of 60–90 °C. Furthermore, Meyer observed significant reductions in the initial spore concentration in low-acid foods after treatments ranging between 700 and 1000 MPa and a product temperature of 70 °C. With those conditions, they obtained foods that were microbiologically stable at room temperature, and in many cases, the quality of the products was higher than that of those processed by heat.

7.12. THERMAL METHODS

7.12.1. Microwave

Microwaves are defined as a part of electromagnetic waves which have frequency range between 300 MHz and 300 GHz corresponding to wavelength from 1mm to 1m. Microwave frequencies of 915 MHz and 2.45 GHz can be utilized for industrial, scientific, and medical applications. Microwaves have been applied in a broad range of food processing such as drying, tempering, blanching, cooking, pasteurization, sterilization, and baking. Microwave heating has considerable advantages over conventional heating methods, especially with regard to energy efficiency. Since heat is transferred from the surface of food to the interior by convection and conduction in conventional cooking method, it may result in a temperature gradient between outside and inside food. In addition, it requires higher energy consumption and relatively long processing time. In microwave heating, on the other hand, heat is generated (volumetric heating) inside the food in a short time when microwave penetrates through it. Microwaves have greater penetration depth, and this property coupled with volumetric heating can lead to rapid heating rate with short processing time; and also contribute to the minimization of temperature difference between the surface and interior of food material.

As mentioned earlier, microwaves generate heat throughout the volume of food material rapidly because of the complete interaction between microwave, polar water molecules and charged ions in food. Molecular friction resulting from dipolar rotation of water molecule can generate heat. Water constitutes a major portion of most food products. Therefore, water is the primary component that interacts with microwaves due to its strong dipole rotation. Furthermore, heat can be generated through ionic migration that positive and negative ions of dissolved salts in food interact with the electric field by moving towards the oppositely charged regions of the electrical field and disrupt the hydrogen bonds with water.

The main ingredient that enables food to be heated by microwaves is water. The higher the water content of food, the faster is the heating rate. Water in molecular level behaves exactly like a magnet. Water has two oppositely charged ends due to presence of two positively charged hydrogen atoms and a negatively charged oxygen molecule. Therefore, water in food behaves like a magnet. If a bar magnet is held above another bar magnet, and you rotate the held magnet, the other one also rotates. In other words, microwave causes polar water molecules in food to constantly rotate and couple with electromagnetic field. Similarly, due to two different poles in water, when microwaves oscillate, the water molecules rotate. This is because the negatively charged end of water is attracted to positively charged end of microwave, while the positive end of water is attracted to the negatively charged end of microwaves. The microwaves rotate at extremely high speed of 2450 time per second. This means for every second a microwave rotates, the water molecule also rotates 2450 times. This extremely high rotation rate causes water molecules to collide with each other at very fast rate. This creates friction between water molecules. This friction generates heat. The heat flows through the food by conduction, convection or radiation. Therefore food warms up.

Figure 7.18: Using microwave technology for food processing, (*Source: https://www.alamy.com/stock-photo/microwave-technology-for-food-processing.html*).

7.12.1.1. Applications

7.12.1.1.1. Thawing and Tempering

Thawing is the process of bringing the frozen product from a frozen state to a state where there is no heat (usually at temperatures above 0°C) otherwise known as defrosting. Thawing is often considered a reversal of the freezing process. Frozen food are thawed before cooking to ensure that there is enough heat to destroy harmful bacteria during cooking. If large frozen foods are not thawed prior to cooking, the center will remain uncooked while it gets burnt or overcooked on the surface. The main problem of normal thawing is the large space requirements and longtime that can cause chemical and bacterial damage. When the frozen food are thawed, the surface area of the food is the first factor to increase in terms of temperature while the reproduction of bacteria that were restricted in the frozen foods can proceed. The use of microwave energy can overcome this challenge when heat is generated with food from the center to the surface thereby making microwave thawing faster than the other techniques. The main drawback of microwave thawing is that it does not occur homogeneously and this condition is known as runaway heating.

Tempering is the process of introducing the product to a temperature where a large quantity of water in the product is in form of ice, but not all water becomes ice. The temperature should be lower than the freezing point and usually between –5 and –2 °C. In this condition, the product is firm but not hard and thus easier to cut. This product can be cooled to this condition from temperatures above the freezing point or warmed by ice temperatures. The concept of microwave tempering, as opposed to dissolving the microwave, has been studied by many scientists. Early researches predicts that full microwave thawing cannot be applied. However, microwave tempering has shown great prospects. Microwave tempering can be used as a substitute to thawing because in most cases, complete thawing is not necessary, and may consume energy, increases processing time, and affects quality. Microwave tempering has several benefits over most thawing methods. The microwave tempering process can handle large quantities of frozen product economically, has a high yield rate, and is accomplished in small spaces with no favorable condition for the growth of bacteria.

7.12.1.1.2. Dehydration

Microwave heating offers distinct benefits in dehydration because of its penetration depth, and the uniform heating results in water vaporizing from throughout the product. This induces an inner pressure that maintains puffed character of the dried product and preserves color, flavor, and nutritional value. Microwave drying is rapid and more energy–efficient compared with conventional hot–air drying. In microwave drying, the removal of moisture is accelerated and, furthermore, heat transfer to the solid is slowed down significantly owing to the absence of convection. The usual practice of applying microwaves to drying of food materials is at the falling rate period where the migration of water from the center of the products is significantly reduced and the drying rate is comparatively slow in conventional drying process.

A two-stage drying process involving initial forced–air convective drying followed by microwave finish drying has been reported to give better product quality with considerable savings in energy and time. The bakery industry conventionally uses the microwaves for finish drying of biscuits and cookies. Moreover, the moist bakery products exhibit higher loss factor, resulting in more heat generation compared to drying at a later stage with minimum water content. Potato chips have been recognized as a global fast food. The first large-scale application of microwave energy in food processing industry was in finish drying of potato chips. Conventional drying technology cannot rapidly achieve the desired low moisture levels of potato chips and, in addition, browning creates another quality problem for the product due to presence of sugars.

Drying of potato chips up to 6%–8% followed by finish drying by microwave overcomes the difficulty of the process industry. Application of microwave energy in vacuum results in an increase in product temperature; however, the temperature rise is limited to the boiling point of the water at the lowered pressure. At a pressure of 3 kPa, free water boils at 22 °C. This maintains a product temperature at a level below the temperature used under atmospheric conditions. Microwave vacuum dehydration was first used for concentration of citrus juice in France. Microwave vacuum drying of agricultural commodities includes various cereal grains, and further this technology has been adapted to grapes for production of Grape PuffsTM using zoned microwave vacuum dehydration. The drying of pasta products and noodles at 915 MHz is a commercial success in many countries.

7.13. FUTURE TRENDS: FROM LABORATORY TO INDUSTRY

The grippers utilized as a part of automated taking care of in the nourishment business today (2010) are for the most part the conventional squeezing, encasing or suction writes. These advances function admirably in applications that are like the ones seen in different businesses. In any case, in dealing with errands that are special for nourishment enterprises it is by all accounts more hard to discover great gripper arrangements. Consequently, there has been an expanded exertion in explore research facilities to examine new option grasping innovations. All the non-customary grasping strategies depicted in this book have been produced in light of necessities identified with food dealing with.

These new kinds indicate promising exhibitions that can tackle a portion of the issues that emerge in mechanical applications in the food business today. One evident issue is the sterile test. The customary gripper arrangements utilized as a part of industry are not the best from a sterile point of view. The Peltier component gripper is a case of an answer for this test.

More critical however is the ability of dealing with delicate material in single-sided grasps. This is a prerequisite that shows up regularly in exchanges over robotization in the preparing of meat, be it fish or land creatures. The delicate quality is frequently matched with uneven surfaces and shapes. For this class of taking care of issues the Peltier component gripper, the Coanda gripper, and the needle grippers can give great arrangements.

In any case, execution in lab tests isn't sufficient. At this stage, a few clients must will to try different things with the new advancements to decide their genuine incentive in the cruel modern conditions. This progression is important to refine the new advancements and increment understanding in grasping innovation for food materials.

REFERENCES

1. Barbosa-Cánovas, G., Pierson, M., Zhang, Q., & Schaffner, D., (2000). Pulsed electric fields, *J. Food Sci. 65*(8), 65–79.
2. Brodnck, H., & Thomas, A., (1978). Radiation Preservation of Subtropical Fruits in South Africa. *In Food Preservation by Irradiation, 1*, pp. 167–178, IAEA, Vienna (1978).
3. Chua, P. Y., Ilschner, T., & Caldwell, D. G., (2003). Robotic manipulation of food products –an overview. Industrial Robot. doi.org/10.1108/01439910310479612
4. Davis, S., Gray, J. O., & Caldwell, D. G., (2008). An end-effector based on the Bernoulli principle for handling sliced food and vegetables. Robotic and computer integrated Manufacturing. Doi: 10.1016/j.rcim.2006.11.002
5. Fellows, P., (2000). Food processing technology: Principles and practice. [ebook] Woodhead Publishing. Available at: *http://www.webpal.org/SAFE/aaarecovery/2_food_storage/Food%20Processing%20Technology.pdf.*
6. Caldwell G. Darwin, (2013). Robotics and Automation in the Food Industry: Current and Future Technologies. Woodhead Publishing.1st edition
7. Jayasooriya, S., Bhandari, B., Torley, P., & D'Arcy, B., (2004). Effect of High Power Ultrasound Waves on properties of Meat. A Review. *Int. J. Food Prop. 7*(2), 301–319.
8. Knorr, D., Zenker, M., Heinz, V., & Lee, D., (2004). Applications and potential of ultrasonics in food processing. *Trends Food Sci. Technol. 15*, 261–266.
9. Kumar, Y., Patel, K., & Kumar, K., (2015). Pulse Electric Field Processing in Food Technology. *Int J Engg Stud Tech Approach 1*, 6–17.
10. McClements, D., (1995). Advances in the application of ultrasound in food analysis and processing, *Trend Food Sci. Technol., 6*, 293–299.
11. Singh, P., Heldman, R., (2014). Introduction to Food Engineering. [ebook] Fifth ed. Academic Press. Available at: http://197.14.51.10:81/

pmb/AGROALIMENTAIRE/Introduction%20to%20Food%20Engineering%20Fourth%20Edition.pdf [Accessed 14 May 2018].

12. Urbain, W., (1978). Food irradiation Advances in Food Research, *24*, pp. 155–227 Academic Press, USA.

CHAPTER 8

WIRELESS SENSOR NETWORKS (WSNS) IN THE AGRICULTURAL AND FOOD INDUSTRIES

"Food manufacturing is an ideal candidate for targeted accelerated depreciation because the food industry, our biggest industry, creates significant flow-on benefits."

—Anthony Pratt.

CONTENTS

This chapter starts by presenting a wireless sensor network (WSN) innovation and its present improvement. It at that point broadly surveys the applications of WSNs in the zones of condition observing, accuracy horticulture, exactness domesticated animals cultivating, nursery administration, and food traceability frameworks. The chapter likewise talks about the snags and difficulties looked in the improvement and arrangement of WSNs in farming and food creation. At long last, the future patterns in the use of WSN innovation in agribusiness and food generation are exhibited.

8.1. INTRODUCTION

Remote advances have been under fast improvement amid the previous 10 years. Sorts of remote innovations being produced extend from basic short-run, point-to-point information transmission through infrared light (Infrared Data Association, IrDA), to remote individual region systems (WPANs) for short-run, point-to-multipoint interchanges, for instance, Bluetooth and ZigBee, to mid-run, multi-bounce remote neighborhood (WLANs), and to long-separate wireless frameworks, for example, GSM/GPRS and CDMA (Wang et al., 2006). Late logical and innovative unrests have prompt development in numerous modern territories, for example, radio recurrence (RF) innovation, coordinated circuits, chip, keen sensors, and smaller scale electro-mechanical frameworks (MEMS) (Mahfuz and Ahmed, 2005). These most recent advancements have made it feasible for large-scale manufacturing of minimal effort, low power utilization, high dependability, multifunctional, wise, smaller than normal sensor and controller hubs with systems administration capacity. These gadgets are called 'hubs' or 'shrewd tidy.' The organized hubs are regularly sent to application conditions to gather constant information and actualize straightforward control methodologies. Through them, more itemized learning about the earth is gained, particularly in unsafe, risky, and remote zones and areas.

Horticultural and food generation are ordinarily in remote and uncontrolled regions. The creation results depend vigorously on condition and product conditions. With the development and union of a few advances, including the worldwide situating framework (GPS), geographic data frameworks (GIS), scaled down PC segments, programmed control, in-field and remote detecting, portable figuring, propelled data preparing, and broadcast communications, point by point data on field and creation inconstancy both spatially and transiently can be obtained and medications can be acclimated to meet each site's one of a kind needs (Zhang et al.,

2002). Different sensors, controllers, and actuators with knowledge have been broadly utilized as a part of field information obtaining, checking, and control. Their real deficiencies are the necessities for broad wiring for various point estimations and controls and continuous on location visits by administrators for information downloading and upkeep because of unavoidable harms and misfortune. Inability to fulfill these necessities may prompt missing information and breakdown of the general generation framework.

A WSN framework comprises of different hubs, every one of which includes RF handsets (bits), sensors, microcontrollers, and power sources. Their selforganizing, self-designing, self-diagnosing and self-mending capacities empower them to frame impromptu single/multi-jump systems. This lessens and rearranges wiring and connectors, gives establishment adaptability to sensors and controllers, and diminishes support many-sided quality and expenses. Most remote hubs have flag molding and preparing units that can change over simple to advanced signs. Subsequently, clamor get turns into a less noteworthy issue. Besides, progressed microelectronic innovation enables the hubs to have little physical sizes and low power utilization while keeping up their usefulness. In farming and sustenance creation, information gathered from a sent WSN can be utilized to advance administration system in a continuous manner. A remote client (e.g., cultivate chief) can likewise send charges to choose hubs in a field to appoint new errands, change arrangements, and analyze issues. With these favorable circumstances, WSN innovation can assume a vital part in acknowledging continuous information procurement, preparing, and controls with fine spatial and fleeting determination.

8.2. CURRENT STATE OF DEVELOPMENT OF WSNS

The advancement of WSNs has experienced its rise, moderate improvement, and fast reception stages. The quick advancement of PC organizes innovation and extension of remote correspondence systems convey WSNs to another period of fast improvement and organization.

Advancement of WSNs is fundamentally in view of three noteworthy parts: equipment, working framework (OS), and system correspondence. The equipment comprises of: (1) numerous hubs conveyed in an objective site to frame a remote correspondence arrange, every one of which has at least one sensors as well as controllers, onboard flag molding and handling units, restricted information stockpiling, a radio handset, and a power unit; and (2) one or different sink or door crossing over units to hand-off estimation information to a remote control focus through remote LAN, cell, or different systems (Boegena et al., 2006). The OSs are attached to the equipment and keep running on the hubs to facilitate on-board segments to finish the doled out assignments, for example, the information obtaining, handling, transmission, and capacity (Gay et al., 2003). The system correspondence characterizes organize topology and correspondence conventions with the thought of steering, power and asset administration, and hub confinement, and so forth. Table 8.1 demonstrates the most widely recognized business remote detecting and controlling frameworks. They are arranged into two classes: 'WSN equipment stages' and 'Information lumberjack stages.' Radio recurrence recognizable proof (RFID) has been generally received in numerous applications. It assumes vital parts in nourishment traceability frameworks. At the point when coordinated with WSN innovation, RFID frameworks can give more item data notwithstanding distinguishing proof.

8.2.1. WSN Hardware Platforms

Numerous broadly useful business stages are right now accessible in the WSN item showcase. The stages are comparable in real segments, for example, hubs with/without onboard sensors, information procurement modules, sinks/doors modules, and power unit. They are regularly open-source and for the most part, have huge client gatherings. Specialized help is given through assets/gatherings shared among framework engineers (Table 8.1).

Table 8.1: Commercial WSN Hardware Platforms and Data Logger Platforms

Platform	CPU	Power	Memory	I/O interfaces	Radio	Max. range	OS	Manufacturer
WSN hardware platforms								
Mica2	Atmega128	3.3V battery, 15μA, Sleep; 8mA, Active	128 kB ROM 512 kB Flash 4 kB EEPROM	Regular I/Os*, 51 pin interface to other extension boards	CC1000	300 m outdoor	TinyOS 1.x/2.x	Crossbow Technology
MicaZ	Atmega128	3.3V battery, 15μA, Sleep; 8mA, Active	128 kB ROM 512 kB Flash 4 kB EEPROM	Regular I/Os*, 51 pin interface to other extension boards	CC2420	75–100 m outdoor 20–30 m indoor	TinyOS	Crossbow Technology
IRIS	Atmega128	3.3V battery, 15μA, Sleep; 8mA, Active	4 kB RAM 128 kB ROM 512 kB Flash 4 kB EEPROM	Regular I/Os*, 51 pin interface to other extension boards	RF230	>300 m outdoor >50 m indoor	TinyOS	Crossbow Technology
Imote2	Intel PXA271	3.3V battery, 390μA, Sleep; 66 mA, Active	256 kB SRAM 32 MB SDRAM 32 MB Flash	GPIO/SPI/UART/ I²S/USB/ AC'97/Camera/ IMB400 multimedia extension board	CC2420	30 m with attached antenna	Embedded Linux or Windows support	Crossbow Technology
TinyNode584	MSP430	3.6V battery 4μA, Sleep; 77mA, Active	10 kB RAM 512 kB Flash	Regular I/Os, factory made extension board for custom interface electronics	XE1205 (915 MHz)	Up to 2000 m	TinyOS	Shockfish SA
TinyNode184	MSP430	3.6V battery 22 μA, Sleep; 2.2A, Active	10 kB RAM 512 kB Flash	Regular I/Os, factory made extension board for custom interface electronics	SX1211 (868/915 MHz)	150 m outdoor 50 m indoor	TinyOS	Shockfish SA
Tmote Sky	TI MSP430	3.6V battery 5.1 μA, Sleep; 23mA, Active	10 kB RAM 48 kB Flash	Regular I/Os, on-board sensors, no manufacture built extension board	CC2420	125 m outdoor 50 m indoor	TinyOS	Moteiv
BTnode rev3	Atmega128L	Sleep 9.9 mW Active with Bluetooth, 198 mW or 102.3 max for CC1000	64+128 kB SRAM 128 kB Flash 4kB EEPROM	Regular I/Os, no manufacture built extension board	ZV4002(Bluetooth) or CC1000	10 m using Bluetooth 300 m outdoor using CC1000	TinyOS or BTnut	ETH Zurich
eyesIFXv2.1	MSP430F1611	36 mW max	10 kB RAM 48 kB ROM 4 MB Flash	Regular I/Os, no manufacture built extension board	Infineon TDA5250 (868 MHz)	200 m–600 m depending on data rates	TinyOS	Infineon Technologies
Wireless data logger platforms								
CR1000+RF4xx	Renesas H8S 2322 for CR1000	12 V battery Sleep, 0.6 mA Active, 27.6 mA/100 mA for CR1000; 9-16 V battery 1 mA Stand-by 75 mA max for RF4xx	4 MB SRAM CF card extension	16 Analog inputs, 3 analog outputs, 8 digital I/Os, communication ports	RF401 (915 MHz) RF411 (922 MHz) RF416 (2.4 GHz)	Up to 1.5 km using short distance antennas Longer distance using high gain antennas	LoggerNet 3.x, PC400 1.2 or ShortCut 2.2	Campbell Scientific
V/SG/G-Link nodes+ 2.4 GHz base station	N/A	0.5 mA sleep 25 mA max	Up to 2 MB (approximately 1 000 000 data points)	Fixed on-board sensors depending on series 'x'	2.4 GHz, 16 nodes maximum for simultaneous streaming	70 m line of sight, up to 300 m with optional high gain antenna	Precompiled system in VB/VC++ / LabView	Micro Strain
Em50R/G	N/A	5 AA batteries	1 MB data storage	4 Analog inputs, 1 serial input port	50R: 900 MHz or 2.4 GHz selectable 50G: using cellular network	N/A	ECH_2O Utility and DataTrac	Decagon Devices
HOBO W-TMB/ SMC	N/A	AC powered or 3 AAA battery	N/A	TMB: Temperature SMC: Soil moisture	2.4 GHZ	420 m with clear line of sight or 300 m typical	HOBO Node viewer	Onset

Source: G. Caldwell Darwin (2013), Robotics and Automation in the Food Industry: Current and Future Technologies, Woodhead Publishing

The business WSN stages are made out of remote handset modules, information procurement interface modules, basic sensor modules, and entryway modules. Clients can stack the modules together to shape a hub in view of the necessity of their applications. A remote handset module comprises of a low-control chip and a RF handset. The DAQ modules give interfaces to outer sensors and controllers, for example, simple I/O, broadly useful advanced I/O, I 2 C interface, and UART ports. The sensor modules incorporate regular sensors, for example, temperature, moistness, light, quickening, and sound on one circuit board with a standard association interface to alternate modules. The door/base station modules are utilized as a sink to permit the total of sensor information from a WSN to a remote PC through different means, for example, Ethernet, cell arrange, high-control radio, and satellite. The pioneer producer in this classification is Crossbow Technology, CA, USA who gives remote handsets (Mica2, MicaZ, IRIS and Imote2), information obtaining sheets (MTS, MDA, ITS and IMB400), entryways (Star-door and MIB), and improvement programming (XMesh and XServer, Crossbow Technology Inc., California, USA).

This product offering furnishes WSN engineers and clients with adaptability on plan application-arranged WSNs. Jennic Technology in the United Kingdom gives a comparable group of remote microcontrollers, modules, and advancement devices that have been utilized as a part of numerous applications. Shockfish (2010) in Switzerland delivers the TinyNodes arrangement principally for 868 MHz band applications. The expansion sheets give interfaces to include outer power supply, sensors, and controls. Sentilla Corporation in the USA gives Tmote Sky and Tmote smaller than expected, which are 'one-board' module coordinating sensors, a chip, and a RF handset. They are shabby underway cost, and powerful because of the coordinated plan, yet with low application adaptability. Coal Corporation in the USA offers a low-control, elite coordinated ZigBee System-on-Chip with a size of just 7 × 7 mm.

Different global principles have been set up for WSN applications as of late. Among them, the guidelines for remote LAN, IEEE 802.11b ('WiFi') and remote PAN, IEEE 802.15.1 (Bluetooth) and IEEE 802.15.4 (ZigBee), are utilized all the more generally for estimation and mechanization applications. Every one of these gauges utilize the instrumentation, logical and therapeutic (ISM) radio groups, including the sub-GHz groups of 902–928 MHz (US), 868–870 MHz (Europe), 433.05– 434.79 MHz (US and Europe) and 314– 316 MHz (Japan) and the GHz groups of 2.400– 2.4835 GHz (overall adequate). When all is said in done, a lower recurrence permits a more drawn out transmission run and a more grounded capacity to enter through dividers and glass. Be that as it may, because of the way that radio waves with bring down frequencies are all the more effortlessly consumed by different materials, for example, water and trees, and that radio waves with higher frequencies are less demanding to disperse, the compelling transmission remove for signals conveyed by a high recurrence radio wave may not really be shorter than that of a lower recurrence bearer of a similar power rating. The 2.4 GHz band has a more extensive transmission capacity (250 kbs) that permits more channels and recurrence jumping and allows reduced reception apparatuses. The ZigBee standard was set up by the ZigBee Alliance that is upheld by in excess of 70 part organizations. It includes system, security and application programming to the IEEE 802.15.4 standard. Attributable to its low power utilization and basic systems administration arrangement, ZigBee is viewed as the most encouraging for remote sensors and has been installed into numerous business items around the world.

One of the regular attributes of the WSN equipment stage is it needs ability to program the hubs and general framework. Framework engineers with a decent foundation of electronic circuit plan and chip writing computer programs are required. Since a large portion of the makers of the WSN stages don't give helpful specialized help, it turns into a major obstacle for application designs in rural and nourishment preparations to create and send the WSN frameworks.

8.2.2. Wireless Data Logger Platforms

In agricultural applications, stand-alone data loggers with various standard interfaces (analog I/O, digital I/O, serial ports, parallel ports, etc.) are widely used. They are easy to use, easy to program, and very rugged under various environmental conditions. For example, data loggers from Campbell Scientific Inc. (2010) have been widely used in weather stations to collect environmental data. Some of the data loggers integrate RF modules to remotely transmit data to a control center. Recently, more and more data logger manufacturers are adding wireless communication modules to their existing data loggers. Campbell Scientific Inc. now provides communication modules of Ethernet, Spread spectrum RF, satellite, cellular (GPS and CDMA), etc. which can easily connect to their data loggers. Decagon Devices, Inc., in the USA (Decagon, 2010) recently provides the Em50G wireless cellular data logger and Em50R wireless radio data logger, which allow long-distance networking and data communication. MicroStrain (2010) has a production line of wireless nodes with various sensors, wireless base stations, and a wireless sensor data aggregator. Onset (2010) manufacture a series of Hobo data loggers.

They have recently started to provide wireless data loggers for temperature and/or humidity for indoor and outdoor applications. These data logger platforms are tied closely with the OSs and development tools from their respective manufacturers. They can be easily networked among modules from the same manufacturer, but it can be difficult to connect with systems from different manufacturers. This reduces the flexibility on both hardware and software selections. Meanwhile, these systems are often expensive. With these shortcomings, they are not suitable for large-scale WSN applications.

In agrarian applications, remain solitary information lumberjacks with different standard interfaces (simple I/O, advanced I/O, serial ports, parallel ports, and so on.) are generally utilized. They are anything but difficult to utilize, simple to program, and exceptionally rough under different natural conditions. For instance, information lumberjacks from Campbell Scientific Inc. have been generally utilized as a part of climate stations to gather ecological information. A portion of the information lumberjacks coordinate RF modules to remotely transmit information to a control focus. As of late, an ever increasing number of information lumberjack makers are adding remote correspondence modules to their current information lumberjacks. Campbell Scientific Inc. presently gives correspondence modules of Ethernet, Spread range RF, satellite, cell (GPS and CDMA), and so on which can without much of a stretch interface with their information lumberjacks.

Decagon Devices Inc. in the USA as of late gives the Em50G remote cell information lumberjack and Em50R remote radio information lumberjack, which permit long-remove systems administration and information correspondence. MicroStrain has a generation line of remote hubs with different sensors, remote base stations, and a remote sensor information aggregator. Beginning (2010) fabricate a progression of Hobo information lumberjacks. They have as of late given remote information lumberjacks to temperature as well as moistness for indoor and outside applications. These information lumberjack stages are tied intimately with the OSs and advancement instruments from their individual makers. They can be effortlessly organized among modules from a similar producer, however, it can be hard to associate with frameworks from various makers. This decreases the adaptability on both equipment and programming choices.

Then, these frameworks are regularly costly. With these deficiencies, they are not appropriate for expansive scale WSN applications.

8.2.3. Radio Frequency Identification (RFID) Systems

A RFID framework comprises of labels or transponders, perusers/cross-examiners, and a PC with a data administration framework. The labels/transponders convey item IDs, which are perused by the peruser/investigative specialist. The PC framework stores and deals with the item IDs and other data. There are two sorts of RFID labels, dynamic and inactive. The dynamic labels are fuelled by an onboard battery, while the detached labels get their energy from a per user or a cross-examiner through electrical attractive field. Thus, dynamic RFID labels are outfitted with bigger memory, more I/O ports, and onboard sensors, and an all the more effective radio. They convey more data notwithstanding item ID. The all the more intense radio in a dynamic tag permits a more extended transmission remove than a latent tag.

Most detached labels work at low frequencies in the vicinity of 125 and 148 kHz. These labels are shabby (around US$2 per tag) and have been embraced for creature labels since the 1980s. ISO 11784/11785/14223 are the measures to direct the creature RFID applications. As of late an ever-increasing number of dynamic labels working at frequencies of 455 MHz, 2.45 GHz, and 5.8 GHz have been accounted for. Their multifunctional highlights make them more like the hubs in WSNs and simpler to be incorporated into WSNs. Z2RFID framework created by ZigBeef (2010) comprises of long-range steers RFID labels, which can transmit one of a kind ID numbers and movement and temperature data along separation (>60 m) utilizing ZigBee gauges and a handheld examiner (PDA) to store the data. These labels have been utilized by farmers and rodeos to track and tally their creatures. Another organization, TekVet, has been promoting a little dynamic 418 MHz RFID tag, TekSensor, which can be settled inside a creature's ear waterway to quantify outer ear trench temperature, which is a marker of creature center body temperature.

A peruser gets the IDs and temperature readings from the TekSensors consistently, 24 h daily. A product bundle, TekAccess, was created to get to and break down the information to give an early sign of anomalous changes in temperature, which might be related with numerous basic sicknesses.

8.2.4. WSN Operating Systems

All the WSN equipment stages share highlights of restricted onboard or on-chip assets, serious memory requirements, and constrained power get to. Thus, an OS running on them should be both little in impression and occasion driven.

Originally developed by University of California-Berkeley, Intel Research, and Crossbow Technology, TinyOS is a free, open-source embedded OS for low-power WSNs. The development group has now grown to be an international consortium, the TinyOS Alliance, which makes TinyOS the de facto standard OS for WSNs. The latest released version, TinyOS 2.1.1 supports most commercial WSN hardware platforms. TinyOS is written in nesC, a specially designed C dialect, and supports many popular microcontrollers and radio chips. It provides a large library of software components related to those hardware elements and has only one stack and a single execution context. Hence, the TinyOS programs are very robust and efficient. TinyOS strongly supports low-power operations including multi-hop networking, network-wide sub-millisecond time synchronization, data collection to a designated root or gateway, reliable data dissemination to every node in a network, and installing new codes over the wireless network. The scheduling mechanism used in TinyOS is specifically designed to fit WSN platforms with limited system resources, low-power consumption, and high concurrency.

Initially created by University of California-Berkeley, Intel Research, and Crossbow Technology, TinyOS is a free, open-source installed OS for low-control WSNs. The improvement amass has now become a universal consortium, the TinyOS Alliance, which makes TinyOS the accepted standard OS for WSNs. The most recent discharged form, TinyOS 2.1.1 backings most business WSN equipment stages. TinyOS is composed in nesC, an extraordinarily outlined C tongue, and backings numerous well-known microcontrollers and radio chips. It gives an expansive library of programming segments identified with those equipment components and has just a single stack and a solitary execution setting. Henceforth, the TinyOS programs are exceptionally hearty and effective. TinyOS emphatically bolsters low-control tasks including multihop organizing, arrange wide sub-millisecond time synchronization, information gathering to an assigned root or door, dependable information dispersal to each hub in a system, and putting in new codes over the remote system. The booking instrument utilized as a part of TinyOS is particularly intended to fit WSN stages with restricted framework assets, low-control utilization, and high simultaneousness.

The fundamental target of WSN applications is to gather, transmit and store a tremendous measure of information. The gathered crude information is normally a blend of helpful data, undesirable data, and clamor. To effectively deal with these information utilizing constrained assets, a query processing framework running on TinyOS, TinyDB (2010), was created to deal with the information and concentrate helpful data. It gives a basic, SQL-like interface to enable clients to effectively ask, channel, total, and course the information through power-productive in-arrange handling calculations without composing NesC code.

Other little impression and high proficiency OSs for WSNs were likewise announced. A limited state machine-based OS, specifically SenOS, was proposed by Kim and Hong. It can work in a greatly asset compelled condition (counting force and equipment assets) while giving wanted assignment simultaneousness and system reconfigurability. A bit class WSN OS, named SOS, was additionally proposed to accomplish dynamic reconstructing. Progressively loadable modules and a typical part were settled to help dynamic expansion, change, and expulsion of system administrations. Table 8.2 abridges the Oss created for WSN applications.

Most OSs for WSNs give barebones usefulness and leave framework improvement to application designers. This turns into a major obstacle for clients with frail programming improvement foundation. Numerous producers are giving system administration instruments to enable clients to set up, to screen, and deal with a WSN. They are coded as firmware in light of the activity frameworks and give PC based graphical UIs (GUI). XMesh from Crossbow Technology is a product library in view of TinyOS which keeps running on bits with pre-manufacture multi-bounce, specially appointed, work organizing conventions and restricted sensor board get to strategies. XSniffer, additionally from Crossbow Technology, is an apparatus to screen bundle transmission, parcel substance, interface quality and hub wellbeing data of all bits in a WSN after sending (Table 8.2).

Table 8.2: Summary of Operation Systems Developed for WSNs

Name	Execution model	Levels of granularity supported in reprogramming	Scheduling	Power management	Supported platform
TinyOS (2009)	Event driven	Application level	Not real-time	Yes	Telos, Mica2Dot, Mica2, TMote Sky, Eyes, MicaZ, iMote
Sensor OS (SOS) (Han ***et at,*** 2005)	Event driven	Modular/component level	Not real-time	No	Cricket, imote2, Mica2, MicaZ, tmote, Protosb, emu
MantisOS (Bhatti ***et al.,*** 2005)	Thread-based	Modular/component level	Not real-time	Yes	Mica2, MicaZ, Telos, Mantis nymph
SenOS (Kim and Hong, 2005).	Finite state machine based	Instruction/variable level	Not real-time	Yes	Not specified
Nano-RK (Rajkumar ***et at,*** 1997)	Reservation-based	Instruction/variable level	Real-time	Yes	Atmel ATMEGA 128 with Chipcon CC2420 transceiver
kOS (Britton ***et al.,*** 2005)	Hybrid (event and object based)	Instruction/variable level	Not real-time	No	User defined

Source: G. Caldwell Darwin (2013), Robotics and Automation in the Food Industry: Current and Future Technologies, Woodhead Publishing

8.2.5. Network Architecture

The remote principles likewise address the system issues for remote sensors. A WSN can be single-jump or different bounce (Figure 8.1). The single-jump systems permit coordinate correspondence between a source (remote sensor) and a sink (door). Because of the power impediment of a radio, the attainable separation of transmission may not be adequate to set up a dependable correspondence between a source and a sink. Multi-jump systems transmit information bundles through a few transfers keeping in mind the end goal to broaden arrange scope, maintain a strategic distance from obstructions between a source and a sink, and give clients adaptability to outline information directing maps to enhance execution and vitality effectiveness.

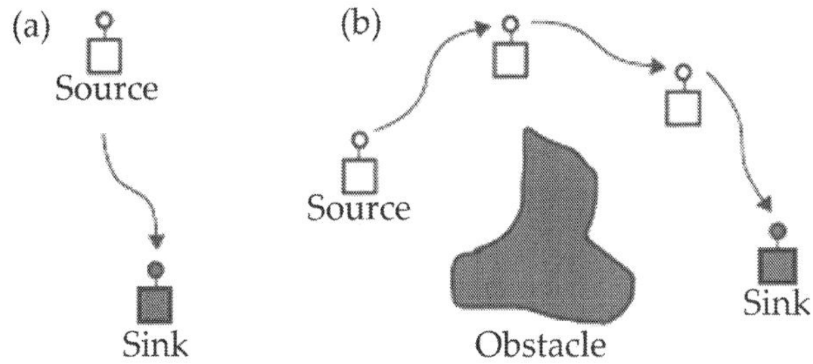

Figure 8.1: (a) Single-hop and (b) multi-hop communication, (*Source: G. Caldwell Darwin (2013), Robotics and Automation in the Food Industry: Current and Future Technologies, Woodhead Publishing*).

Three kinds of systems: star organize, half breed system and work arrange, are normally utilized. The star topologies primarily rely upon single-jump correspondences (Figure 8.2a). The hubs just converse with their folks or children. It has favorable circumstances in effortlessness, simple establishment, simple administration and support, and seclusion of gadgets. In any case, the system execution very relies upon the center, and the system scale is constrained by the reachable separation of the hubs to the center. The most effective system utilizes shared, work systems, in which every one of the hubs in the system have steering capacity (Figure 8.2b). Work systems enable hubs to self-collect into the system, sensor data to engender over the system with a high unwavering quality and over an expanded range, and time synchronization and low power utilization for the 'audience members' in the system to broaden battery life.

At the point when countless sensors should be organized, a few levels of systems administration might be joined. For instance, an 802.11 (WiFi) work arrange including top of the line hubs, for example, passage units, can be overlaid on a ZigBee sensor system to keep up an abnormal state of system execution. A remote application server (RAS) can likewise be sent in the field near a confined sensor system to deal with the system, to gather limited information, to have electronic applications, to remotely get to the cell arrange through a GSM/GPRS or a CDMA-based modem and, thusly, to get to the Internet and remote clients.

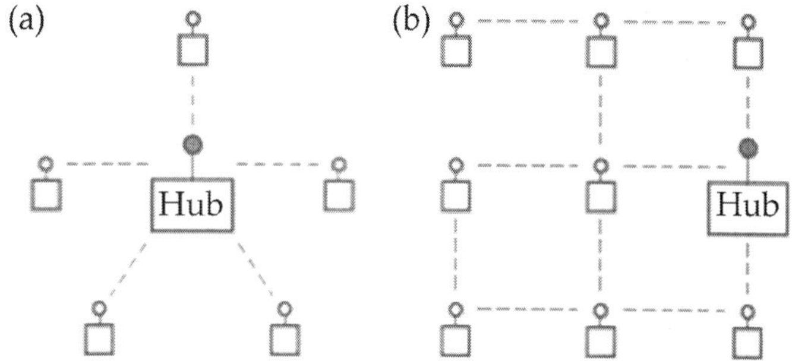

Figure 8.2: Examples of wireless network topology: (a) star network (b) mesh network, (*Source: G. Caldwell Darwin (2013), Robotics and Automation in the Food Industry: Current and Future Technologies, Woodhead Publishing*).

8.2.6. Energy Harvesting

Most WSN stages are controlled by batteries, for example, basic, cell, and lithium batteries, which can be supplanted when required. Albeit numerous makers assert long battery life in their items, clients regularly confront significantly shorter battery life from the power sources, particularly when outside sensors, high examining rate, and high information transmitting rate are utilized and outrageous condition conditions are experienced. To grow durable and genuinely self-sufficient remote WSNs, predictable and stable power sources are pivotal. Presently the ultra-low power utilization plan on WSN equipment and programming reduce the need of vitality for tasks. Miniaturized scale vitality collecting (at smaller scale or milli-watt level) from encompassing condition is turning into an attainable way to deal with be joined with WSN applications. In the 2010– 2011 Energy Harvesting Report, IDTechEx, a counseling firm in printed hardware, RFID, and vitality reaping, figures a market of more than two billion dollars in 2016 only for the gathering components, barring power stockpiling and electronic interfaces and almost 10 billion vitality collecting gadgets will be sold in 2020.

There are two noteworthy undertakings in vitality collecting, that is, vitality catch and capacity. The regular focused on surrounding vitality sources incorporate photovoltaics, temperature slopes, vibrations, weight varieties, stream of air/fluid, and electromagnetic (RF). Sun oriented vitality gathering frameworks for WSN applications have been generally promoted and conveyed. Other vitality gathering frameworks for WSNs are generally new. A few makers began to advertise their original power reapers from spring 2010. Progressed Cerametrics Inc. in the USA gives Energy Harvester which utilizes piezoelectric fiber composite material to gather vibration and give milliwatts vitality. Powercast and Perpetuum in the USA have created control gatherer recipients to chase for vitality in the RF field.

Micropelt GmbH in Germany have advertised their sans battery TE-POWER NODE, which harvests warm vitality in light of a warmth transition through a thermogenerator. The organization guaranteed that the thermogenerators can deliver voltages in the scope of 0.5– 5 V on only a couple of square millimeters of impression and can drive low power remote sensors. Then, different sellers are giving rechargeable vitality stockpiling gadget that will supplant as of now utilized batteries. Cymbet Corp in the USA is building up a thin-film battery innovation, EnerChip, which can be utilized to store vitality gathered from eight distinct sources, including vibration, warm, RF, stream and sun oriented based frameworks. Boundless Power Solutions in the USA creates a group of strong state, rechargeable, miniaturized scale vitality cell (MEC) items named Thinergy.

There is presumably that gathering vitality from encompassing condition offers an energizing future for long-haul WSN organizations. Be that as it may, more innovative work and field testing on vitality collecting and capacity advances are as yet required. In rural and food creation applications, numerous sensors and actuators require higher current to drive, consequently devour generally more power amid tasks in correlation with those on-board semiconductor sensors. Henceforth, effective power reaping, administration, and capacity methodologies should be created.

8.2.7. Data Management

The achievement of WSNs is controlled by the nature of information produced, which should be prepared, separated, deciphered, put away, and showed to end clients. From one viewpoint, WSNs give clients nitty gritty learning of target situations; then again, they make a colossal load on information preparing and taking care of. The restricted assets regarding WSN segments likewise challenges conventional techniques for information administration. In WSN applications, the information are ordinarily utilized as a part of two ways: (1) inquiries on current information; (2) questions on chronicled information. The present information are regularly utilized for basic leadership to decide control activities. For instance, if a dirt dampness level is underneath a predefined limit, a water pump ought to be fuelled on. To spare vitality on correspondence, some push-down channel strategies are utilized to preprocess crude information before transmitting. This is particularly valuable for mixed media WSNs with expansive informational collections including pictures, sound as well as video streams.

TinyDB, BBQ and Direct Diffusion give instruments to consistent inquiries to the present information. Another strategy, acquisitional question handling (AQP), offers usefulness to figure out which hubs, which parameters, and at what time to gather information. In numerous WSN applications, information gathered by WSNs are spilled to a remote conventional database through different long-separate correspondence systems. End clients can inquiry put away 'authentic' information whenever. Information mining, counterfeit consciousness, and multivariate factual investigation strategies can be utilized to process, examine, and question the information. Some new vitality proficient inquiry techniques and database administration frameworks are still a work in progress that view the WSNs as a database that backings authentic question handling.

8.3. WSN APPLICATIONS IN AGRICULTURE AND FOOD PRODUCTION

WSN innovation has quickly entered into numerous applications including mechanical process, restorative gear, condition checking, development, agribusiness, military, food following, stock administration, stock taking care of and recording, and so forth. It has been utilized as a part of numerous new applications that were not beforehand thought to be conceivable. Agribusiness and food creation are altogether different from different businesses. For instance, trim fields are regularly remote and extensive however with fluctuation in soil properties, edit conditions, and condition. Absence of power get to is extremely basic in the fields, while the product covering can regularly create obstruction to radio engendering. In creature observing applications, remote hubs introduced on creatures are versatile. Henceforth, the WSNs have a dynamic system topology. In particular, aside from esteem included deliver, advancements received by rural generation should be minimal effort, dependable, simple to utilize, and simple to keep up. Consequently, WSN applications in agribusiness and food generation are behind the advance in different enterprises and territories.

8.3.1. Environment Monitoring

Numerous field estimations of condition factors, for example, climate information and geo-referenced water quality information, still rely upon stationary sensors and information lumberjacks. They require standard visits by administrators to download the information and change setups and activities, if necessary. Henceforth, they are work serious and powerless to recording blunders amid translation.

WSNs are composed and created to defeat these hindrances by giving vast scale, unsupervised, constant, short-interim condition checking. Remote gadgets are not new in condition checking. Numerous information lumberjacks for remote condition observing have been outfitted with radios or land-line phone modems to bring clients field estimation information. As of late, with the extension of cell systems, Internet, and general parcel radio administration (GPRS), an ever-increasing number of information lumberjacks have an inward cell modem introduced keeping in mind the end goal to stack the information to a web server and be gotten to by clients from the Internet. The Oklahoma Mesonet is an ecological checking system with 120 computerized stations covering 77 provinces in Oklahoma. All the estimation information are transmitted to a focal office each 5 min, 24 h for every day year-round, and can be gotten to from Internet. Each station has an information lumberjack with radio correspondence ability and sensors to take estimations including air and soil temperatures, wind speed and bearing, rainfall, relative moistness, sunlight based radiation, soil properties, and so on. With an on-line device, AgWeather, the Mesonet has been giving the most recent climate information to assist farming makers with making choices on tasks.

In Italy, a propelled vineyard arrange framework i.e. advanced vineyard network system (NAV) was produced and sent for remote, ongoing observing and small-scale meteorological information gathering in a vineyard. The framework incorporated a base agrometeorological station (ace unit) and a few remote hubs (slave units). The ace unit was an entryway that connected the slave units to a remote server through cell administrations. Different sensors, including temperature, weight moistness, sun-powered radiation, rainfall, and anemometer, were appended to the ace unit and every one of the slave units through information obtaining sheets. A GUI was created to enable clients to effectively get to the information and screen the field conditions. Test outcomes demonstrated that the NAV framework could give constant and ongoing natural observing with adaptability on arrange plan, design, and support.

Crossbow innovation declared the eKo TM Pro arrangement remote yield observing framework in January 2008 that comprised of numerous sensor hubs with work organize engineering. The highlights of self-arrangement and self-recuperating ability made the framework sending considerably simpler for clients. The framework could gather data, for example, soil dampness, surrounding temperature, and mugginess, leaf wetness, soil water content, sun based radiation. A web-interface, eKoView, gave an instrument to clients to picture and examine the gathered information and settle on significant choice on horticultural activities.

Vivoni and Camilli (2003) built up a remote model framework to secure, store, show and transmit constant, geo-referenced ecological information between various field groups and remote areas. Each field group had a handheld information gathering unit which could convey to different groups or a field station server through a WLAN. The field station server joined data got from every one of the groups and intermittently answered to a remote web/information server through a double recurrence cell phone (GSM/GPRS benefit at 900 MHz and 1.9 GHz). Field tests directed in Maryland in the USA and New South Wales in Australia showed an incredible potential to enhance proficiency and accuracy on field estimation.

A WSN framework was accounted for by Zhang et al. for constant remote checking of residue overflow at a low-water crossing. The optical soil silt sensor was submerged into the water while whatever is left of the framework was situated on the bank. The in-site network utilized straightforward but dependable star topology. Information was first transmitted from sensor hubs to a door in light of ZigBee convention and afterward from the passage to the Internet server through a cell network.

8.3.2. Precision Agriculture

Precision agriculture (PA) innovation has been advanced and actualized the world over in late decades. The authentic base of PA is the spatial and worldly inconstancy of soil and product factors between and inside fields. Generally, crops have been dealt with under the suppositions of 'uniform' soil, supplement, dampness, weed, creepy crawly, and development conditions. This has prompted over-or under-utilization of herbicides, pesticides, water system, manures, and different medicines. The most extreme objective of

PA is to apply suitable medicines at opportune place and correct time to seek after a low-input, high-effectiveness, maintainable rural generation.

8.3.2.1. Spatial Data Collection

A versatile field information obtaining framework was created by Gomide et al. (2001) to gather information for edit administration and spatial-changeability thinks about. The framework comprised of an information accumulation vehicle, a supervisor vehicle and information obtaining and control frameworks on cultivate machines. The framework could lead nearby field reviews and to gather information on soil water accessibility, soil compaction, soil fruitfulness, biomass yield, leaf territory file, leaf temperature, leaf chlorophyll content, plant water status, neighborhood atmosphere information, creepy crawly malady weed invasion, grain yield, and so forth. The information accumulation vehicle recovered information from cultivate machines by means of a WLAN and broke down, put away and transmitted the information to the director vehicle remotely. The director and designers in the chief vehicle observed the exhibitions of the ranch machines and the information procurement frameworks, and troubleshot issues in light of got information.

Lee et al. (2002) built up a silage yield mapping framework, which incorporated a GPS, stack cells, a dampness sensor and a Bluetooth remote correspondence module. The dampness sensor and the Bluetooth transmitter were introduced on a testing truck. Signs from the dampness sensor were sent to a Bluetooth recipient on a host PC at an information rate of 115 kbps and were utilized to revise the yield information.

Li et al. (2009) revealed a crossbreed soil sensor arrange (HSSN) composed and conveyed for in-situ, continuous soil property checking (Figure 8.3). The HSSN incorporated a nearby remote sensor arrange (LWSN), shaped by numerous sensor hubs introduced at pre-chosen areas in the field to obtain readings from soil property sensors covered underground at four profundities and transmit the information remotely to an information sink introduced on the edge of the field; and a long-removed cell correspondence organize (LCCN). The field information were transmitted to a remote web server through a GPRS information exchange benefit gave by a business cell supplier. The information sink worked as a portal which got information from all sensor hubs; repacked the information, cradled the information as per cell correspondence plan, and transmitted the information bundles to LCCN. A web server was executed on a PC to get, store, process and show the ongoing field information. Information parcels were transmitted in view of a vitality mindful self-sorted out steering calculation. The information parcel conveyance rate was over 90% for a large portion of the hubs (Figure 8.3).

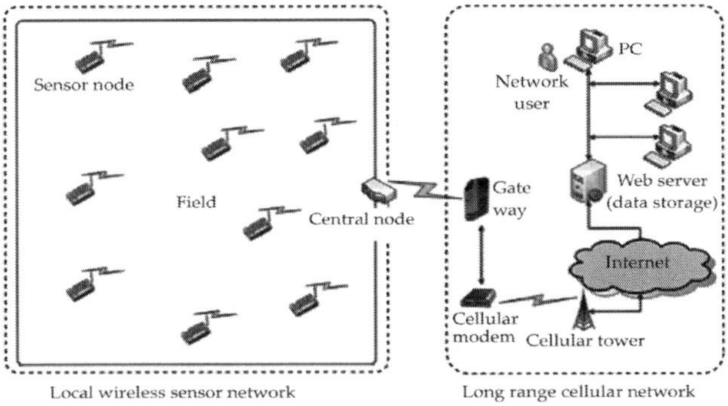

Figure 8.3: WSN for soil property monitoring, (*Source: Li et al., 2009*).

8.3.2.2. Precision Irrigation

Damas et al. created and tried a disseminated, remotely controlled, programmed water system framework to control a 1500 ha inundated territory in Spain. The territory was isolated into seven sub-districts with 1850 hydrants introduced. Each sub-district was observed and controlled by a control division. The seven control areas spoke with each other and a focal controller through a WLAN arrange. Field tests indicated 30– 60% sparing in water utilization.

Evans and Bergman drove a USDA inquire about gathering to examine exactness water system control of self-moved, straight move and focus rotate water system frameworks.

Remote sensors were utilized as a part of the framework to help water system planning in view of on location climate information, remotely detected information and producer inclinations. O'Shaughnessy and Evett set up a test stage including two WSNs of infrared thermometers to screen shelter temperature, which was ordinarily utilized as a pointer of harvest water pressure. One of the WSNs was introduced on the sidelong arm of an inside rotate water system framework and used to screen trim covering temperatures while moving. Another WSN shaped by stationary hubs was introduced in the field beneath the rotate arm to give the stationary reference covering temperature. RF modules, XBee (MaxStream, Logan, Utah, USA), were utilized for interchanges. An inserted PC was utilized for recording and dissecting the gathered information and settling on water system choices.

Kim et al. (2006) created and tried a shut circle robotized water system framework. The framework comprised of in-field detecting stations, a water system control station, a climate station, and a base station. The detecting station and the climate station gave in-field status readings. The base station gathered and handled the in-field information, settled on choices on water system booking, and sending the control orders to the water system control station to control the tasks of sprinkler spouts. The water system control station additionally refreshed GPS areas of the direct water system framework consistently and sent them to the base station. The remote correspondences among the stations were through Bluetooth innovation.

Ruler et al. (2005) built up a shut circle, circulated control and information obtaining framework for site-particular water system administration for a middle turns framework. The framework was shaped by a gathering of stationary field detecting hubs sent in a field and a gathering of hubs, called P2K hubs, introduced on the horizontal arms of the focal rotate framework. The field sensor hubs spoke with the P2K hubs utilizing low power RF signals. The P2K hubs gathered the information from the field and the focal rotate framework and transmitted them to an ace controller through an electrical cable bearer correspondence. The framework execution was contrasted and a regular uniform water system treatment in a potato field. Results demonstrated that, with basically measure up to water utilization, the tuber yield under site-particular water system administration was altogether more noteworthy, by around 4%, than the yield from uniform water system.

8.3.3. Machine and Process Control

An ever-increasing number of innovations are being created to help wired or remote (WPAN, WLAN, cell systems, and so forth.) interchanges from machine to machine, from machine to versatile or from portable to machine, and from human to machine or machine to human. These innovations enormously improve the robotization of machines, the proficiency and viability of machine administration frameworks through a mix of discrete segments inside the framework.

Yang et al. (2010) built up a farm horticulture hardware data administration framework, that coordinated undertakings including data administration, cultivate apparatus work planning, apparatus execution assessment, work-stack circulation, fuel and extra parts supply data, upkeep booking, measurable examination on specialized and monetary pointers, money related administration and online meeting emotionally supportive networks. The framework was conveyed in farms in Heilongjiang, China. Each agri-machine in the tried homestead was outfitted with a GPS unit and a cell gadget. It detailed its areas and execution parameters consistently to a focal administration focus through GPRS benefit gave by cell bearers. Some hardware had camera frameworks on board, which enabled the administrator to speak with the focal chief through video conferencing administrations. The field tests demonstrated an awesome change on assignment planning for agri-apparatus employments.

More agri-hardware producers incorporate correspondence modules to their frameworks to enable the apparatus to remotely report current data, for example, topographical area, fuel condition, framework conditions, mishaps, and so forth. The JDLink product offering from John Deere enables the hardware supervisor to get today-by-day information through the Internet to a focal server secretly. Breaking down caution messages can be conveyed to the director through messages and short instant messages. Some New Holland join collectors have organize cards created by EIA Electronics in Belgium that enable a chief to remotely screen their tasks by means of cell system or satellite correspondence. A German organization, RTS Rieger, have built up a framework to record different working parameters of hardware and transmit the information to a PDA (without benefit membership charge) through Bluetooth or to a web server by means of cell arrange.

A WLAN-based, ongoing, vehicle-to-vehicle information correspondence framework was set up by Guo and Zhang to trade data between vehicles on vehicle state and task control factors. Research center and field tests showed the achievability of ongoing, remote information correspondences between vehicles in self-sufficient, ace slave vehicle direction. Charles and Stenz actualized a self-ruling tractor for splash tasks in fields. Amid showering, the tractor drove completely self-sufficiently no less than 90% of the time. This tractor could likewise be exactly controlled by a chief through a radio connection. Ribeiro et al. built up a self-ruling direction tractor for splash activities in citrus and olive trees fields in Spain. An easy to use perception specialist was produced for human administrators to remotely control and regulate unmanned tractors in a field through WLAN. Stentz et al. built up a remote connection amongst tractors and a human administrator in an armada of semi-self-ruling tractors. Every tractor had the ability to identify individuals, creatures, and different vehicles in its predefined way and to stop before hitting such impediments until the point that it got control charges from a boss over a remote connection.

8.3.4 Precision Livestock Farming

Exactness domesticated animals cultivating intends to enhance creature efficiency and generation maintainability by encouraging the correct creature at the perfect place with appropriate sum at the opportune time. To accomplish that objective, creatures should be deliberately checked and made do with particular fleeting and spatial determination. Be that as it may, creatures are moving articles, which makes wired detecting techniques infeasible. WSN innovation and smaller than usual electronic gadgets open an entryway for accuracy domesticated animals administration. As of late, much research on the improvement of creature observing frameworks have been accounted for.

8.3.4.1. Animal Behavior Monitoring

A WSN framework was produced to ponder the periods of time dairy animals spent almost a water trough. Each cow in the investigation wore a neckline with a sensor hub including a GPS collector, a chip, and a radio handset. The sensor hub recorded GPS information at predefined time interims and transferred the putaway information to a base station remotely, utilizing 802.15.4 standard convention, at whatever point it entered the correspondence scope of the base station. A steer's administration WSN framework was shown to screen dairy animals' quality and field time in a segment of new grass continuously. Micaz sensor bits from Crossbow Technology were utilized and encased in collars worn by dairy animals. A door was introduced in the zone of new segment of grass. It got and put away information parcels when a bovine entered to its correspondence region. The creature field time was then figured.

Observing dairy cattle eating action isn't just imperative to know their eating conduct, but on the other hand is an advantage to field administration. Reed built up a remote sensor to screen unfenced brushing movement. The sensor comprised of a microcontroller, flag molding, and a radio handset incorporated on a PCB board with measurements of 19.6 mm × 71.8 mm × 11 mm. A small-scale GPS lumberjack was additionally incorporated into the framework. Information handling calculations were produced to describe touching movements against ordinary developments. The field tests demonstrated that the framework could give geo-referenced steers nibbling maps.

Attributable to concerns with respect to creature welfare and nourishment security, non-confine lodging frameworks are winding up increasingly prominent in the poultry business. In the United States, numerous egg makers are currently progress from conventional activities to non-confine tasks. The progression of hens' practices under such noncage condition should be considered and deliberately observed with a specific end goal to influence a fitting administration to control system and find unfortunate creatures at a beginning period. Quwaider et al. (2010) built up a stage to screen hens' movement. The stage included remote sensors mounted on hens and stationary remote sensors. The stage gave data including hens' relative areas to stationary sensors and different hens, speed of developments, demonstrating of headway (flying or ambulating), and particular practices (eating, drinking, and egging). The Mica2Dot bit from Crossbow Technology was utilized as the remote sensor with onboard light, quickening, temperature, and sound sensors. The relative area of the hens was resolved in view of radio-flag quality marker (RSSI). Preparatory outcomes demonstrated that the stage could find hens with a precision of 80– 90%. This outcome exhibited that the created framework could identify when and where hens were dynamic. This data could be important to the makers to screen hens progressively, record action designs, for example, egg laying, eating, drinking, and tidy showering, and so on. and alter tasks continuously.

8.3.4.2. Animal Health Monitoring

Pastell et al. (2009) announced a remote detecting framework created to gauge stride includes in dairy bovines. Every sensor hub with a three-dimensional accelerometer was connected to an appendage proximal to fetlock joint of a bovine. Three-dimensional developments of appendages were logged to the sensor hub and transmitted to a PC remotely utilizing the 869 MHz radio channel to decide the faltering level of dairy animals. A remote drifting base sensor organize framework was produced in Brazil to screen creature physiological reactions. The framework included two modules, creature module, and base module. The creature modules were introduced on tried creatures to gather EEG flags and transmit the signs to the base module utilizing 433 MHz bearer recurrence. The base module was utilized to get and store the EEG signals which could be utilized to screen ox-like cerebrum electrical action.

Deep body temperature (DBT) was normally utilized as a pointer of warmth worry of ovens. Remote sensors were created to consistently quantify DBT of ovens. Each feathered creature was embedded with a remote sensor at a profundity of 2 cm into the stomach depression, which transmitted DBT estimations to a stationary collector in an oven house. The outcomes demonstrated that the remote sensors worked well and were responsive the progressions of DBT.

8.3.4.3. Animal Environment Monitoring

Exact controls of atmosphere related factors inside a creature house can enhance creature welfare and keep up great wellbeing. Pessel and Denzer built up a convenient, portable instrument to quantify temperature, relative stickiness, commotion, brilliance and alkali content noticeable all around inside a creature house and exchanged the information remotely to a PC through an infrared information interface. Darr and Zhao (2008) built up a remote sensor and control framework to screen temperature varieties and control ventilation in a swine house. Nadimi et al. (2008) built up a WLAN sensor system and instant message conveyance framework to watch and ready atmosphere changes in an oven house. Remote temperature sensor hubs and web-cameras were introduced in various grill houses and organized through WLAN. The information were spilled into a remote server which sent instant messages when anomalous conditions were identified.

8.3.5. Greenhouse Management

Greenhouse creation needs numerous sensors and controls to guarantee efficiency. They have been utilized to screen indoor atmosphere changes, soil conditions, trim wellbeing, and nuisance invasions and control the utilization of assets, for example, light, water, chemicals, and so forth. Because of the huge measure of wired detecting and control frameworks utilized, nurseries have a notoriety for being 'houses brimming with wires and links.' A major exertion is expected to deal with the wires and links amid framework establishment and upkeep. At the point when remote innovation showed up, nursery makers were pioneers in embracing it. They supplanted the old frameworks with cleaner, more dependable, more adaptable, and more effective remote frameworks. Most nurseries have power and Internet gets to, which additionally make the organization of WSNs substantially less demanding than open-air applications.

Morais et al. actualized a remote information securing system to gather open air and indoor atmosphere information for nurseries in Portugal. A few sun oriented controlled information securing stations (SPWAS) were introduced inside and outside to quantify and screen the atmosphere information. RF joins were set up among various (up to 32) SPWASs and a base station, which was utilized to control the SPWASs and to store the information. Serodio et al. created and tried a dispersed information obtaining and control framework for dealing with an arrangement of nurseries. A few correspondence procedures were utilized for information interchanges. At a lower supervisory level, inside every nursery, a WLAN connect with a RF of 433.92 MHz was utilized to interface a sensor system to a neighborhood controller. A controller zone organize (CAN) was given to interface an actuator system to the neighborhood controller. Through another RF interface (458 MHz), a few neighborhood controllers were associated with a focal PC. Abnormal state information correspondence was given through Ethernet to associate the focal PC to a remote system. Liu and Ying announced a nursery observing and control framework utilizing the Bluetooth innovation. The framework gathered ecological information from a sensor arrange in a nursery and transmitted this information to a focal control framework. Mizunuma et al. sent a WLAN in a homestead field and nursery to screen plant development and actualized remote control for the creation framework. They trusted that this kind of remote control procedure could extraordinarily enhance profitability and decrease work prerequisite.

Systems based on ZigBee technology are widely used in greenhouse management to form mesh networks, which makes the system configuration and maintenance more convenient. Zhu *et al*. (2006) established a WSN in a greenhouse to collect temperature, humidity, and light intensity periodically. The sensor nodes communicated with a gateway using ZigBee protocol, while the gateway node communicated with a web server through Ethernet. Park et al. (2009) developed an automatic control system for greenhouse use to collect temperature and humidity of crop leaves. Sensor nodes were installed inside a greenhouse to collect air temperature, humidity, leaf temperature, and leaf humidity. Through ZigBee compliance radio in the sensor nodes, the collected data were transmitted to a base station and stored in a database. The base station periodically sent the data to a remote control system via the Internet, which analyzed the data and made decisions on the control strategy. The control commands were then transmitted to a control station in the greenhouse to implement relevant operations on ventilation, irrigation, and heating, etc. Lea-Cox et al. (2007) established a WSN to measure substrate water, temperature, electrical conductivity, daily photosynthetic radiation and leaf wetness in real-time in nursery houses. These data could be used for the control of irrigation. A web-based graphical reporting system was also developed to provide easy access of the data via the Internet. Growers who adopted the technology could benefit from improved plant health, more efficient water and fertilizer applications, and a reduction in disease problems related to over-watering.

Frameworks in view of ZigBee innovation are broadly utilized as a part of greenhouse management to shape work systems, which makes the framework setup and support more helpful. Zhu et al. set up a WSN in a nursery to gather temperature, mugginess, and light force intermittently. The sensor hubs spoke with a passage utilizing ZigBee convention, while the portal hub-spoke with a web server through Ethernet. Stop et al. built up a programmed control framework for nursery use to gather temperature and moistness of yield clears out. Sensor hubs were introduced inside a nursery to gather air temperature, moistness, leaf temperature, and leaf stickiness. Through ZigBee consistence radio in the sensor hubs, the gathered information were transmitted to a base station and put away in a database. The base station occasionally sent the information to a remote control framework by means of the Internet, which investigated the information and settled on choices on the control procedure. The control charges were then transmitted to a control station in the nursery to execute important activities on ventilation, water system, and warming, and so forth. Lea-Cox et al. built up a WSN to gauge substrate water, temperature, electrical conductivity, day-by-day photosynthetic radiation and leaf wetness progressively in greenhouse management. These information could be utilized for the control of water system. An online graphical announcing framework was likewise created to give simple access of the information by means of the Internet. Cultivators who received the innovation could profit by enhanced plant wellbeing, more productive water and compost applications, and a lessening in sickness issues identified with over-watering.

8.4. FUTURE TRENDS IN WSN TECHNOLOGY IN AGRICULTURE AND FOOD PRODUCTION

In the past, excessive wiring was considered a sign of technological advancement in many sensing and control systems. In 1955, the total length of electrical wires on a state-of-the-art automobile was about 45 m. This number grew to more than 5 km on a high-end vehicle in recent years. The thick wiring harness is often hard to manage and difficult in diagnosing problems. In field applications, a lot of efforts are given to protect the wiring harness from damages of weather, animals, and theft. Many greenhouses are full of wires on the floor and ceilings which needs to be carefully protected to avoid shortage due to high moisture and humidity.

WSN technology shows great potential to overcome these shortcomings and be economically viable to replace wired networks. The autonomous massive data collection provides user detailed information in a target environment. Combining historical data and agronomic knowledge, more effective management strategy can be made to improve the production. As the demand for food quality, health benefits, and safety increases, more stringent scrutiny on the inspection of agricultural and food products have become mandatory. Also being increasingly demanded is 'traceability,' which requires not only rigorous inspections, but also systematic detection, labeling, and recording of quality and safety parameters while archiving the entire agricultural production chain, from farms to consumers' tables. RFID has shown a booming trend with adoption by producers, food processing and handling industry, and merchants to establish an effective 'traceability system.' RFID allows an 'intelligent tag' assigned to each individual product to be read at any checkpoint. The intelligent tag can be updated along the entire supply chain to provide complete archives of information on the growth, processing, packaging, transportation, and distribution, storage, shelving, and recycling. Active RFID systems can also record environmental parameters and specific quality/safety attributes of the product along the chain.

Communication has changed significantly over the past decade. Various new devices including smart cell phones, Internet, Tablet PCs, etc., become imperative in human life. These new technologies greatly shorten the distances between people, between nature and human beings, and between goods and consumers. There is no doubt that WSNs will soon penetrate all aspects of agriculture production, food process, handling, and storage, transportation, and marketing and trading.

REFERENCES

1. Caldwell G. Darwin, (2013). Robotics and Automation in the Food Industry: Current and Future Technologies. Woodhead Publishing.1st edition

CHAPTER 9

INTELLIGENT QUALITY CONTROL SYSTEMS IN FOOD PROCESSING BASED ON FUZZY LOGIC

"Agricultural sustainability doesn't depend on agritechnology. To believe it does is to put the emphasis on the wrong bit of 'agriculture.' What sustainability depends on isn't agri- so much as culture."

—Raj Patel, an academic, journalist, activist, and author of The Value of Nothing: How to Reshape Market Society and Redefine Democracy

CONTENTS

In the food business, final results must accomplish a bargain between a few properties, including tangible (surface or shading for instance), sterile, and mechanical properties (measurements of a bread or mass loss of a cheddar for instance). Dealing with these properties ideal from the creation phase with the point of controlling them is no simple undertaking. At present, numerous generation forms are to a great extent dependent on the expertise and experience of the administrator, something that no framework will be equipped for supplanting within a reasonable time-frame. The upshot is that the nature of a food item can't be depicted solely through strict advances between the fresh qualifiers 'great' or 'awful,' yet rather requires an interminable arrangement of possibly unclear evaluations to be legitimately characterized and, eventually, controlled. This chapter diagrams the key elements of fluffy sets and proposes probability speculations to coordinate aptitude and experience for control purposes inside the food business.

9.1. INTRODUCTION

In the food business, final results must accomplish a trade-off between a few properties, including tactile, sterile, and mechanical properties. Tactile and clean properties are fundamental as they impact buyer decision and inclination. In any case, dealing with these properties appropriate from the creation phase for consequent control is no simple undertaking, for a few reasons:

1. The food industry works inside numerous parameters that must be considered accordingly. A solitary tactile property, for example, shading or surface, can be connected independently to a few measurements enrolled by the human mind. For instance, the shade of a biscuit can be assessed online by administrators utilizing their senses which is the union of three instrumental estimations spoke to as L, a, b.
2. The food business operates with non-uniform variable crude materials that, when handled, need to prompt an item that fulfills a settled standard.
3. The marvels engaged with food handling are exceedingly non-straight, and factors are coupled. For instance, the surface development of a milk gel, for example, cream cheddar amid handling, is a non-direct and circuitous consequence of the structure advancement of the gel at a nanosomic scale

4. The food industry works with a various arrangement of items and forms and has necessities as far as the compactness and versatility of the frameworks created
5. Less information is accessible in conventional assembling plants creating, for instance, hotdog or cheddar, and this circumstance is reflected all through the food business. Moreover, notwithstanding when databases do exist, it isn't generally conceivable to utilize them for controlling food item quality.

In this specific circumstance, in spite of the way that the plan of guidelines and solid methods for controlling item quality is a noteworthy goal for the food business, mechanization is restricted:

- Few sensors are accessible to complete the estimations required like tangible surface or scent. Albeit new sensors, for example, counterfeit noses have been produced, the street is long, troublesome, and unfeasible for SMEs.
- For a few procedures, it is hard to set up models that are adequately illustrative of the wonders included, notwithstanding for control purposes.
- Classical mechanized methodologies stay constrained, for the reasons specified previously.

At present, numerous creation forms are to a great extent dependent on the expertise and experience of the administrator, something that no framework will be fit for supplanting within a reasonable time-frame. Thus, by and by, administrators frequently assume an imperative part and coordinate with robotization in order to: (1) make online assessments of the tactile properties of the item, and additionally (2) change the on-line process in like manner. Also, experienced administrators make naturally visible elucidations of physicochemical marvels happening amid handling, which can act in cooperative energy with established building learning on the procedure.

Incorporating administrator and master aptitude in a control structure is an applicable course, particularly for conventional procedures. In any case, it prompts outlining scientific devices that need to incorporate: (1) thinking in view of the utilization of semantic images, for example, 'finished covered,' 'great shading,' and so forth., which are communicated not on a numerical scale but rather on a spasmodic graduated scale alluding to an assessment of a deviation in contrast with a set point; (2) vulnerability on these images, which is interpreted after combination in a particular activity; and (3) an activity that is the consequence of a verifiable or unequivocal addition between two particular states recorded by the administrator over the long run.

Choices identified with the administration of food forms progressively depend on scientific models that (1) speak to the accessible best in class information about the wonders included, and (2) can reproduce the diverse transient and balance states after some time. Be that as it may, because of time limits, budgetary limitations, and logical and innovative obstructions, data for show parameters is frequently inadequate and uncertain. This prompts vulnerability, which should be represented in the basic leadership process. Coordinating the vulnerabilities intrinsic to genuine levels of learning accessible on the sustenance forms is a critical advance in the improvement of process portrayal models. Given that specialists need to battle with an absence of valuable information for building such models, they may here and there pick to make their own assessments of inadequately known amounts given in certainty interim arrangement by applying their own particular experience and instinct.

Accessible learning about sustenance preparing may show (1) arbitrariness (otherwise called 'stochastic vulnerability') because of the common inconstancy of perceptions, or (2) imprecision (otherwise called 'epistemic vulnerability') because of an absence of data. Haphazardness and imprecision may coincide, particularly in a foundation of a few heterogeneous learning sources, for example, measurable information and master conclusions. When in doubt, fractional numbness is spoken to by a uniform circulation in view of what is known as the 'lack of concern guideline' (all that is similarly conceivable is similarly plausible). This decision, despite the fact that guided by standards, remains profoundly easy to refute, since the investigator is bringing additional data (equiprobability) to information that was not sourced from specialists or information and does not, in this way, exhibit any psychological esteem. Practically speaking, the decision is managed by the way that the main feasible alternative is a traditional probabilistic model, for example, uniform, typical, Beta or log-ordinary likelihood conveyances for instance, which can be contended unequipped for representing epistemological vulnerability and imprecision when all is said in done. While data with respect to fluctuation is best passed on utilizing likelihood appropriations, data fixing to imprecision might be all the more precisely passed on utilizing groups of likelihood dispersions encoded by probability conveyances.

Fluffy sets and probability speculations were presented by Zadeh in 1965 and assist in 1978 as an expansion of the set hypothesis by supplanting the trademark capacity of a set by a participation work whose qualities run from 0 to 1. Plausibility hypothesis furnishes fluffy sets with the full settings required for a sound correlation with likelihood and is important for speaking to consonant uncertain information, for example, certainty interims given by specialists relating to sick known amounts. A plausibility dispersion can demonstrate loose data with respect to a settled obscure parameter, however, can likewise fill in as an inexact portrayal of deficient perceptions of an arbitrary variable.

The central idea is the likelihood circulation depicting the pretty much conceivable estimations of some dubious variable. Probability hypothesis is currently a wide field of concentrate that, in the course of the most recent 20 years, has seen the improvement of a scope of devices. Connected to sustenance designing, it has been considered as relevant by a few creators for various applications, particularly to take into account the thinking procedure, communicated in etymological terms, of administrators and specialists where its concentrations include: (1) portrayal of the illustrative tactile assessment performed by a quality group, an administrator, or a customer; (2) aberrant estimation of the properties of a nourishment item, for example, the demonstrating of the master tangible hotdog quality assessment; (3) analysis, supervision, and control of sustenance quality, for example, the control of procedures, for instance broiling peanuts, the working of choice emotionally supportive networks in cooking procedures, or cheddar aging procedures. Illustrations will be portrayed in this chapter.

9.2. STANDARDS OF SHREWD CONTROL FRAMEWORKS USING FLUFFY RATIONALE

Albeit Boolean rationale discovered across the board applications in the twentieth century (from double rationale to rationale incorporated circuits), it remains underequipped to model or utilize human thinking in circumstances of vulnerability or imprecision, since it just arrangements in parallel data that can't decipher the image dialect or the many-sided quality of our observations. Numerical apparatuses, for example, fluffy rationale and probability hypothesis, seem better adjusted to managing or utilizing human thinking, close by other hopeful methodologies, for example, subjective material science.

In customary rationale, the 'universe of talk' (U) is shaped of components (u) portrayed by add up to enrollment (trademark work: μ E (u), equivalent to 1) or aggregate rejection (trademark work: μ E (u), equivalent to 0) of a given set E. This prompts strict between-set advances forcing a structure that can demonstrate excessively fragile, making it impossible to catch the multifaceted nature of various occasion forms. The upshot is that the nature of a nourishment item can't be portrayed solely through strict advances between the fresh qualifiers 'great' or 'terrible,' yet rather requires a vast arrangement of possibly ambiguous evaluations to be appropriately characterized and, at last, controlled. Fluffy rationale hypothesis and plausibility hypothesis take an alternate position grounded in dynamic consistent between-set changes. The enrollment work in fluffy rationale is a decent delineation of these advances, as it reaches out to the customary trademark work. Any component u found in the universe of talk U can be described by its level of participation of a set E (μ E (u)), esteemed in the genuine nonstop interim [0,1]. A fluffy set E in the universe of talk U can be characterized by:

$$E = \{(u, \mu_E(u)) \backslash u \in E\}$$

$$\mu_E : U \rightarrow \lfloor 0,1 \rfloor \tag{9.1}$$

μE is in this way the enrollment capacity of set E. It speaks to the arrangement of participation degrees (μE (u)) of a variable u mapped to fluffy set E. To represent, an enrollment capacity can be utilized to portray the connection between the numerical variable sogginess and the etymological label utilized by an administrator for a role that is 'excessively sodden' (Figure 9.1).

By and by, biscuits of estimated dampness (x) lower than 3.5 g/100 g of the dry issue (gDM) are not viewed as wet, though scones with a dampness content going from 3.5 to 4.5 g/100 gDM are sodden to an expanding degree that is given by the enrollment work. In this way, a bread of dampness 4 g/100 gDM will be considered 'excessively wet' with a degree (μ "excessively soggy" (4)) of 0.5. Scones exhibiting a dampness more prominent than 4.5 g/100 gDM will be considered absolutely wet with a level of enrollment to semantic tag "excessively clammy" of 1.

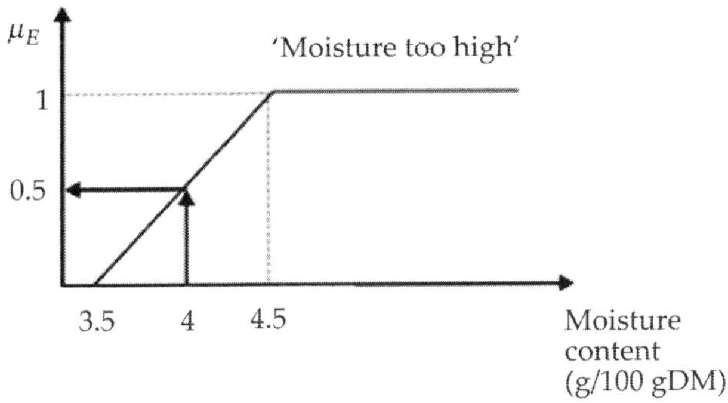

Figure 9.1: Case of a fluffy enrollment work connected to qualify the dampness of a biscuit, (*Source: G. Caldwell Darwin (2013), Robotics and Automation in the Food Industry: Current and Future Technologies, Woodhead Publishing*).

$$\mu(x) = \begin{cases} 0 & (x < a_1) \\ \dfrac{x - a_1}{a_2 - a_1} & (a_1 \le x < a_2) \\ \dfrac{a_3 - x}{a_3 - a_2} & (a_2 \le x < a_3) \\ 0 & (a_3 \le x) \end{cases} \quad (9.2)$$

Enrollment capacities can be communicated through different portrayals. The portrayals most generally utilized are triangular (Equation (2)) for a given triplet arrangement a 1, a 2, a 3, diagramed in Figure 9.2, trapezoidal, or sigmoidal. Certain applications (counting charge controls) likewise utilize what is named a fluffy singleton, where μ (x) is 1 for a given x esteem and 0 in every single other point.

A fluffy set can be portrayed by its limit parameters–a 1, a 2, a 3, for instance, describing a triangular capacity. This implies a scope of impacts and clamors can be incorporated by making the correct changes in accordance with the set attributes–for instance, the enrollment capacity can be made to coordinate a level of affectability or exactness, that is, trueness and precision, of the estimation.

In light of this key meaning of fluffy subsets and fluffy participation capacities, distinctive devices can be created for the points of the control framework. The accompanying segments will investigate these apparatuses further.

This chapter will create two lines of approach. The primary tends to the inquiry 'by what means can fluffy rationale hypothesis be utilized to assemble a control framework for sustenance designing?' while the second tends to the inquiry 'in what capacity would probably be able to and plausibility hypothesis be utilized to speak to and together engender vulnerabilities in respect to the learning about the procedures dealing with?' The point is to gauge the vulnerability in show results and feature that a joint treatment of vulnerabilities can be valuable for controlling the procedure, even with few instruments as well as little data accessible.

9.2.1. Building a Control Framework Utilizing Fluffy Rationale

Diverse advances are vital keeping in mind the end goal to fabricate an insightful control framework utilizing fluffy rationale (Figure 9.3). A first level, named 'adjusted foundational investigations is a key issue in fluffy insightful control frameworks. It is an information extraction step, completed in view of a foundational examination of the systems executed by specialists to determine a procedure control issue. A moment level, named scientific information formalization,' speaks to the numerical formalization of the human learning took care of in the initial step. The outcome came to is a calculation that copies the master handling a procedure, that is, a numeric model of the master systems for taking care of control issues. It is worked by the objective of the investigation: circuitous estimation, state determination, or process control. The systems utilized are profoundly reliant on whether the procedure compose included is a clump procedure or a consistent procedure. Information is then composed in space or in time. The third and last level, named "approval instruments," speaks to chosen methods for approving these methodologies and preparing emblematic information.

9.2.1.1. Adjusted Precise Investigation (Level 1)

The point is to characterize the cutoff points of the framework to be spoken to and the theory, to characterize and formalize unified the factors of intrigue and the sort of associations between the procedure examined and the factors controlled, and to distinguish and formalize unified the systems specialists utilize to take care of the control issue. It coordinates the non-straight and coupled connections in certain settled composed groups. Figure 9.4 portrays the procedure for dealing with master learning. In the wake of talking the chose specialists (when in doubt, there won't be too much), an initial step is the formalization of the estimations and the related estimation granularity. In the event that it is approved, at that point the second step, formalizing and organizing the group's specialists have at the top of the priority list is completed, and ought to clarify the non-straight connections between the sources of info and yields.

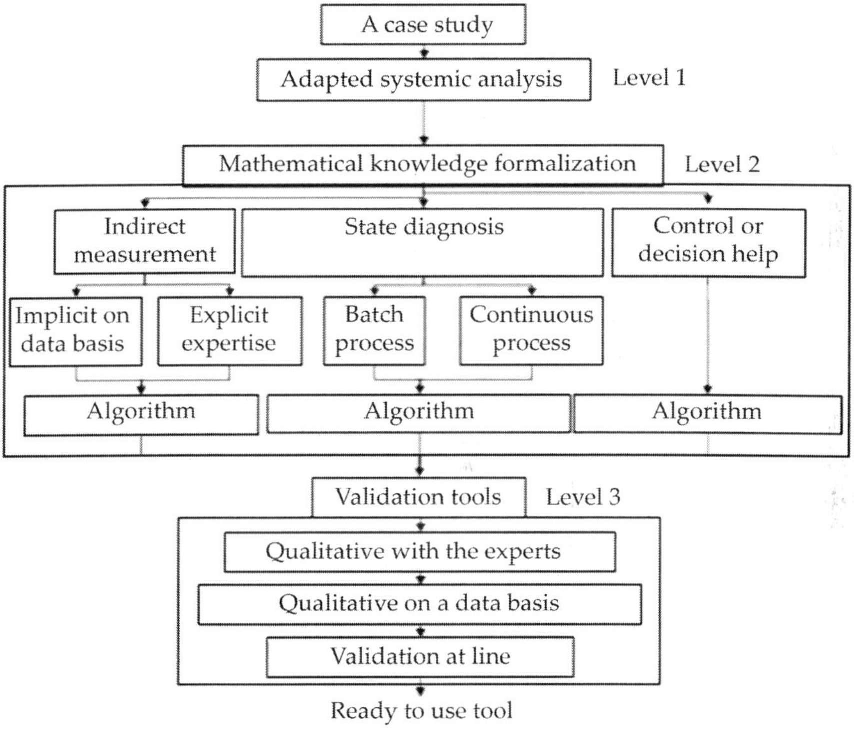

Figure 9.2: Steps to build a control system using a fuzzy logic, (*Source: G. Caldwell Darwin (2013), Robotics and Automation in the Food Industry: Current and Future Technologies, Woodhead Publishing*).

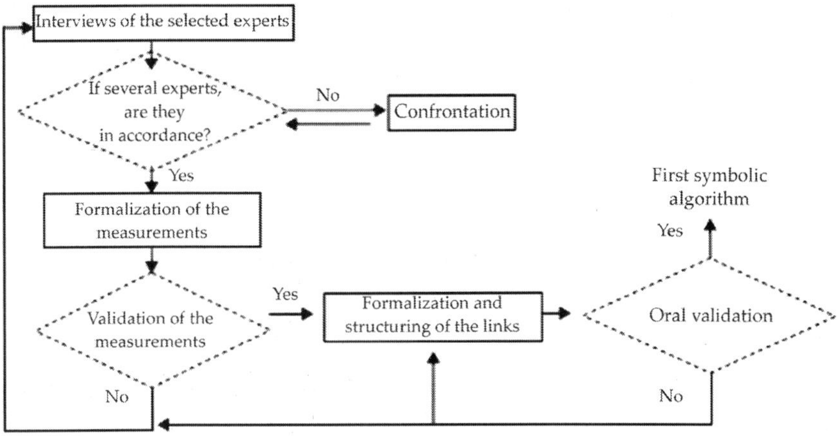

Figure 9.3: Methodology for expert knowledge handling, (*Source: G. Caldwell Darwin (2013), Robotics and Automation in the Food Industry: Current and Future Technologies, Woodhead Publishing*).

Table 9.1: Type of Formats for the Variables of Interest

	Variables		
Variable	Sensor measure	Sensory measurement on a structured scale - sensory indicators (Curt, 2001)	Sensory measurement on an ordinated linguistic scale
Format	Numeric continuous	Numeric discontinuous	Linguistic
Example	Luminance (camera)	Number of spots for a browning process (Ioannou *et al.*, 2004) 	Color of a cheese: line scale from 1 to 6 with three anchors: ivory, yellow and brown (Perrot et *al.*, 2004)

Source. G. Caldwell Darwin (2013), Robotics and Automation in the Food Industry: Current and Future Technologies, Woodhead Publishing

The foundation of this progression includes characterizing and selecting specialists. The specialist's selected will to a great extent shape (1) the outcomes acquired, (2) the extent of induction versus learning infused in the models, and (3) the generalizability of the models constructed, whose properties are themselves fixing to '(2).' In spite of the fact that ability has been broadly broken down in brain science and in manmade brainpower, there is still no operational and all around acknowledged meaning of skill perse. In this examination, specialists were selected based on reliable practice, elite levels, and associate acknowledgment. This approach of demonstrating groups set up by specialists on a given procedure control issue made it conceivable to deal with numerical-design factors as well as various sorts of emblematic arrangement factors communicated by the specialists. Table 9.1 recaps these arrangement composes.

9.2.1.2. Numerical Learning of Formalization (Level 2)

We have selected the scientific devices accessible as indicated by the kind of study attempted: aberrant estimation, state conclusion or choice help frameworks.

Diverse devices adjusted from fluffy set hypothesis can be executed relying upon the sort of information accessible.

With no master learning accessible, material ideas incorporate directed bunching, for example, the fluffy k-closest neighbor technique or the supposed fluffy multicomponent participation capacities strategy, which sums up the enrollment capacities idea given by Benoit and Foulloy.

A guideline is an approach ready to naturally make the connection between master tangible assessments and instrumental estimations.

With express master information accessible, the idea of the fluffy emblematic sensor, in view of the fluffy enrollment work idea exhibited above, can be created. Fluffy methods are utilized to characterize a dialect, for example, a connection between an arrangement of words (L) and a numerical set (N). This connection is described by an enrollment work, μ RF, which speaks to the degree or the quality of the connection amongst images and numbers. This fluffy connection can be portrayed by two projections that take their qualities from the arrangement of fluffy numerical subsets F (N) and from the arrangement of the fluffy emblematic subsets F (L): Meanings and Descriptions.

Fluffy significance speaks to the images utilized by administrators in word design. To outline, fluffy significance can be utilized, after examination of master dealing with, to speak to a projection in numeric space of the representative way administrators qualify a cheddar regarding item dampness content: 'low, center, high' (Figure 9.5). It implies for instance, that in the vicinity of 0 and 1.5 g/100 gDM we are certain that the item is low in term of master assessment. In the vicinity of 1.5 and 2.5 g/100 gDM, the item is amongst low and center with a given direct law.

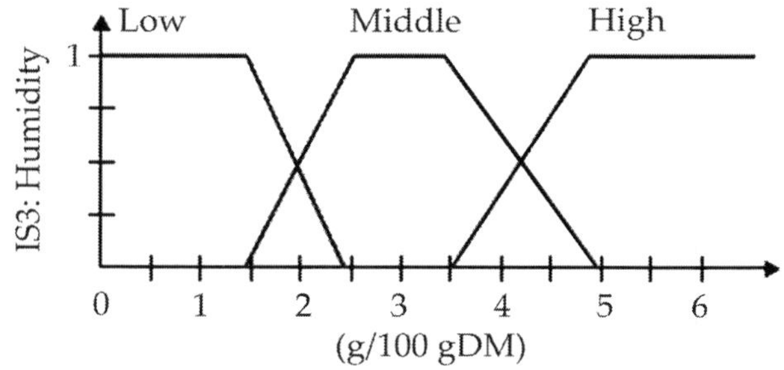

Figure 9.4: Case of a fluffy importance depicting a dampness substance of a cheddar, (*Source: G. Caldwell Darwin (2013), Robotics and Automation in the Food Industry: Current and Future Technologies, Woodhead Publishing*).

Fluffy portrayal is a straightforward method for depicting an estimation with words. For instance, in Figure 9.5 a dampness substance of 4 g/100 gDM can be portrayed as μ D(4) ('dry') = 0; μ D(4) ('typical') = 0.5; μ D(4) ('clammy') = 0.5. We can likewise document the idea as D(4) = 0/dry + 0.5/ ordinary + 0.5/soggy.

To finish this device, it is important to manufacture approaches that have the capability to club few images together. This is the situation in a control structure where choices for activities depend on a finding of the consolidated deviations of the diverse measurements of item quality and the condition of the procedure. For instance, the at line administrator needs four fundamental tangible attributes to assess the direction of a cheddar amid maturing and to suspect his control activities: shading, coat, consistency, dampness content. The fluffy capacity utilized for blending is made based on a specialist clarification of the coherent connections between images. This framework, tried in the business, was utilized to foresee the quality development of the cheddar creation with a decent precision and heartiness by correlation with a specialist assessment (around 85% of the comparative answer). This framework is utilized as a help choice framework when the master isn't accessible. These between-image joins consolidate the worldwide point of the combining. The surmising is made by Zadeh's compositional govern of derivation, connected to the fluffy representative depictions of the contributions, as clarified beneath.

9.2.1.3. Process Control or Choice Guide Frameworks

For process control or choice guide framework purposes, the yield of the fluffy controller ought to by and large be a numerical yield, the set point of a physical controller. The most generally utilized fluffy controller is a steady kind Takagi– Sugeno fluffy controller, where the yield gi, j of each manage is acclimatized to a consistent (e.g., a temperature setpoint of a maturing chamber for a given run the show). Numerical yields are straightforwardly figured utilizing the consequence of Equation [9.3], participation review μ ExF(An I, B j) for each run l explain α l, and the related consistent yield clarified P lk, for each run 1 and every steady K.

$$D_{lk} = \frac{\sum_l \alpha_l^* P_{lk}}{\sum_l \alpha_l} \qquad (9.3)$$

9.2.1.4. Validation Tools (Level 3)

The third and last level is committed to approving the instruments that could feasibly be utilized for the calculations created (see Figure 9.3). An initial step includes working with specialists to secure the subjective approval. This initial step requires an expert– eyewitness input trade to check that the information communicated by the master coordinates the learning executed in the calculation. In this progression, the main conceivable approval is subjective. Quantitative approval must be accomplished on test purposes of a specially assembled database. This quantitative approval is the base pattern approval required to have the capacity to consider the instrument as perfect with the goal focused on, which is either (1) to accomplish a roundabout measure or state analysis coordinating the administrator's reactions, or (2) to accomplish a level of process control that is at any rate tantamount to the administrator's and similarly as vigorous, solid, and stable. In the event that (1), the cogency of the approach is checked in light of test focuses that give the most ideal representativity of the scope of various framework states. The model's yield is contrasted and the administrator's reaction. The quantity of good reactions (C) is given by Equation [9.4], where Rm is the model reaction comparing to the info estimations signed in the approval database, Ro is the reaction of the relating master administrator, and d is augmented by 1 when D is not exactly or equivalent to the resilience limit (St). A percent similarity esteem is created by ascertaining the proportion of number of perfect reactions to number of items making up the approval database.

$$C = \sum_{i=1,2,3}^{n} d$$
$$d = 1 \;\; if \;\; D_i \leq St \;\; with \;\; D_i = | Rm - Ro |$$
$$d = 1 \;\; if \;\; D_i > St \qquad (9.4)$$

A key variable in this progression is the resilience edge. It is, for the most part, indicated pair with the specialists for tangible estimations. In situations where the yields are communicated as etymological qualities, the approval pointers depend on conditions created in Bouchon-Meunier et al. (1996).

On the off chance that the model is approved in light of recorded information, the approval stage proceeds under genuine conditions; generally, the main arrangement is to backpedal to the model plan planning phase: either the learning was severely detailed or the questioner missed some essential data. Each progression in this stage should be rehashed through and through. For approval of a procedure control, the criteria utilized are the same as those utilized under great computerized preparing. Supporting this progression is the need to arrange a notable database. This database should be a delegate as conceivable of the full arrangement of potential framework states and the most widely recognized framework unsettling influences. In the event that the factors incorporated in the database are surveyed by specialists, it is essential to control the nature of these estimations. This can be accomplished utilizing philosophies adjusted from standard NF X 07-001 to test the reproducibility and repeatability of human estimations communicated on a representative scale (numeric irregular or semantic), along these lines summing up prior works drove by Curt et al. (2004) on tactile pointers.

9.2.2. Representation and Propagation of Uncertain and Imprecise Knowledge

Give T a chance to be a scientific model speaking to a wonder that happens amid food handling whose contentions are spoken to by the fitting formalism (probability– plausibility) as per the character of accessible information (changeability, imprecision, inadequacy, and so forth.).

9.2.2.1. Information Portrayal by Methods for Likelihood Dispersions

Looked with significant information, it might be conceivable to recognize a likelihood dispersion portraying the fluctuation. For instance, Figure 9.6 demonstrates the dissemination of the cumulated relative frequencies of overwhelming precipitation (depicting the likelihood that the genuine estimation of substantial rain is lower than a specific limit) assessed from 30 years of information gathered from a climate station. This circulation appears, for instance, that there is 95% of chance that the overwhelming precipitation does not surpass 300 mm but rather surpasses 150 mm. In a hazard evaluation setting, the hazard supervisor could utilize the likelihood law fitted from information (see the case in Figure 9.6 demonstrating the total ordinary appropriation with mean 222.2 mm and standard deviation 39.5 mm fitted from crude frequencies).

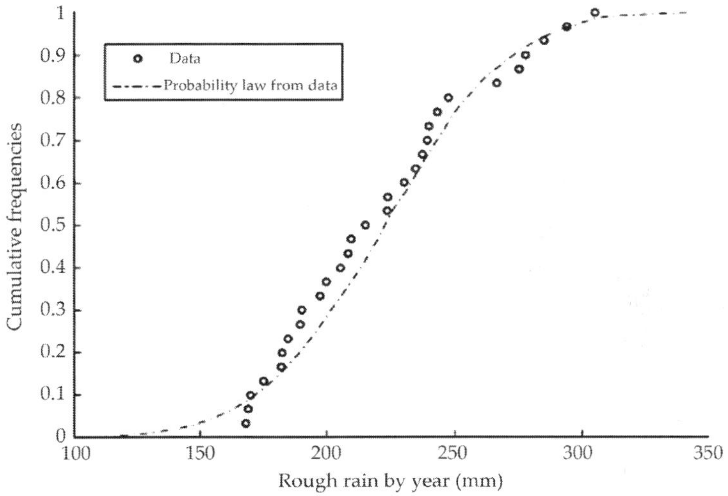

Figure 9.5: Cumulated relative frequencies and likelihood law from information about yearly overwhelming precipitation. Rain history more than 30 years, (*Source: G. Caldwell Darwin (2013), Robotics and Automation in the Food Industry: Current and Future Technologies, Woodhead Publishing*).

Probability distributions can be also used for theoretical reasons, or where there is *a priori* knowledge on the variability involved. Likelihood conveyances can be additionally utilized for hypothetical reasons, or where there is from the earlier information on the inconstancy included.

9.2.2.2. Information Portrayal by Methods for Plausibility Circulations

Looked with data along the lines of 'I am certain that amount x exists in an interim [a, b] yet values situated inside the interim [c, d] (included inside [a, b]) are more probable,' it is intelligible to propose a trapezoidal probability conveyance (formally proportional to a fluffy interim–see Figure 9.7). For instance, the likelihood circulation π characterized in Figure 9.7 implies that the in all likelihood (hazardous yet instructive) warm mass exchange esteems are situated inside the interim [3, 3.2] (alluded to as the 'center' of π) with no inclination inside this interim. We locate the preservationist yet not extremely enlightening qualities inside [2.86, 3.33], otherwise called the 'support 'of π. Numerical probability conveyance may likewise be seen as a settled arrangement of certainty interims (data ready to be given in part by specialists), known as cuts of π noted π θ = {e/π (e) ≥ θ}. The level of conviction that interim π θ contains the genuine estimation of warmth mass exchange is 1 − θ, implying that the likelihood that the warmth mass exchange exists in π θ is more prominent than 1–θ.

Plausibility appropriation empowers the portrayal of sets of probabilities (Dubois et al., 2000) constrained by an upper likelihood bound known as the likelihood measure (see the upper aggregate likelihood bound on Figure 9.7) and a lower likelihood bound known as the need measure (see the lower total likelihood bound on Figure 9.7). Plausibility dispersion in this way makes it conceivable to speak to inadequate probabilistic learning and get a sectioning of probabilities.

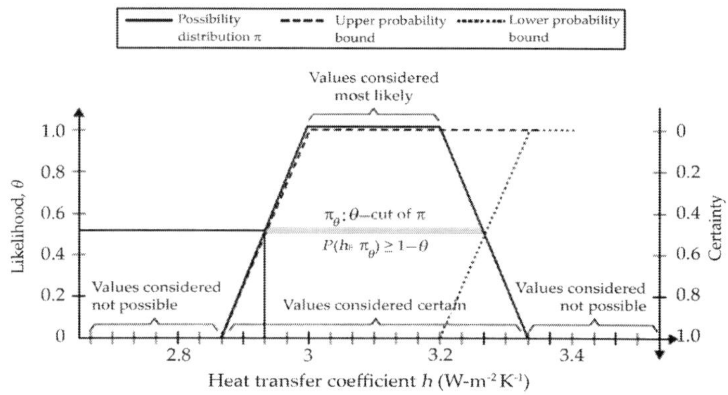

Figure 9.6: Case of a probability circulation initiated from a learning master on a warmth exchange coefficient utilized as a parameter in a cheddar maturing model, (*Source: Baudrit et al., 2009*).

9.3. CURRENT APPLICATIONS IN THE FOOD INDUSTRY

Fluffy rationale has just been connected in the food business for quality estimation, state determination, and process control. It will be created in this segment.

9.3.1. Indirect Estimation of Representative Factors Utilized By Specialists for Estimation of Food Product Quality

Circuitous estimation can be utilized alone for review purposes, as in Chao et al. (1999) for after death poultry order of shading pictures of viscera to recognize air sacculitis livers from ordinary livers, or as in Nielsen and Paul (1997) for tomato, quality evaluating. It can likewise be combined with a controller or implanted in a choice guide framework to control nourishment item quality, as in Perrot et al. (1996) to control the nature of rolls in a modern passage broiler. The prerequisites as far as estimation accuracy and unwavering quality will in this manner vary. In the two cases, it is frequently an 'element level combination' as characterized in Valet et al. (2000). The combination is performed on particular highlights separated from crude data that is for the most part given by sensors, for example, cameras, electronic noses, NMR sensors, and so on. For instance, Yea et al. (1994) utilized three business gas sensors to distinguish four sorts of fragrant odors with a segregation rate of 99.2%. In the initial step, fluffy thinking identifies the fragrant scents, at that point in the second step, smell is segregated by a neural system. De Silva et al. (1996) utilized PCA-separated highlights from acoustic video pictures to decide the immovability of herring roe. A fluffy basic leadership framework tried on 160 examples gave around 84% great separation. Sundic et al. (2000) utilized highlights separated from an electronic tongue combined with an electronic nose to impersonate human tangible view of potato chips and cream.

A delineation of such an approach in light of the ideas introduced in Section 9.2 is produced in the case underneath connected to the assessment of the tactile properties of a crusting frankfurter. It is a wiener imperfection that shows up under particular states of drying. The deformity of crusting is one of the two essential imperfections that can show up on the item and ought to be stayed away from. The assessment of this deformity in the business is made near the assembling line (Mauris et al., 2000). The level of hotdog crusting assessed by the administrators is then repeated utilizing fluffy implications (Section 9.2.1) in view of adjusted picture preparing and express master runs about crusting (Figure 9.8). Crusting is assessed from the human visual impression of the shade of the focal point of the cut, length of dark zones (LB), and length of dim red territories (LD) on frankfurter cuts. Figure 9.9 gives a case of fluffy significance for LB. Crusting classes are characterized by master information by an arrangement of fluffy emblematic standards connecting the terms of dark lengths and dull red lengths, and classes of crusting, together with master guidelines, for example, 'if LB is non-existent and LD is non-existent, crusting degree is nonexistent, class 1.' The approach was actualized with an extremely set number of tenets: 16 rules handled in this application. It takes 2 months to deal with those guidelines. The outcome is emblematic data as a level of enrollment to crusting classes 1 to 4. The outcome for a limit of resilience (see Section 9.2.1) limited in a first guess by the esteem 0.5 is 88% for perfect answers tried on two databases. This framework is valuable as respects to the hotdog crusting assessment that is hard to accomplish for the administrators. This device is utilized as a choice emotionally supportive network in the business to accomplish the correct assessments.

In applications where specialized or financial reasons direct that no sensors are accessible, the main contributions to the backhanded estimation module are human administrators. For this situation, the representative space utilized by the administrators to assess the item is utilized straightforwardly by the framework, as in Ioannou et al. (2004) for general sautéing appearance amid a warming procedure. This application is illustrative of circumstances ordinarily experienced in the food business, where it is imperative to propose numerical methodologies able to do straightforwardly adapting to the representative information took care of by the administrators. The caramelizing of sustenance items is gotten through warming the item surface, which prompts biochemical responses called Maillard responses. In this application, we have built up a determination model to help the assessment of the carmelizing variable. It is a vital tactile variable surveyed by an administrator to control the searing procedure, which impacts purchaser decision and inclinations. All things considered, in the business, it is hard to gauge and is accomplished by the quality group of the processing plant through a visual quality rule purported the sautéing worldwide appearance (BGA). The conclusion show comprises in amassing the enlightening factors evaluated by the administrators into the BGA (V1, level of spotted zones; V2, shade of the spotted zone; V3, shade of the item surface aside from the spot territories), see Table 9.2 for V2. Fluffy tenets are in the shape, for instance, of the enlightening factors being evaluated as takes after V3 is beige, V2 is dull darker, and V1 is non-existent with a participation degree to 0.75 and is little with an enrollment level of 0.25. Thusly, the standards having a participation degree unique in relation to 0 are R1 and R2.

R1: if V1 is non-existent, V2 is dim darker and V3 is beige then the BGA is Low searing LB.

R2: if V1 is little, V2 is dim darker and V3 is beige then the BGA is No searing NB.

The finding was approved on 40 items with a level of similarity of 92.5% at an edge of one-half image (Figure 9.10).

The chose number of fluffy enrollment capacities, and thusly the quantity of images portrayed an etymological variable, is a urgent factor in this approach. It gives a sign of the significance of every factor in the master's methodology. It is additionally connected to the affectability of the administrator's estimation on this variable, which is itself attached to its significance, if not to challenges in playing out the estimation. This is the thing that can be named the 'granularity' of the estimation, which is connected to the graduality chose for the fluffy implications and portrayals. For instance, in the application created underneath on the worldwide appearance of carmelizing (BGA). A test was keep running on the repeatability of the master estimations for these three factors (Table 9.3), uncovering that repeatability can be connected to the quantity of images related to every one of these factors (V1 and V3 were clarified by 5 images, and V2 by 4 images).

Table 9.2: Tangible Level Pointer of Spotted Territory (V2)

Name	The percentage of spot areas
Definition	Proportion of spots in comparison with the total area
Protocol	Observe the product area *at* the output of the browning oven On the product area, observe the areas which are different from the background color of the product Assess the rapport between these areas in comparison with the total area
Assessment scale	Non-existent, small, medium, large, very large
References	
Time	At the output of the browning *oven*
Location	The center of a set of products

Source: Ioannou et al., 2004

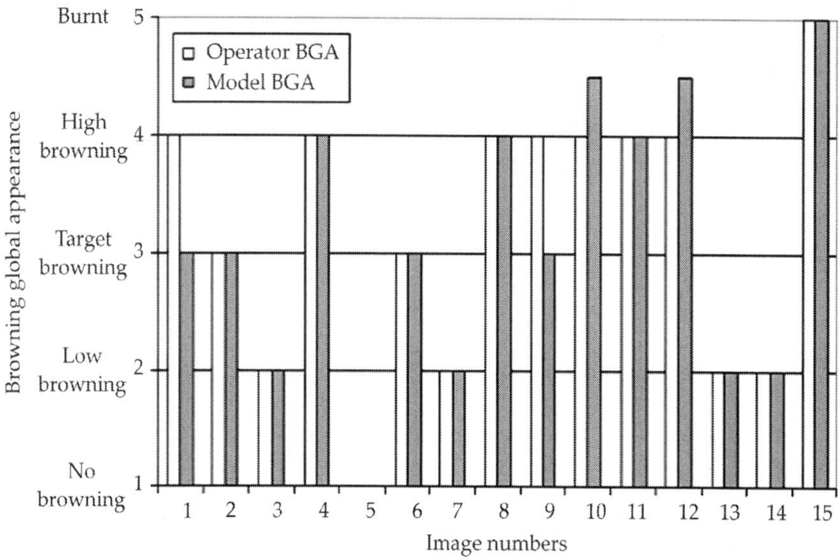

Figure 9.7: Correlation between the administrator and the determination show for the assessment of the BGA, (*Source: Ioannou et al., 2004*).

Table 9.3: Repeatability for the Master Estimations Required to Anticipate the Worldwide Cooking Appearance amid a Warming Procedure

	Tolerance threshold BGA (%)		VI (%)	V2 (%)	V3 (%)
Operator repeatability	0.5	91.67	98	92	90
	0.25	82	80	62	83

Source: G. Caldwell Darwin (2013), Robotics and Automation in the Food Industry: Current and Future Technologies, Woodhead Publishing

This graduality incited by the fluffy implications is vital to overseeing fluffy control, as it offers an approach to clarify some portion of the resilience of the framework constructed. Backpedalling to the wiener crusting application, on the off chance that we twofold the incline of the line connecting the image 'somewhat long' and 'long' for LB for a given LD settled to an enrollment level of 1 for the 'long' image and 0 somewhere else, it prompts a difference in item characterization from class 3 to class

4, which implies the item is then delegated inadequate. This graduality could be unequivocally set up while altering the parameters of verifiable characterization strategies. An outcome is displayed in Perrot et al. (1996) for an issue including a backhanded estimation of treat shading.

9.3.2. State Analysis and Food Process Control

This subject has been the focal point of a lot of grant in the 'fluffy and food' field (51% of the aggregate papers for a question on FSTA; Perrot et al. (2006)). The greater part of these papers are traditional utilization of the Takagi–Sugeno controller (see area Mathematical learning formalization (level 2, for example, Alvarez et al. (1999) for controlling isomerized bounce pellet creation, Honda et al. (1998) for controlling the purpose fermenting procedure, or O'Connor et al. (2002) for controlling the lager fermenting process. The creators utilize two particular ways to deal with build up the analysis, control and supervision modules: (1) 'information-driven' methodologies, and (2) 'expert learning driven' methodologies.

For 'information-driven' methodologies (around 30% of papers), the framework is distinguished naturally utilizing a database and an enhancement calculation. For instance, Honda et al. (1998) built up a fluffy neural system to control the temperature of the Ginjo purpose pounding process, O'Connor et al. (2002) built up a fluffy PID to control the preparing procedure, Guillaume and Charnomordic (2001) upgraded a fluffy lead premise utilizing a hereditary calculation to build up a choice guide framework for the cheddar making process. For this situation, the hazard is that the models created should adjust with countless (50–100 or more) and therefore have countless to distinguish. For instance, in Honda et al. (1998), 283 information indicates are utilized recognize the model. Taking care of such extensive amounts of information can prompt troubles in setting every one of the parameters, and is a noteworthy restriction for applications utilized as a part of sustenance forms. Also, the instruments manufactured must be utilized to insert yet not extrapolate new cases. In any case, in these cases, the additional esteem brought by utilizing fluffy rationale depends on (1) an easy to use administrator interface that handles information in a space near the emblematic space utilized by the administrators; (2) the likelihood of naturally adjusting the module to new preparing conditions, (for example, another recipe for a bread) if a model or a set with enough information to portray the procedure end up accessible.

For 'master information-driven' methodologies, fluffy rationale is utilized as a method for giving a numerical formalization of master learning and implanting it in choice guide calculations and controllers. For instance, Ioannou et al. (2004) proposed a fluffy rationale way to deal with control the sautéing procedure amid modern warming, Curt et al. (2002) created five Takagi– Sugeno modules to control frankfurter quality amid maturing.

Fluffy rationale is likewise utilized for supervisory undertakings, as in Acosta-Lazo et al. (2001) for the supervision of a sugar production line, or Goelzera et al. for choice help programming including a physiological model of yeast movement and fluffy rationale to give quick and effective answers for encourage winery administration on a mechanical scale in different winemaking conditions. For this situation, the modules are utilized specifically, without database adjustment, and the datasets are just used to approve the methodologies.

The control bases are by and large conservative, including just a set number of standards (around 20 by and large). It is exceptionally intriguing for two reasons: right off the bat, the working of such a fluffy master framework is less tedious than for traditional master frameworks building; besides, it is anything but difficult to rectify and adjust starting with one process then onto the next on a similar item.

We might outline this chapter utilizing the outcomes created in Perrot et al. (2000). The guideline was to build up a choice guide framework for the heating procedure. The whole procedure was non-direct and time-variation (the working states of a stove change after it has been utilized for expanded periods), with solid collaborations between factors, for example, the effect of the temperature up and down a mechanical passage broiler on the dampness substance, shading, and thickness of the scone.

The administrator assumes an essential part in this procedure, and the nature of the rolls prepared depends on the administrator's estimations and activities. Working in this application, we assembled and approved a shut circle quality control input framework on a backhanded constant pilot burrow broiler (Figure 9.11).

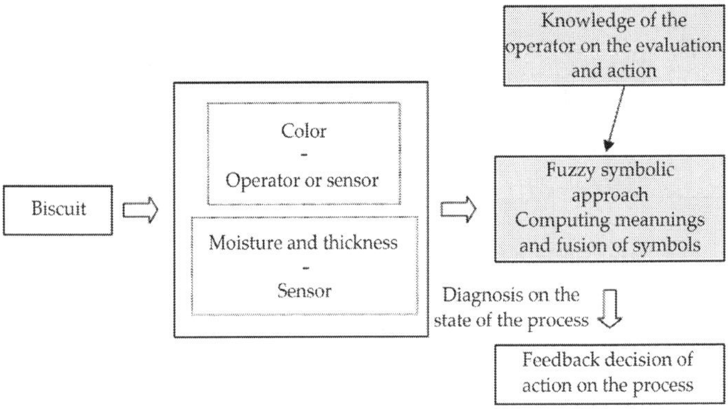

Figure 9.8: The fluffy representative approach connected to biscuit baking, (*Source: G. Caldwell Darwin (2013), Robotics and Automation in the Food Industry: Current and Future Technologies, Woodhead Publishing*).

Numerical contributions for dampness and thickness are designed in a phonetic space utilizing the fluffy implications depicted in Section 9.2.1. The space organized along these lines can be accumulated at an indistinguishable level from the administrators' representative portrayal of the shading. The second step is to register the combination between the images communicated by the administrators as indicated by the fluffy formalism. For instance: 'broiler excessively hot in area c of the stove = (dampness typical OR dampness dry) AND shading over-cooked' or 'stove not sufficiently hot in segment c of the broiler = (dampness ordinary OR a lot of dampness) AND shading marginally under-cooked.' Finally, a steady sort Takagi– Sugeno fluffy controller is executed to help choices on activity on the yield factors, that is, temperature setpoints for the air in the distinctive assemblies of a consistent broiler. This controller depends on the enrollment grades ascertained for framework finding. Figure 9.10 represents a case of an analysis on a roundabout ceaseless pilot broiler, giving the reaction of the criticism controller utilizing the yields of the emotionally supportive network and the procedure state finding.

Great direction exhibitions were accomplished for all scone quality attributes amid preparing, regardless of the unsettling influences as affirmed after solid aggravation connected to the procedure at 9000 and 1200s. The administrator recognizes the quality float after around 8 min and gives his/her assessment to the emblematic module. In parallel, a similar difference in shading was checked by the sensor. At long last, the representative module totals the administrator estimation with the instrumental estimations of bread dampness substance and thickness and builds up an analysis (displayed Figure 9.10b) that is utilized by the programmed control framework to alter its activity on the broiler. Along these lines, the three shading deviations are communicated regarding process deviation, that is, 'stove excessively hot in segment C' or 'broiler insufficient hot in area C,' with an evaluating between [0, 100]%. These judgments are transmitted to the programmed controller, which at that point settles new setpoints to keep up roll quality. Similar outcomes are come to at the same time for item dampness substance and thickness. Both control precision and dependability are approved as attractive.

By and large, learning is bunched on the sort of process experienced. For bunch forms, information groups are composed by time groups. For instance in Perrot et al. (2004), where the creators manage a utilization of cheddar aging, information is sorted out in four phases of maturing connected to the aging time. For consistent procedures, for example, the case exhibited above, information bunches are sorted out in space. What is fascinating to underline is the soundness of the specialists' learning organizing communicated under fluffy guidelines, with the logical information that is in some cases accessible. For instance, in the scone heating application depicted above, there were robotic models accessible.

In these unthinking models, the wonders were guided by the temperature came to at the roll surface. Underneath 100°C, drying speed expanded, while over 100°C it stayed stable, and water movement met 1. For the persistent stove considered in this investigation, this key temperature was come to in the second third of the broiler. The information groups increased after master taking care of were intelligent with this, and sorted out, for dampness analysis, in the supposed 'segment C' of the broiler which relates to the second third of the stove. For the energy of item tinge, both learning portrayals prompted activity in the last third of the stove. It has prompted a more customary control that has prompted staying away from tangible imperfections on the bread rolls.

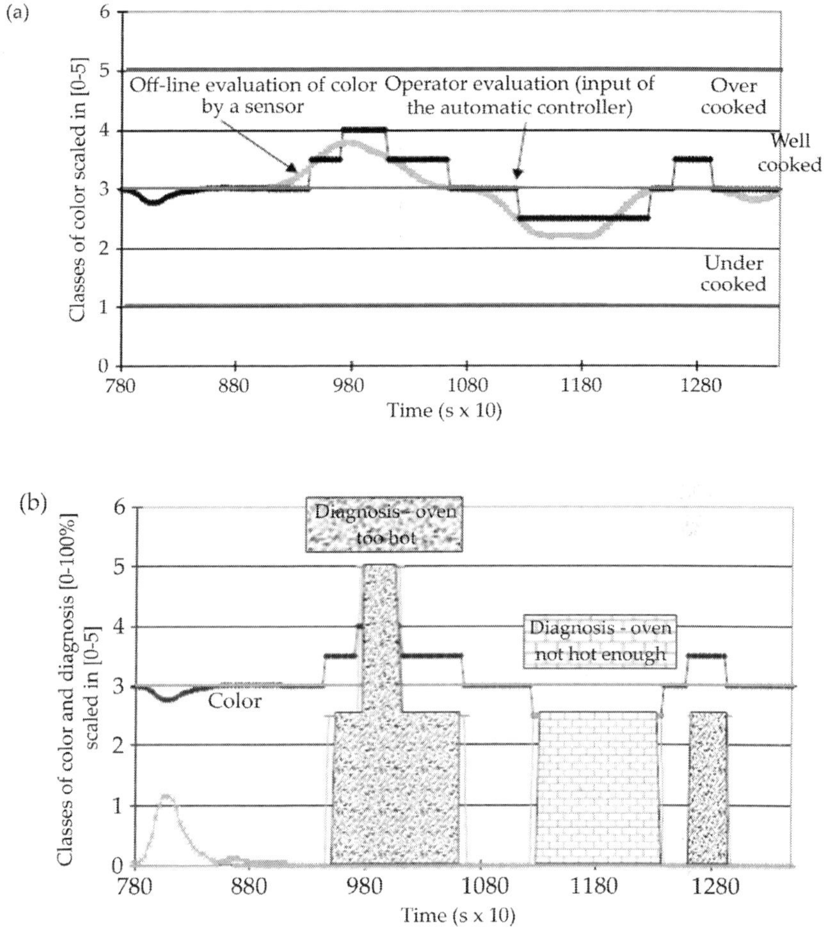

Figure 9.9: Criticism control (a) and determination of the shade of the biscuit (b) utilizing the fluffy representative module amid a test for the cooking procedure, (*Source: G. Caldwell Darwin (2013), Robotics and Automation in the Food Industry: Current and Future Technologies, Woodhead Publishing*).

9.3.3. Joint Preparing of Imprecision Changeability in the Displaying of Cheese Mass Misfortune amid Ripening

The progression of cheddar mass misfortune is a key factor in the ripening procedure, with outcomes for profitability and a hazard that the subsequent item might be dropped in status. These progressions are subject to climatic conditions in the maturing chambers where ventilation is utilized to empty warmth and moistness created by cheeses. In a modern setting, it isn't practical to introduce sensors at each point in the ripening chamber keeping in mind the end goal to control barometrical conditions. Because of privacy issues, the writing contains few investigations on the connection between climatic conditions and wind stream. Warmth and mass exchanges have been broadly examined in connection to cooking and drying forms, however couple of information have been distributed in the field of cheddar maturing, and exchange coefficients amongst cheeses and the climate are not correctly depicted. The point here is to consider the loose and inadequate learning of aging chamber attributes in the cheddar mass misfortune demonstrate. Each information variable and model parameter is spoken to by methods for likelihood and probability circulations as per the idea of the accessible learning, and we executed the joint spread plan for each time-step. Figure 9.13 (left) demonstrates the lower and upper aggregate likelihood limits at day 14 preceding wrapping for an underlying mass of 0.33 kg coming about because of vulnerability spread. The hole between these two points of confinement is expected essentially to the loose idea of accessible data.

We would thus be able to outline the aggregate vulnerability on cheddar weight before the wrapping step, utilizing the interim [0.260, 0.284]. That implies we are 95% certain that cheeses will surpass 0.26 kg at day 14. Figure 9.13 shows the developments of vulnerability edges of 5% and 95% percentiles for the mass misfortune at each timestep amid the two aging procedures. From one perspective, we are 95% sure that the mass loss of cheeses all through the aging room won't surpass 70 g for the principal trial at day 14 of maturing. Then again, we are 95% certain that cheeses have lost no less than 46 g before they are wrapped. Based on a model adjusted to pilot scale, our approach could give key data to enhance the control of cheddar mass misfortunes in mechanical conditions.

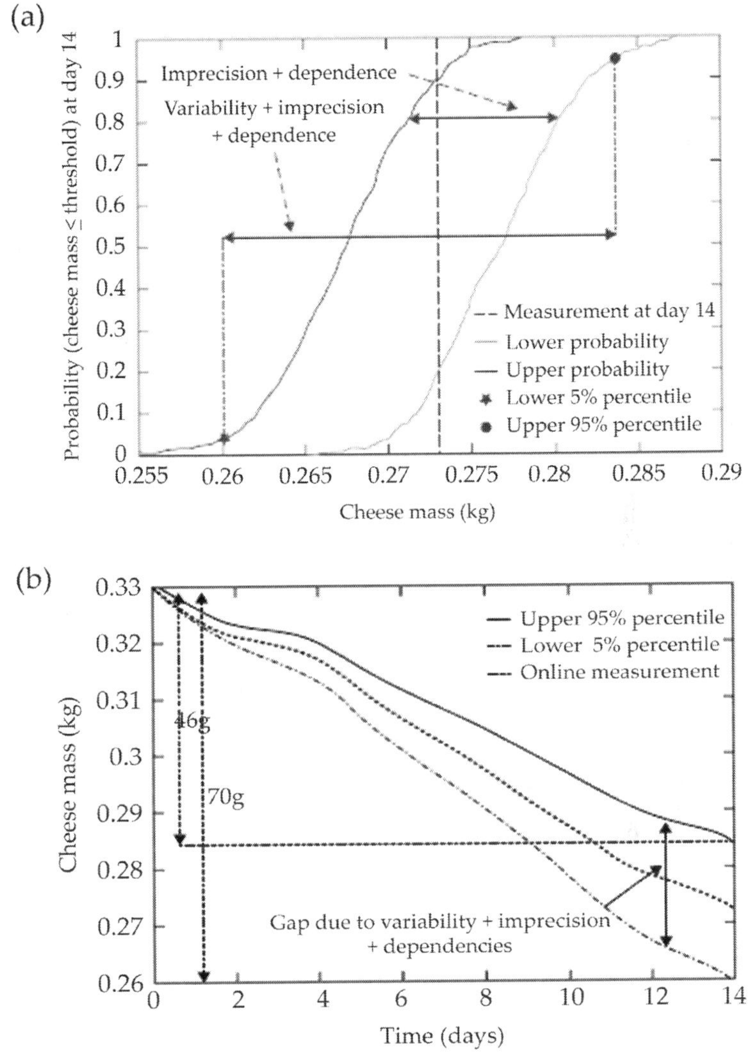

Figure 9.10: (a) Lower and upper probabilities that cheddar mass is lower than a specific weight at day 14. (b) Uncertainty edges of the fifth and 95th percentiles relating to the cheddar mass misfortune through the maturing procedure from an underlying mass of 0.33 kg, (*Source: G. Caldwell Darwin (2013), Robotics and Automation in the Food Industry: Current and Future Technologies, Woodhead Publishing*).

9.4. ADVANCES IN RESEARCH AND FUTURE PATTERNS

As we have seen, fluffy rationale could have a promising future in food applications to (1) catch and formalize the spellbinding tactile assessment performed by a quality group, an administrator or a buyer, (2) convey a state finding, and (3) control or help choices in food designing.

The cutting edge of learning accessible for displaying food forms instigates vulnerability on specific marvels and, subsequently, on certain model info factors and parameters. Plausibility conveyances are all around outfitted to portrayals of master learning as the specialists can be required to be steady with their own particular choices: the interim of qualities that they consider doubtlessly is fundamentally incorporated into the interim outside which they consider values are impractical. Incorporating halfway numbness in the numerical model set up at pilot scale can help transpose this information to a modern scale. Spreading imprecision can help enhance the control procedure. To round up, it will end up conceivable to break down the commitment of imprecision or potentially deficiency to demonstrated reaction yields keeping in mind the end goal to explain whether a questionable reaction is because of an absence of data or to a wellspring of erratic inconstancy, and along these lines decide the key factors as well as impact occasions where facilitate bits of knowledge are basically required.

A few inquiries emerge when assembling a control framework utilizing master learning converted into fluffy capacities.

Question 1 fixates on master learning and the challenges in dealing with it, despite the fact that subjective science has built up some helpful instruments for this reasons.

One point to underline is the significance of picking 'great' specialists, which will generally shape the outcomes acquired. More top to bottom work with the intellectual science group could demonstrate significant in advancing this inquiry.

Question 2 brings up a progression of issues focused on the decision of graduality for fluffy capacities and fluffy implications. In what manner should the correct graduality be picked by goals focused on? In what manner should this graduality be approved? What is the best similar strategy as indicated by objective focused on? How is it conceivable to approve demonstrate comes about that go past human impression of graduality, however, are as yet important when utilized for process control purposes? Which numerical instruments should be sent in order to actualize a hypothetical approach on a given circuitous estimations display with a specific end goal to survey the graduality accomplished on the yield variable as per bends built upon input factors?

Question 3 is fixated on the optimality of the models constructed and their ability for enhancement through auxiliary change without annihilating the model-installed semantics. Connected to this inquiry is the issue of the generalizability of the models manufactured.

Question 4 concerns the dynamic properties of the models in light of the expertise handled. The psychological idea of the specialists' assets makes it to a great degree hard to extricate mental portrayals of the progression of a framework. This is a critical bottleneck for the motivations behind controlling complex frameworks.

To finish up, the advantages that fluffy rationale can give to the nourishment business are:

- An simple approach to formalize and profit by the master learning that speaks to the memory of the processing plant.
- Improvements in the administration and traceability of the nature of generation with frameworks that assistance the administrators in their estimation or control undertakings.
- Management of the vulnerability with apparatuses that can adapt to this vulnerability and propose process administration changes.

REFERENCES

1. Alvarez, E., Cancela, M. A., Correa, J. M., Navaza, J. M., & Riverol, C., (1999). 'Fuzzy logic control for the isomerized hop pellets production. *'Journal of Food Engineering,'* 145–150.
2. Caldwell G. Darwin, (2013). Robotics and Automation in the Food Industry: Current and Future Technologies. Woodhead Publishing.1st edition, page 432-448

CHAPTER 10

ADVANCED METHOD FOR THE CONTROL OF FOOD PROCESSES: THE CASE OF BIOCONVERSION IN A FED-BATCH REACTOR

"Omega-3s occur naturally in food like fish, chicken and eggs, and plants to a lesser extent. Why do we need to get it from bread?"

—Marion Nestle, Professor of Nutrition, Food Studies, and Public Health at New York University, author of Food Politics

CONTENTS

Novel ways to deal with ensure finish and shrewd control of whole food preparing plants are required to guarantee item quality, adaptable and effective activity, and to diminish the effect of procedures on the earth. The goal of this section is to give a few issues related, and propelled techniques for, displaying, checking and control that might be suitable for the treatment of on-line process enhancement with regards to nourishment forms. The part, for the most part, focuses on a bioconversion procedure in an encouraged bunch reactor.

10.1. INTRODUCTION

The food business is settled and numerous procedures in task these days are the subject of concentrated work as to methods for conceiving better activity modes as far as item quality and security (how to work keeping in mind the end goal to guarantee quality and conform to wellbeing requirements) and regarding task costs and ecological effect. There is additionally escalated advancement work went for reacting to shopper requests by planning new items and outlining and working the most fitting mix of unit tasks expected to create them.

Notwithstanding, and in spite of the way that the fundamental physical, biochemical and microbiological standards are sensibly surely known, food are intricate frameworks with properties that, since they are associated with quality and security, are typically exceptionally hard to gauge, assess or even speak to through solid models. Such properties may incorporate physico-concoction parameters related with quality, for example, supplement content, surface, shading or rheology, or microbiological attributes generally associated with sustenance security. Also, and from a procedure building point of view, the sustenance business incorporates a rich assortment of obviously exceptionally various procedures and advances, in this manner hampering the look for bringing together ideal models that would be helpful for managing extraordinary yet comparable to forms. Such procedures have just as of late been grouped into a sensibly modest number of classes, to be specific bioconversion, division, safeguarding, and organizing.

Bioconversion incorporates likely the biggest class of procedures. It is focal in the creation of matured foods, with some outstanding cases from the dairy or winemaking industry, however, it is likewise the reason for cooking in its more broad origination. Initially, it depends on the utilization of a naturally dynamic component, for example, microorganisms or yeast.

Furthermore, the biochemical change of food is frequently performed through cooking. In the two cases, with the suitable mix of fixings and added substances, various applicable responses enacted by temperature or bioactive operators happen. These responses not just change the structure of proteins (denaturalization) to make them more absorbable, yet additionally include seasons and advance blends between lipids, proteins and starches that expansion the nature of sustenance as far as appearance, surface, and taste, notwithstanding making a few supplements or biochemical mixes of intrigue enhance the bioavailability. Understanding the basic dynamic components and creating models of utilization in detecting innovations and control are vital parts of these kinds of procedures. Detecting innovations would permit the precise estimation of value parameters that should be under control. Models and reenactment apparatuses will likewise be required to create proficient control calculations to augment quality for this kind of process.

Division and cleansing tasks are broadly utilized in the food industries, for the most part as downstream handling after bioconversion, or as a method for broadening the timeframe of realistic usability of the item by decreasing the water accessible in the foodstuff. Run of the mill illustrations incorporate microfiltration, dissipation, and drying. Extraction tasks, (for example, in gelatine or gelatin generation) are likewise partition activities broadly utilized as a part of the nourishment business to recoup high-esteem bio-compounds. Other division forms, for example, refining in its renditions of glimmer, sub-atomic or vacuum refining, are likewise cases normal for, for example, the oil-refining industry. Stop drying, cryoconcentration of juices or different liquids of enthusiasm for the food business, or super-basic extraction forms, which are commonly utilized to recuperate very significant bioactive mixes, finish the not insignificant rundown of tasks associated with detachment.

Preservation processes and technologies are those aimed at increasing the product life and ensuring health and safety by reducing the growth of microorganisms (in freezing or refrigeration, for instance) or by eliminating them (as in pasteurization or sterilization) with thermal treatments such as steam heating, microwave or ohmic heating. The so-called emerging technologies, based on high-voltage electric pulses or high pressures, fall into the same category. Other preservation technologies are those having a chemical or biochemical basis, such as smoking, salting, or those aimed at reducing water activity and thus reducing degradation kinetics or growth of microorganisms, as in the case of freeze-drying.

Organizing forms coordinate various advancements went for altering the food item microstructure, for example, in expulsion, emulsification or crystallization. Various tasks are worried, in which the structure of the item is altered or advanced. The vast majority of them are thermo-mechanical activities. For example, the presentation of expulsion, amid the 1960s, advanced the plan of new-properties-based items. Likewise warming (or cooling, solidifying), in view of a superior comprehension of consolidated warm and mechanical systems, turned into an approach to control the structure at various levels. Then again, frothing and emulsion tasks are the premise of new items, regardless of whether control is as yet troublesome because of absence of sensors. From a control and mechanization perspective, the accompanying shortcomings have been identified in most food organizations.

Most plant control plans decrease to nearby and decentralized control circles following up on a normally modest number of states (ordinarily temperature or weight) not straightforwardly associated with item quality, and as a rule not associated either with basic parts of the task, for example, water or vitality utilization. In spite of the way that the execution of PID control can be in numerous occurrences more than worthy (as is additionally the case in different process businesses), the control circles ought to be joined among the distinctive procedures to feature cooperative energies and not to drop them. What's more, this administrative layer isn't instructed or incorporated on higher supervisory levels.

Albeit by and by numerous food plants advantage from progresses in information securing and checking of the full creation line to accumulate and store enormous measures of information, the utilization of such data is very constrained, typically not productively utilized, and decreased to arrange cautions (frequently took care of at a low level) or to help in delivering basic generation choice standards and disconnected control of inventories. Substantially more compelling utilization of such data could be made conceivable, when appropriately joined with process models and guess devices, to have the capacity to appraise unmeasured yet applicable plant states and to anticipate future situations, even in a constant setting.

Regularly perceived as a specificity of food forms, the absence of sensors for applicable item attributes is as yet an issue. Regardless of whether various calculations are accessible for cutting-edge control purposes, clearly online, continuous solid data is essential. Normal sensors (temperature, weight pH, flow rates) give data just in a roundabout way identified with the item properties of intrigue, for example, surface, smell content, biomass, contaminants, vitamins, and so forth. Sensors for such item properties are either missing totally or extremely costly and not hearty or sufficiently dependable to be utilized as a part of regular mechanical practice. Enhancing the unwavering quality of detecting gadgets and growing new hardware–programming detecting strategies for on-line estimation of troublesome quality item parameters, are basic in creating savvy control applications for sustenance production lines.

It is essential to take note of that, for the most part talking, there is an absence of control outline and task ideal models to ideally work plants, either learned or acquired from assorted yet comparative procedures or situations. One such worldview is that sought after by the designing group concentrating on similar ideas, for example, reproduction, optimality or advancement as those containing a genuine deliberate way to deal with the issues of conceiving shrewd control for nourishment forms.

Keeping in mind the end goal to conquer the present confinements recognized in food creation plants, it is critical to offer novel and unique indispensable ways to deal with ensure finish and savvy control of the entire plant in light of changes in the request and the supply of materials, to guarantee item quality, adaptability and proficient task, agreeable with nature.

In this chapter, for clear space reasons, we should focus on one class of food forms (bioconversion) and present a few issues and propelled techniques in displaying, observing and control that might be proper to deal with the online process improvement with regards to food forms.

The chapter is sorted out as follows. We should begin by quickly presenting in Section 10.2 the mass adjust demonstrate that will fill in, as a rule, all through this part. In Section 10.3 we might focus on demonstrating issues, by considering the significance today of thinking about the particulate measurement (by means of the circulation of size, mass or age) in sustenance forms, quickly presenting the utilization of populace adjust to supplement mass adjust demonstrating and address a portion of the unpredictability identified with these fractional differential condition models. In Sections 10.4, 10.5 and 10.6, we might for the most part focus on a few key issues identified with sustained group forms, which are assuming a focal part in the bioconversion procedures of the sustenance business today. Sustained clump bioreactors speak to a vital class of bioprocesses, for the most part in the sustenance business (e.g., yeast creation or alcoholic aging) and in the pharmaceutical business, (for example, the generation of the antibody against hepatitis B) yet additionally, for instance, for biopolymer applications (poly-beta-hydroxybutyric corrosive, PHB). It is additionally particularly engaged with the field of catalyst creation, which has been produced over the previous decade because of the recombinant DNA innovation and by means of the utilization of filamentous microorganisms.

In a food group reactor, the tank is just in part topped before the beginning off of the procedure activity, and the responsive issue is logically included until the point when the tank is full (this is the sustaining stage compressed by the word bolstered in the division of the task). One sits tight for the distinctive responses to be finished (bunch stage) before purging the tank. In Section 10.4, we should address the checking of bolstered clump forms by considering the tuning of onlooker based estimators for obscure parameters that can deal with the time-differing nature of the fed-cluster process progression. This issue is additionally the protest of Section 10.5, which will focus on the specificities of the outline of PID controllers for sustained cluster reactors. Segment 10.6 will at long last focus on the plan of continuous streamlining calculations, to be specific versatile extremum-chasing controllers, for encouraged clump forms.

10.2. THE PRINCIPLES OF OHMIC HEATING

Ohmic heating is a thermal process in which heat is internally produced by the passage of alternating electrical current (AC) through a body such as a food structure that serves as an electrical resistance. Ohmic heating is otherwise called resistance heating or direct resistance heating. The principles of ohmic heating are very simple. During ohmic heating, AC voltage is applied to the electrodes at both ends of the product body. The degree of heating is directly proportional to the square of the electric field strength, the electrical conductivity, and the type of food being heated. The electric field strength can be adjusted by adjusting the electrode gap or the applied voltage, while the electrical conductivities of foods fluctuate critically, but can be adjusted by the addition of electrolytes. Adequate heat is generated to pasteurize or sterilize foods. Generally, pasteurization involves heating high-acid (pH < 4.5) foods to 90–95 °C for 30–90 seconds to inactivate spoilage enzymes and microbes (vegetative bacteria, yeasts, molds, and lactobacillus organisms). Low-acid (pH > 4.5) foods can support the growth of Clostridium botulinum and depending on the actual pH and other properties of the food, require heating to 121 °C for at least 3 minutes (lethality Fo = 3 min) in order to attain sterility (12D colony reduction). Within the past two decades, new and improved materials and designs for ohmic heating have become available. The particular interest in this technology stems from the food industry's ongoing interest in aseptic processing of low–acid liquid-particulate foods. In the case of particulates suspended in viscous liquids, conventional heating transfers heat from the carrier medium to the particulates, and the time required to adequately heat the center of the biggest particulate (the designated 'cold-spot') results in overprocessing. On the other hand, ohmic heating is volumetric and heats both phases at the same time. Ohmic heating is a high-temperature short-time method (HTST) that can heat an 80% solids food product from room temperature to 129 °C in about 90 seconds, allowing the likelihood of decreasing the magnitude of high temperature over processing. A sharp contrast between ohmic heating and conventional heating is that ohmic can heat particulates faster than the carrier liquid, called the heating inversion, which is not possible by conservative, conductive heating.

10.2.1. The Effect of Ohmic Heating on Nutrition

10.2.1.1. Thermal Destruction

Since systematic research on ohmic heating has a much shorter history than has conventional heating, food scientists and technologists might look to microwave heating for information on nutrient changes. In general, many improvements in nutritional quality were found using microwaves (cooking in a minimum of water retained more K, vitamin B12, and vitamin C, and the absence of surface browning retained more amino acid availability, especially lysine), and microwave heating induces no significant effects different to those induced by conventional heating.

The benefit of attaining food safety with less nutrient degradation using HTST processes such as ohmic heating or microwave heating is based on differences in the kinetic parameters (k, z, Ea) for bacterial spores compared to those for biochemical reactions. First, rate constants for microbial destruction are usually much larger than those for the chemical reactions responsible for nutrient degradation, and second, rate constants for microbial destruction are usually more sensitive to temperature increases (z(thiamin)=48, z(peroxidase)=36.1, and z(C. botulinum)=10 °C). Methods for rapidly reaching the target temperatures, therefore, tend to destroy microorganisms while giving less time to compromise the nutrient content and other quality attributes. In fact, the slow heating rate associated with conventional retorting can activate protease to degrade myofibrillar proteins before the protease is eventually heat-inactivated. Tests for conventional heating showed that heating large (25 mm) particulates in a liquid medium at 135 °C to achieve Fo=5 at the particulate center required extensive over processing of the liquid phase (Fo = 150 for the liquid).

For this reason, the common process conditions for scraped surface heat exchangers are maximum particulate sizes of 15 mm and sterilization temperatures of 125–130 °C (producing liquid Fo = 25) while limiting particulates to 30–40 % so that there is enough hot liquid available to heat particulates. For ohmic heating, direct heating sterilization temperatures can reach 140 °C (the temperature limit of plastics in the machinery) without grossly overheating the liquid phase and can support greater particulate loading suspended in highly viscous carrier liquids. For comparative purposes, conventional heating at 130 °C to produce a lethality of Fo = 8 produced a cook value Co (based on thiamin degradation) of Co = 8, whereas ohmic heating at 140 °C produced Fo=24 and Co=4.

Vitamin losses in foods are determined by the temperature and the moisture of the applied heating method. Vitamin C is particularly temperature sensitive and destroyed at relatively low temperatures, so heating foods must be for as short a time as possible to retain the vitamin C. Thiamin and riboflavin are unstable at higher temperatures such as those used in rapid grilling. Vitamin C is also water soluble and can be lost when cooking with moist heat or by auto-oxidation with dissolved oxygen in the food or cooking water. This reaction is catalyzed by adventitious iron and copper ions. By comparison, thiamin is the most water-soluble vitamin, and vitamin A and vitamin D are water insoluble. Unfortunately, studies on food nutrients affected by ohmic heating are barely discussed bare in the literature.

Ohmic heating is an effective method to pasteurize milk (220 V, 15 kW AC, C electrodes–70 C for 15 seconds) and has been used successfully to produce quality viscous products and to foods containing various combinations of particulates such as meat, vegetables, pastas, or fruits in a viscous medium, including a wide variety of high acid (ratatouille, pasta sauce and vegetables, vegetables Provençale, fruit compote, strawberries, applesauce, sliced kiwi fruit) and low acid (tortellini in tomato sauce, cappelletti in basil sauce, tagliatelle a la crème, beef bourguignonne, Beijing lamb, beef and vegetable stew, lamb Wala Gosht, vegetable curry, minestrone soup concentrate) food products.

10.2.1.2. Diffusion

Studies have shown that compared with conventional heating, ohmic heating enhanced diffusion of charged species between solid particles and the surrounding liquid, which could have some impact on loss of nutrients from solid particles to carrier liquid. This becomes undesirable only if the carrier liquid is not to be consumed together with the solid particles.

Ohmic heating is superior over conventional heating in the case of blanching of plant tissues such as vegetables. The loss of soluble solids in water blanching of vegetables affects both the quality and nutrition of the products. In addition, blanching water containing a large amount of soluble solids cannot be discharged without proper treatment. Mizrahi compared hot water blanching and ohmic heating blanching (1996). Hot water blanching was carried out by placing sliced or diced beet into boiling water and taking water samples every 30 seconds; blanching by ohmic heating was done by immersing whole, sliced or diced beets in an aqueous salt solution and passing an AC voltage through the medium. Betaine and betalamic acid concentration in the samples were determined. Solute leaching with both methods followed a similar pattern, and was proportional to the surface to volume ratio and the square root of the process time. By removing the need for dicing and shortening the process time, ohmic heating blanching considerably reduced by one order of magnitude the loss of solutes during blanching of vegetables.

10.3. PASTEURIZATION

This is a heating regime (generally below 105 °C) that primarily aims to achieve commercial sterility by virtue of additional factors that contribute towards preserving the food. The actual degree of heat process required for an effective pasteurization will vary depending on the nature of the food and the types and numbers of microorganisms present. In certain cases, an extended pasteurization treatment may be required to inactivate heat-resistant enzymes. Milk is the most widely consumed pasteurized food in the UK, and the process was first introduced commercially in the UK during the 1930s, when a treatment of 63 °C for 30 min was used. Modern milk pasteurization uses an equivalent process of 72 °C for 15 s. Pasteurization is used extensively in the production of many different types of food, including fruit products, pickled vegetables, jams and chilled ready meals.

Food may be pasteurized in a sealed container (analogous to a canned food) or in a continuous process (analogous to an aseptic filling operation). It is important to note that pasteurized foods are not sterile and will usually rely on other preservative mechanisms to ensure their extended stability for the desired length of time. Chilled temperatures are often used, but some foods have a sufficiently high salt, sugar or acid content to render them stable at room temperature (e.g., pasteurized canned ham). Whereas heat-preserved packaged foods for ambient storage have shelf lives normally measured in years, cook-chill foods typically have shelf-lives measured in weeks or months, and in those cases where only chilling is the preservation factor, they are significantly less than this.

Pasteurization is one of the most important steps in preservation and is essential for food safety. It greatly improves the product's storage quality by effectively destroying virtually all disease-producing and most other bacteria. Pasteurization is a process of heat treatment to inactivate enzymes and kill relatively heat-sensitive microorganisms that cause spoilage, with minimal changes in food properties, for example, sensory and nutritional properties. The primary objective of pasteurization is to free the food of any microorganism that might cause deterioration or endanger the health of the consumer. The severity of the heat treatment and the resulting extension of shelf life are determined mostly by the pH of the food. In low-acid foods (pH 4.5), the main purpose is destruction of pathogenic bacteria, whereas below pH 4.5 destruction of spoilage microorganism or enzyme inactivation is usually more important. Pasteurization does not aim at killing the spore-bearing organisms, such as thermophilic Bacillus subtilis, but these organisms and most other spore-bearing bacteria cannot grow in acidic fruit juices and consequently their presence is of no practical significance. Pasteurization of carbonated juices needs only to be conducted at such a temperature and for such a time that yeasts and molds are destroyed.

Yeast is killed by heating at 60 °C to 65 °C and the resistant mold spores, negative in most cases, are heated up to 80 °C for 20 min. But molds require oxygen for growth, and for this reason, heavily carbonated juice can be pasteurized safely at 65 °C, which destroys yeast cells. Most still (noncarbonated) juices must be pasteurized at 80 °C. Juices of high acidity may be pasteurized at lower temperatures of 60 °C to 65 °C. Processing containers of food that have a naturally low pH (e.g., fruit pieces) or in which the pH is artificially lowered (e.g., pickles) is similar to canning. Yeasts, molds, and bacteria (both thermophilic and mesophilic) thrive in acidic products such as tomatoes, mangoes, and bananas because of their pH which usually range between 4.0–4.4.

The main risk of spoilage is from spore-forming species other than Clostridium botulinum, especially B. coagulans among the aerobes, and C. pasteurianum and C. thermosaccharolyticum among the anaerobes. In high-acid foods (pH 3.9) such as pineapple juice, the spoilage is generally caused by non-spore-forming bacteria (Lactobacillus and Leuconostoc), yeast, or molds. Fruits with a pH lower than 4.5 contain enzyme systems such as catalase, peroxidase, polyphenol oxidase, pectinesterase, etc. in addition to spoilage organisms. Unless inactivated, these enzymes are likely to cause undesirable changes in the canned products. Some of these enzymes, particularly peroxidases, have higher heat resistance than the spoilage organisms and have been used in evaluating the thermal processing of canned fruits (Figure 10.1).

Figure 10.1: Pasteurization: Milk bottles on conveyor line, (*Source: http://www.milkpasteurizermachine.com/gently-pasteurized-milk*).

10.3.1. EFFECTS ON FOODS

10.3.1.1. Color, Flavor, and Aroma

In fruit juices, the main cause of color deterioration is enzymic browning by polyphenol oxidase. This is promoted by the presence of oxygen, and fruit juices are therefore routinely deaerated prior to pasteurization. The difference between the whiteness of raw milk and that of pasteurized milk is due to homogenization, and pasteurization alone has no measurable effect. Other pigments in plant and animal products are also mostly unaffected by pasteurization. A small loss of volatile aroma compounds during pasteurization of juices causes a reduction in quality and may also unmask other 'cooked' flavors. Volatile recovery may be used to produce high-quality juices but this is not routinely used, due to the high cost. Loss of volatiles from raw milk removes a hay-like aroma and produces a blander product.

10.3.1.2. Vitamin Loss

In fruit juices, losses of vitamin C and carotene are minimized by deaeration. Changes to milk are confined to a 5% loss of serum proteins and small changes to the vitamin content.

10.3.1.3. Sterilization

It means the destruction of all viable organisms as may be measured by an appropriate plating or enumerating technique. Canned foods are sometimes called "commercially sterile" to indicate that no viable organisms can be detected by the usual cultural methods employed or that the number of survivors is so low as to be of no significance under the conditions of canning and storage. Also, microorganisms may be present in canned foods that cannot grow in the product by reason of undesirable pH, oxidation-reduction potential (Eh), or temperature of storage.

10.4. FREEZING

Refrigeration (cooling) is the total process of reducing the temperature of a food and maintaining that temperature during storage, transport, and retailing. If the temperature of the food does not fall below one where ice is formed in the food, the food is considered chilled and the temperature reduction process is called chilling. Freezing is a temperature reduction process that occurs when ice is formed and the food is considered frozen. Freezing provides a significant extended shelf life and has been successfully employed for long-term preservation of many foods. Freezing is still one of the most widely used methods of food preservation even though several new technologies, such as high pressure, infrared irradiation, pulsed electric field, and ultrasound, are gaining importance. Freezing processes are continuing to emerge. Freezing changes the physical state of a substance by changing water into ice when energy is removed in the form of cooling below freezing temperature. Usually, the temperature is further reduced to storage level (e.g., 18 °C).

Storage life of frozen foods tends to be dictated more by consumer handling than the effectiveness of the freezing processes. Typical domestic freezers operate at temperatures much higher than those used for production and distribution. Repeated freeze-thaw cycles damage the food structure around the edges and promote chemical and physical breakdown. Ice cream is an example of a high-quality food that is manufactured to contain small (invisible) ice crystals within a complex matrix. However, the ice crystals grow in size as the food is abused, with the result that the smooth structure gradually breaks down and is replaced with a harder, coarse-textured food with visible ice crystals (Figure 10.2).

Figure 10.2: Storing foods in freezer bags, (*Source: https://www.ourprovidenthomestead.com/2017/08/food-preservation-series-1-freezing-food/*).

10.4.1. Effects of Freezing on the Quality of Food

10.4.1.1. Microbiology and Food Safety

Temperature is a major factor affecting microbial growth. Microbial growth is described in terms of the lag phase and the generation time. When a microorganism is introduced to a particular environment, there is a time (the lag phase) in which no increase in microbial numbers is apparent, followed by a period when microbial growth occurs exponentially. The generation time is a measure of the time it takes for the population. Microorganisms have an optimum growth temperature at which a particular strain grows most rapidly, i.e. the lag phase and generation times are both at their shortest time. They also have a maximum growth temperature above which growth no longer occurs. Above this temperature, one or more of the enzymes essential for growth are inactivated and the cell is considered to be heat-injured. However, in general, unless the temperature is raised to a point substantially above the maximum growth temperature, the injury is not lethal and growth will recommence as the temperature is reduced. Attaining temperatures substantially above the maximum growth temperature is therefore critical during cooking and reheating operations. Of most concern during storage, distribution, and retail display of food is a third temperature threshold, the minimum growth temperature for a microorganism. As the temperature of an organism is reduced below that for optimum growth, the lag phase and generation time both increase. The minimum growth temperature can be considered to be the highest temperature at which either of the growth criteria, i.e., lag phase and generation time, becomes infinitely long. The minimum growth temperature is not only a function of the particular organism but also the type of food or growth media that is used for the incubation. Although some pathogens can grow at 0 °C or even slightly lower, from a practical point of view the risks to food safety are considerably reduced if food is maintained below 5 °C. However, food spoilage may still be an issue. Providing that the food does not rise above –12 °C during frozen storage and display, there should be no issues of food safety with frozen storage and display.

10.4.1.2. Chemical Changes during Freezing

Fresh fruits and vegetables, when harvested, continue to undergo chemical changes which can cause spoilage and deterioration of the product. This is why these products should be frozen as soon after harvest as possible and at their peak degree of ripeness. Fresh produce contains chemical compounds called enzymes which cause the loss of color, loss of nutrients, flavor changes, and color changes in frozen fruits and vegetables. These enzymes must be inactivated to prevent such reactions from taking place.

Enzymes in vegetables are inactivated by the blanching process. Blanching is the exposure of the vegetables to boiling water or steam for a brief period of time. The vegetable must then be rapidly cooled in ice water to prevent it from cooking. Contrary to statements in some publications on home freezing, in most cases, blanching is absolutely essential for producing quality frozen vegetables. Blanching also helps to destroy microorganisms on the surface of the vegetable and to make some vegetables, such as broccoli and spinach, more compact.

The major problem associated with enzymes in fruits is the development of brown colors and loss of vitamin C. Because fruits are usually served raw, they are not blanched like vegetables. Instead, enzymes in frozen fruit are controlled by using chemical compounds which interfere with deteriorative chemical reactions. The most common control chemical is ascorbic acid (vitamin C). Ascorbic acid may be used in its pure form or in commercial mixtures with sugars.

Some directions for freezing fruits also include temporary measures to control enzyme-activated browning. Such temporary measures include soaking the fruit in dilute vinegar solutions or coating the fruit with sugar and lemon juice. However, these latter methods do not prevent browning as effectively as treatment with ascorbic acid.

10.4.1.3. Textural Changes during Freezing

Water makes up over 90% of the weight of most fruits and vegetables. This water and other chemical substances are held within the fairly rigid cell walls which give support structure, and texture to the fruit or vegetable. Freezing fruits and vegetables actually consists of freezing the water contained in the plant cells.

When the water freezes, it expands and the ice crystals cause the cell walls to rupture. Consequently, the texture of the produce, when thawed, will be much softer than it was when raw. This textural difference is especially noticeable in products which are usually consumed raw. For example, when a frozen tomato is thawed, it becomes mushy and watery. This explains why celery and lettuce are not usually frozen and is the reason for the suggestion that frozen fruits, usually consumed raw, be served before they have completely thawed. In the partially thawed state, the effect of freezing on the fruit tissue is less noticeable. Other textural defects that may occur during improper storage conditions include loss of juiciness in meats, loss of fluidity in egg yolk, dryness of fish muscle that makes it crumbly, and aggregation of starches in sauces. Textural changes due to freezing are not as obvious in products which are cooked before eating because cooking also softens cell walls. These changes are also less noticeable in high starch vegetables, such as peas, corn, and lima beans.

10.4.1.4. Weight Loss

Some foods exhibit particular quality advantages as a result of rapid cooling. In meat, the pH starts to fall immediately after slaughter and protein denaturation begins. The result of this denaturation is a pink proteinaceous fluid, commonly called "drip," often seen in prepackaged cuts of meat. The rate of denaturation is directly related to temperature and it, therefore, follows that the faster the chilling rate, the less the drip. Investigations using pork and beef muscles have shown that faster rates of chilling (e.g., reducing the deep leg temperature of a pig in 10 hours to 5 °C compared with 14 °C) can halve the amount of drip loss (Taylor, 1972). Another quality and economic advantage of temperature control is a reduction in weight loss, which results in a higher yield of saleable material. Meats, fruits, and vegetables, for example, have high water contents and the rate of evaporation depends directly on the vapor pressure at the surface. Vapor pressure increases with temperature and thus any reduction in the surface temperature will reduce the rate of evaporation. The use of very rapid chilling systems (3–4 h compared with overnight) for pork carcasses has been shown to reduce weight loss by at least 1% when compared with conventional systems (James et al., 1983).

10.4.2. Principles of Refrigeration Systems

A detailed analysis of refrigeration cycles and systems may be found in numerous refrigeration textbooks. Nevertheless, a basic understanding of the principles of how a refrigeration system works is necessary for all users of refrigeration equipment. There are two main types of refrigeration system: total loss refrigeration systems, and mechanical refrigeration systems. The rate of heat removed from the food ($Q_{product}$, J/s) accounts for the majority of the refrigeration load and can be determined from:

$$Q_{product} = \frac{mc_p(\theta_{initial} - \theta_{final})}{\Delta t}$$ Where,

m = total mass, kg
c_p = specific heat capacity, J/kg/C
$\theta_{initial}$ = initial temperature of the food, C
θ_{final} = final temperature of food, °C, and Δt is the time taken to remove the heat

10.5. CHILLING AND COOLING

Chilling may be referred to as the process that lowers the food temperature to a safe storage temperature of between 0 and 5 °C, whereas cooling is a more general term applied to the lowering of a food temperature. Chilled foods can potentially present a greater risk to public safety than frozen foods. Keeping products at a low temperature reduces the rate of microbiological and chemical deterioration of the food. In most processed chilled foods, it is the microbial growth that limits the shelf life; even the slow growth rates that occur under chilled conditions will eventually result in microbial levels that can affect the food or present a potential hazard. This microbial growth can result in the spoilage of the food (it may go putrid or cloudy or show the effects of fermentation), but pathogens, if present, may have the potential to grow and may show no noticeable signs of change in the food. Under the UK's Food Protection Act of 1990, a use by date must be declared on labels for packaged chilled dairy, meat, egg, seafood and poultry products which must be distributed, stored and retailed at 5 °C or below (Figure 10.3).

Figure 10.3: Keeping foods for chilling in the refrigerator, (*Source: https://greatist.com/eat/ultimate-way-organize-your-fridge*).

10.6. THE BASIC DYNAMIC MODEL

Consider the following dynamical model of a simple microbial growth process:

$$S \rightarrow X \tag{10.1}$$

In a completely mixed fed-batch reactor:

$$\frac{dX}{dt} = \mu X - DX \tag{10.2}$$

$$\frac{dS}{dt} = k_1 \mu X + D(S_{in} - S) \tag{10.3}$$

$$\frac{dV}{dt} = DV \tag{10.4}$$

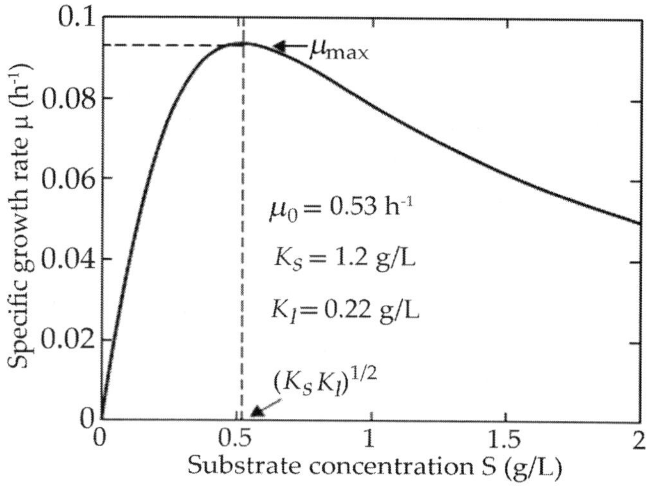

Figure 10.4: Haldane Model, (*Source: G. Caldwell Darwin (2013), Robotics and Automation in the Food Industry: Current and Future Technologies, Woodhead Publishing*).

where X (g/L) and S (g/L) are the biomass and substrate focuses, separately. μ (h −1) is the particular development rate, D (h −1) is the weakening rate, S in (g/L) signifies the grouping of the substrate in the encourage, k 1 is a yield coefficient, and V (L) is the volume of fluid medium in the tank (Figure 10.4).

An average circumstance where bioprocesses are worked in nourished clump is when substrate restraint is conceivably present. The Haldane dynamic model is then regularly used to describe such hindrance wonder. This model (see Figure 10.1) is given by:

$$\mu = \frac{\mu_0 S}{K_S + S + (S^2 / K_I)} \tag{10.5}$$

where μ_0 is a parameter related to the maximum value of the specific growth rate as follows: $\mu_0 = \mu_{max} =(1 + 2(K_S K_I))$. The coefficients K_S and K_I denote the saturation constant and the inhibition constant, respectively.

10.7. MODELING ISSUES: POPULATION BALANCE MODELING IN FOOD PROCESSES

One of the key issues by and by in process control is to suitably address the way that generation quality isn't only a worldwide property of the mass delivered inside the procedure yet additionally of the circulation of various factors, for example, size, mass or age. In crystallization forms, clearly, the primary target is to deliver precious stones of a predefined estimate. As said in the presentation, crystallization assumes a noteworthy part in the food business. For instance, the EC venture CAFÉ bargains specifically with the control of frozen yogurt crystallization. A populace adjust show has been considered to accentuate the procedure flow. Such a model has been distinguished and the model properties have been examined, with a specific end goal to create and actualize a fitting controller of the procedure. A similar concern applies additionally to the creation of added substances, for example, against froth particles for washing powders, where the extent of added substances should be like the powder to guarantee that the blend is adequately homogeneous. When thinking about living life forms, obviously the age of the phones assumes an imperative part in the proficiency of the procedure, and along these lines, the learning of the age dissemination (ordinarily by means of mass conveyance estimations) is a key issue to enhancing the effectiveness of natural procedures.

The common method to portray the particulate measurement of mechanical procedures is to consider populace adjust. There has been as of late a lot of research exercises concerning the displaying and investigation of the properties of the cell populace adjust show. Such models depict the elements of cell development and consider the natural actuality that the cell properties (e.g., mass or age) are conveyed among the cells of a populace. It comprises of a non-direct incomplete integrodifferential condition with a non-straight limit condition combined with a conventional integrodifferential condition. Another trouble with such a model is because of the inborn physiological capacities, to be specific the development rate work, the cell division likelihood thickness work, and the apportioning likelihood thickness work, whose determination may give off an impression of being an intricate undertaking in numerous examples.

For instance, let us consider a cell populace developing in a consistent blended tank reactor. The cells are recognizable from each other regarding their mass or some other property of the cell that complies with the protection law. Let N (m, t) be the quantity of cells that have a mass amongst m and m + d m at time t. The cells are considered to develop at a rate r (m, S) that relies upon their mass and on the centralization of the restricting substrate S. We additionally accept that the estimation of the mass is institutionalized and that m ϵ [0,1]. The cell division and the birth procedures of the cell populace are depicted by the division rate Γ(m, S) characterized as takes after:

$$\aleph(m,S) \quad \frac{f(m)}{1-\int_0^m f(m')dm'} r(m,S) \quad \gamma(m)r(m,S)$$

(10.6)

where *f (m)* is the division likelihood thickness work, which is expected to depend just on the cell mass, and is taken to be a left-hand side truncated Gaussian conveyance with mean of μf and standard deviation of σf.

We additionally accept that the likelihood *p* that a mother cell of mass will bring forth a little girl cell of mass *m* is autonomous of substrate focus. This capacity should additionally fulfill the accompanying standardization conditions:

$$\int_0^{m'} p(m,m')dm = 1$$

and furthermore, ought to likewise be to such an extent that biomass is saved at cell division, that is:

$$p(m,m') = p(m'-m,m)$$

We at long last accept that the likelihood work is asymmetrical Beta appropriation with a parameter of q characterized by the accompanying condition:

$$p(m,m') = \frac{1}{B(q,q)}\frac{1}{m'}\left(\frac{m}{m'}\right)^{q-1}\left(1-\frac{m}{m'}\right)^{q-1} \tag{10.7}$$

We accept additionally that no cell demise happens and that cells develop in a persistent propagator from which they exit with a weakening rate of D. Under these suspicions, the cell populace progression are portrayed by the accompanying integrodifferential conditions, see case:

$$\frac{\partial N(m,t)}{\partial t} + \frac{\partial}{\partial m}[r(m,S)N(m,t)] + \Gamma(m,S)N(m,t)$$

$$= -DN(m,t) + 2\int_m^1 \Gamma(m',S)p(m,m')N(m',t)dm' \tag{10.8}$$

Subject to the underlying condition:

$$N(m,0) = N_0 \tag{10.9}$$

Framework [10.8]–[10.9] is finished by the accompanying limit conditions

$$r(1,S)N(1,t) = r(o,t)N(0,t) = 0 \tag{10.10}$$

The cell populace Equation [10.8] comprises of five terms: the aggregation term, the development terms, the division term, the weakening term and the birth term.

The conduct of the cell populace relies upon the substrate fixation. The mass adjust condition for the substrate communicates specifically that the substrate utilization is corresponding to the aggregate biomass generation r (m, S) N(m, t)dm. The substrate focus mass adjust at that point peruses as takes after:

$$\frac{dS}{dt} = D(S_n - S) - k_1\int_0^1 r(m,S)N(m,t)dm \tag{10.11}$$

The principle highlight of the above populace adjust show when contrasted with the fundamental mass adjust demonstrate [10.2]–[10.3] is that the model now incorporates a halfway differential condition and basic terms in both mass adjust conditions. Such a class of incomplete differential conditions has not so far been widely considered in the control group. Because of the mind-boggling structure of the dynamical model, the current outcomes on the examination of the framework, including the investigation of the framework arrangement, the investigation of the presence, variety, and security of balance profile or on the control of this framework, have generally been acquired numerically. However, in reference 14, the issue of breaking down and settling of the unfaltering condition of the phone mass appropriation in a consistent bioreactor is examined in an unending dimensional framework structure.

The advancement of control laws for a phone populace adjust show has turned into a dynamic region of research amid the previous decades. A few non-straight controllers or non-direct versatile controllers have been proposed for this class of frameworks, for instance, to guarantee the control of the procedure efficiency (for a bioprocess with two cell cycle organize 13), or to control the diverse minutes (zeroth minute (illustrative of the cell thickness), first minute (illustrative of the biomass fixation) and the second minute) by utilizing a non-direct linearizing controller.

Give us a chance to consider parameter estimation by means of the utilization of the spectator based estimator and stress the specificities of its tuning for cluster and nourished group forms. Consider that the procedure elements are portrayed by the accompanying equations:

$$\frac{dx}{dt} = F_1(x)\theta + F_2(x) \tag{10.12}$$

where x is the state vector (dim(x) = n), θ is the vector of (obscure) parameters (diminish(θ) = p), and F 1 (x) and F 2 (x) are (network), for the most part non-direct, elements of the state vector x . A run of the mill case is the general dynamical model:

$$\frac{dx}{dt} = Kr(x) + F - Q - Dx \tag{10.13}$$

where x, K, r (x), F, Q hold for the segment focus vector, the yield coefficient network, the response rate vector, the feedrate vector, and the vaporous surge rate vector, separately. In the present setting, the above model can be reworked as takes after:

$$\frac{dx}{dt} = KG(x)\theta + F - Q - Dx \tag{10.14}$$

The response rate r (x) is currently reworked as the result of a component of the express (the askew lattice G (x)) and the obscure parameter θ. In the fundamental dynamical model case [10.2]–[10.3], a common decision for G (x) and θ could be:

$$G(x) = \frac{SX}{K_S + S + (S^2 / K_1)}, \quad \theta = \mu_0$$

Expect that:

- H1: the p parameters θ are obscure and perhaps time-differing (with limited time varieties dθ d < M;
- H2: p state factors are accessible for on-line estimation

From the suspicion H2, we can characterize a state parcel:

$$x = \begin{bmatrix} x_1 \\ x_2 \end{bmatrix} \tag{10.15}$$

With x_1 the deliberate factors, and x_2 the unmeasured ones. The dynamical equations would then be able to revise as follows:

$$\frac{dx_1}{dt} = F_{11}(x)\theta + F_{21}(x) \tag{10.16}$$

$$\frac{dx_2}{dt} = F_{12}(x)\theta + F_{22}(x) \tag{10.17}$$

Further assume that:

H3. F_{11} can be made as the outcome out of two p × p systems: F_{11} = F_3 F_4 (x) with F_4 a to one side structure (F_4 = diag{f_4, i}, I = 1 best) and F_3 being full rank for each and every satisfactory estimation of x (for (bio) chemical shapes for instance, simply positive estimations of the state factors (i.e. obsessions and maybe temperature) are seen as); H4. F_{11} (x) and F_{21} (x) are known components of x.

Under suspicion H3, it might be conceivable to construct an asymptotic eyewitness 15 to remake the time development of x_2 freely of the obscure parameters θ. In the accompanying, we expect that states incorporated into x_2 are either open for on-line estimation or accessible by means of such spectators, or that the flow of x_1 are autonomous of x_2.

With regards to mixed tank reactors (STR), $F_3 = K_1$, F_4 = G (x) with K 1 the yield coefficient framework related to x1. Note that by developing and expecting H2, the lattice G (x) is a p × p framework. Suspicion H4 at that point implies that the feedrates F1 (related to x1), the vaporous surge rates Q1 (related to x_1), the influent stream rate q and the volume V are known (by means of on-line estimations or client's decision), and in addition the yield coefficients in K_1 and the capacity G (x). The outline of the spectator construct estimator is based with respect to Equation [10.16] and takes after surely the line of thinking for the plan of Luenberger eyewitnesses. This gives the accompanying estimator conditions:

$$\frac{d\hat{x}_1}{dt} = F_{11}(x)\hat{\theta} + F_{21}(x) - \Omega(x_1 - \hat{x}_1) \tag{10.18}$$

$$\frac{d\hat{\theta}_1}{dt} = [F_{11}(x)^T]\Gamma(x_1 - \hat{x}_1) \tag{10.19}$$

The essential inspiration of the above structure for the estimator is the accompanying. With respect to a traditional spectator, the estimator conditions are the mix of the procedure demonstrate (F_{11} (x) $\hat{\theta}$ + F_{21} (x)) and remedy terms ($-\omega(x_1 - \hat{x}_1)$) and ([F_{11} (x)]T $\Gamma(x_1 - \hat{x}_1)$) on the deliberate factors. In the above eyewitness based estimator, the parameters θ are acclimatized as states without progression. The weighting factor [F_{11} (x)] T in Equation [10.19] is without a doubt the term increasing the obscure parameter in the model condition: its presentation in the estimator condition gets from established estimator outline and it is generally called the regressor. A critical contrast as for the broadened Luenberger or Kalman spectators is that the deliberate factors don't show up as appraisals in the onlooker conditions yet with their 'actual' (estimated) values.

The hypothetical strength investigation of the above eyewitness based estimator is accessible in reference 15: the principle prerequisites are the negative definiteness of Ω T Γ + Γω and the constancy of excitation of F_{11} (x). Be that as it may, its tuning might be troublesome and unpredictable practically speaking, due to the nearby association of the obscure parameters in the estimator conditions, and in light of the fact that its progression relies upon the procedure factors. The last might be of minor significance if the framework is worked around a relentless state, yet it will end up essential if the framework covers a substantial scope of working conditions (concerning nourished clump and group reactors, or process new companies and grade changes) perhaps with extensive varieties of the state factors, and as a result, the network F11(x). Great following abilities of the parameters varieties are especially fundamental in these conditions. Nonetheless, in the above shape Equations [10.18] and [10.19] of the onlooker based estimator, tuning may give extremely traditionalist qualities for the plan parameters Γ and Ω, and result in terrible following execution in a portion of the working locales.

The above goals for the tuning of the onlooker based estimator can be accomplished by considering first the accompanying two stages in the reformulation of the calculation:

1. A state change;
2. The adjustment of the state vector

Think about the accompanying state change:

$$z = F_3^{-1} x \tag{10.20}$$

At that point the dynamic conditions of the framework can be changed as takes after:

$$\frac{dz}{dt} = F_4(x)\theta + F_3^{-1} F_{21}(x) \tag{10.21}$$

Because of the above change, just a single state variable is related with every obscure parameter θ. In the particular instance of STR, the invertibility of F_3 (= K_1) comes about because of the freedom of the p responses and of the p estimated factors, as of now specified previously. The eyewitness based estimator would now be able to be updated based on Equation [10.21]:

$$\frac{dz}{dt} = F_4(x)\theta + F_3^{-1} F_{21}(x) \tag{10.22}$$

$$\frac{d\theta}{dt} = \Gamma(z - \hat{z}) \tag{10.23}$$

Because of the change Equation [10.20], the onlooker based estimator is presently reformulated in a decoupled arrange for the obscure parameters θ_i (I = 1 to p). In light of the decoupled estimation plan, a conspicuous decision for the networks Ω and Γ are corner to corner grids:

$$\Omega = diag\{-\omega_i\}, \Gamma = diag\{-\gamma_i\}, \omega_i > 0, \gamma_i > 0,\ i = 1\ to\ p \tag{10.24}$$

In the above detailing of the estimation plot, we have expelled the regressor term F_4 (x) (G(x) in the STR case) from the estimation condition of θ (Equation [10.23]); since one of its principal parts is to expressly exchange the coupling between the obscure parameters and the deliberate factors, its quality is not any more basic.

The last advance before the detailing of the tuning principle comprises of a revision of the estimator's conditions. Give us a chance to assemble every factor z_i with its related parameter θ_i and adjust the passages of the vector [z, θ] T in the accompanying request in a vector ζ :

$$\varsigma = \begin{bmatrix} z_1 \\ \theta_1 \\ z_2 \\ \theta_2 \\ \cdots \\ z_p \\ \theta_p \end{bmatrix} \tag{10.25}$$

Give us initial a chance to characterize the estimation mistake:

$$e = \varsigma - \hat{\varsigma} \tag{10.26}$$

The estimation mistake elements are promptly gotten from Equations (10.21 – 10.23):

$$\frac{de}{dt} = Ae + b \tag{10.27}$$

with square corner to corner grid A with 2 × 2 blocks:

$$A = diag\{A_i\}, A_i = \begin{bmatrix} -\omega_i & f_{4,i}(x) \\ -\gamma_i & 0 \end{bmatrix}, i = 1 \ to \ p \tag{10.28}$$

What's more, b equivalent to:

$$b = \begin{bmatrix} 0 \\ \frac{d\theta_1}{dt} \\ 0 \\ \frac{d\theta_2}{dt} \\ \cdots \\ 0 \\ \frac{d\theta_p}{dt} \end{bmatrix} \quad [10.29]$$

The trademark equation of the framework A, $\det(\lambda I - A)$, is equivalent to:

$$\det(\lambda I - A) = \prod_{i\_1}^{p} (\lambda^2 + \omega_i \lambda + \gamma_i f_{4,i}(x)) \tag{10.30}$$

The key thought of the tuning principle comprises of picking every y_i conversely relative to the comparing term f_4, i (x):

$$\gamma_i = \frac{\gamma_i}{f_{4,i}(x)}, \overline{\gamma}_i > 0, i = 1\, to\, p \tag{10.31}$$

With the decision above, the trademark Equation [10.30] is changed as follows:

$$\det(\lambda I - A) = \prod_{i=1}^{p} (\lambda^2 + \omega_i \lambda + \overline{\gamma}_i) \tag{10.32}$$

What's more, the onlooker based estimator flow are currently autonomous of the state factors. Such a decision compares to a Lyapunov change. It is clearly legitimate for estimations of f_4, i (x) = 0: this condition is typically met effectively in (bio)process applications, as will be delineated in the accompanying area.

The estimations of the outline parameters would then be able to be set to subjectively settle the estimator's progression for every obscure parameter θ_i. Since the estimator lessens by means of the changes to an arrangement of free second-arrange direct frameworks, the traditional standards for allotting the elements of second-arrange direct frameworks apply clearly here. The peruser is in this manner alluded to established programmed control course books for additional data regarding the matter. However, the accompanying fundamental rules are recommended.

One critical rule is to pick genuine shafts:

$$\omega_i^2 - 4\overline{\gamma}_i \geq 0 \tag{10.33}$$

The goal is then to abstain from initiating motions in the estimation of the parameters that don't compare to any physical marvel identified with the assessed response rates.

Pomerleau and Perrier 17 propose picking twofold shafts, that is:

$$\overline{\gamma}_i = \frac{\omega_i^2}{4} \tag{10.34}$$

Oliveira et al. propose as another option to pick complex shafts with a damping factor equivalent to 0.7 to expand the speed of meeting of the estimator with a diminished overshoot. (As a rule, the damping element can be unreservedly picked; the decision may then rely upon the sort and nature of the application, of the time varieties of the parameter to be evaluated, and of the commotion on the deliberate information. This implies at that point there are two plan parameters for each assessed parameters).

At that point, the tuning of the estimation calculation lessens to the decision of one plan parameter, wi, per assessed parameter. This permits having an outline technique that has the twofold favorable position of being basic (one plan parameter) and adaptable (every parameter estimation can be tuned contrastingly if necessary, for instance, if the time varieties of the parameters are extraordinary).

As an issue of outline, let us demonstrate genuine aftereffects of the eyewitness construct estimator actualized in light of a dough puncher's yeast generation process.

$$\mathrm{Re}\,spirative\ growth\ on\ glu\cos e : S + C \rightarrow X + P \tag{10.35}$$

$$Fermentative\ growth\ on\ glu\cos e : S \rightarrow X + E + P \tag{10.36}$$

$$\mathrm{Re}\,spirative\ growth\ on\ ethanol : E + C \rightarrow X + P \tag{10.37}$$

where S, C, X, P and E hold for glucose, oxygen, yeast, CO_2, and ethanol, individually. The eyewitness base has been planned by two sub-models (one of every an oxygen reductive administration with initial two development responses, and one out of a respirative administration with the first and the third responses) and the online estimations of broke up oxygen and CO_2. A few outcomes are given in Figure 10.2. Figure 10.2a gives the estimation of the yeast focus by means of an asymptotic spectator. The three after boards (Figure 10.2b, 10.2c, and 10.2d) give the estimation comes about for the particular development of the three development responses, to be specific μ_o, μ_r, and μ_e. The approval of the estimations of the assessed particular development rates has been performed by contrasting the deliberate estimation of the respiratory remainder (RQ) with the esteem registered from the estimation of μ_o, μ_r and μ_e (Figure 10.2e).

10.8. DESIGN OF PID CONTROLLERS FOR FED-BATCH PROCESSES

The outline and adjustment of established controllers, for example, PID are commonly in view of a direct time-invariant model of the framework. However, most procedures are described by a non-direct dynamical conduct. The basic presumption is that the non-direct progression can be genuinely all-around approximated by a straight model. On the off chance that this can regularly generally be supported on account of control (i.e. at the point when the goal is to keep up the procedure sufficiently close to some characterized working conditions) when a linearized show around the procedure unfaltering state can give a decent estimate progression, it is in all probability that this won't be the situation when the procedure is worked in non-consistent state conditions as in clump and sustained group working modes. The goal of this area is to delineate with a straightforward sustained group case how to deal with this particular part of the procedure flow (Figure 10.5).

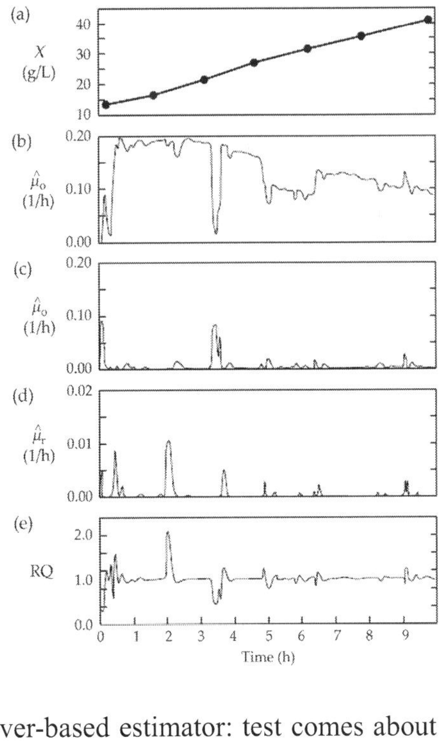

Figure 10.5: Observer-based estimator: test comes about on a baker's cook's yeast creation process: For clarification of boards (a) to (e), (*Source: G. Caldwell Darwin (2013), Robotics and Automation in the Food Industry: Current and Future Technologies, Woodhead Publishing*).

Give us a chance to begin by considering a bolstered bunch process whose progression are portrayed by the mass adjust display Equations [10.2]–[10.4] with Haldane energy Equation (5).

On the off chance that the goal is to expand the biomass generation, instinctively it appears to be sensible to trust that this target would become to if the substrate fixation S is kept at an esteem that amplifies the particular development rate, that is, S KK S I K.

This is to be sure what can concluded by deciding the ideal control input by means of the utilization of the Pontryagin Maximum Principle. Give us a chance to consider two cases relying upon the control input. To start with let us expect that the gulf substrate focus S_{in} is the control input and that the delta stream rate q (= DV) is settled (and e.g., consistent) (this suggests the sustaining time frame is settled). This decision is principally spurred by the straightforwardness of the ideal control calculation.

The ideal control issue comprises of finding the sustaining rate profile S in (t) that boosts the accompanying target work:

$$J = V(t_f)X(t_f) \tag{10.38}$$

It can without much of a stretch be demonstrated that the solitary ideal control input is composed as takes after:

$$S_{in}(t) = \sqrt{K_S K_I} + \frac{V(t)}{q} k_I \frac{\mu_0 X(t)}{1 + 2\sqrt{K_S K_I}} \tag{10.39}$$

Where V (t) and X (t) are arrangements of the differential Equations [10.2], [10.4] in which S takes the steady esteem K_K S_I. It is acquired by first taking note of that the Hamiltonian of the framework is equivalent to:

$$H = \lambda_1 \left(\frac{q}{V}(S_{in} - S) - k_1 \mu X \right) + \lambda_2 (\mu X - DX) + \lambda_3 q \tag{10.40}$$

Which can be rewritten as follows:

$$H = \phi + \psi u \tag{10.41}$$

With u = S_{in} and

$$\phi = \lambda_1 \left(-\frac{q}{V} S - k_1 \mu X \right) + \lambda_2 \left(\mu X - \frac{q}{V} X \right) + \lambda_3 q \tag{10.42}$$

$$\psi = \lambda_1 \frac{q}{V} \tag{10.43}$$

In the above connections λ_1, λ_2, λ_3 are the co-expresses whose flow are given by the accompanying conditions:

$$\frac{d\lambda_1}{dt} = -\frac{\partial H}{\partial S} = (\lambda_1 k_1 - \lambda_2) X \frac{\partial \mu}{\partial S} + \lambda_1 \frac{q}{V} \tag{10.44}$$

$$\frac{d\lambda_2}{dt} = -\frac{\partial H}{\partial X} = (\lambda_1 k_1 - \lambda_2)\mu + \lambda_2 \frac{q}{V} \tag{10.45}$$

$$\frac{d\lambda_3}{dt} = -\frac{\partial H}{\partial V} = \frac{q(\lambda_1 (S_{in} - S) - \lambda_2 X)}{V^2} \tag{10.46}$$

Give us a chance to process the ideal particular control contribution (for which $\psi = 0$). In the first place take note of that:

$$\psi = 0 \Rightarrow \lambda_1 = 0 \tag{10.47}$$

The time subordinate of ψ is then equal to:

$$\frac{d\psi}{dt} = -\lambda_2 X \frac{q}{V} \frac{\partial \mu}{\partial S} \tag{10.48}$$

It is equivalent to zero if $\lambda_2 = 0$ or if $\partial\mu/\partial S = 0$. Just the second choice is of intrigue. It relates to the (consistent) estimation of the substrate fixation S that boosts the development rate μ, that is:

$$S_{\sin g} = \sqrt{K_I K_S} \tag{10.49}$$

The second-order derivative of ψ is written as follows by considering $(\partial\mu/\partial S) = 0$:

$$\frac{d^2\psi}{dt^2} = -\lambda_2 X \frac{\partial^2 \mu}{\partial^2 S^2} \left(\frac{q}{V} (S_{in} - S) - k_1 \mu X \right) \tag{10.50}$$

It is identical to zero on the off chance that we consider the control input [10.39].

In this simple case, the optimal control will be either of the 'bang-singular' type (if the initial substrate concentration S (0) is different from K_K I_S), or purely 'singular' (if $S(0) = K_I K_S$).

In this basic case, the ideal control will be both of the 'blast particular' sort (if the underlying substrate focus S (0) is not quite the same as K_K I_S), or absolutely 'solitary' (if $S(0) = K_I K_S$).

Note that the sustaining rate profile Equation [10.39] is exponential. Undoubtedly under the speculation that q and S (and along these lines μ) are steady, the aggregate biomass V (t) X (t) is equivalent to:

$$V(t)X(t) = V_0 X_0 e^{\mu_{\max} t} \tag{10.51}$$

With

$$\mu_{\max} = \frac{\mu_0}{1 + \sqrt{K_S / K_I}} \tag{10.52}$$

It is significant this relates to a critical measurement in the control (sustained)-group bioreactors: the ideal biomass profile is exponential, and in a perfect world the control activity that enables it to have this exponential profile, and accordingly the augmentation of the biomass creation, should be time-fluctuating (exponential in the current circumstance) (Figure 10.6)

The ideal control of the fed-group bioreactor is delineated in Figure 10.3 in the particular control case [10.39]; that is, when S (0) = $K_1 K_S$ with the accompanying arrangement of parameter and starting qualities:

$$\mu_{\max} = 5h^{-1}, K_S = 10\,g / L, K_I = 0.1g / L, k_1 = 1, q = 0.5277L / h$$
$$S(0) = 1g / L, X(0) = 0.1g / L, V(0) = 10l, V(t_f) = 20l$$

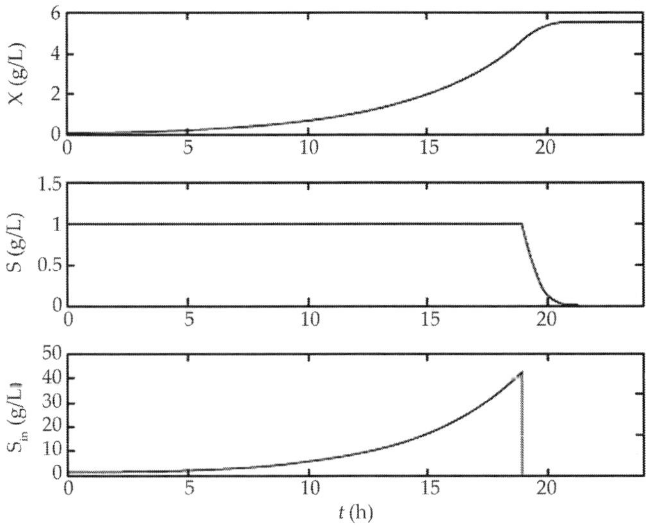

Figure 10.6: Optimal control of a fed-batch bioreactor, (*Source: G. Caldwell Darwin (2013), Robotics and Automation in the Food Industry: Current and Future Technologies, Woodhead Publishing*).

The ideal control stops when the tank is filled (t_f= 18.95 h), at that point the group stage begins (q = 0) when the substrate S is drained.

The illustration is specifically noteworthy since it stresses that controlling the substrate focus S at a (consistent) esteem (that expands the particular development rate) is sufficient to amplify the aggregate biomass generation. This implies we decrease a direction following issue to a straightforward control issue. Anyway, the issue is this is in fact just a 'control.' Truly, while the substrate fixation S stays steady (or is expected to stay consistent), alternate process factors, and specifically the biomass, are covering a wide range of qualities. In the above basic case, we have seen that under ideal conditions, the biomass amount V (t) X (t) is following an exponential profile. It is hence rather clear that not considering this marvel may prompt poor control execution. Furthermore, it is very straightforward that the tuning of the controller might be productive in some working condition run while being less proficient, if not in any way, shape or form sad, in another working condition run. This is outlined in Figure 10.4 for the genuine instance of a modern sustained group reactor where the PI controller (which controls the disintegrated oxygen focus), which has been tuned for conditions comparing to the early piece of the maturation, gives a loud flag after a sharp starting transient and after that shows an oscillatory conduct toward the finish of the aging, basically in light of the fact that the framework flow are then altogether different from those saw toward the start of the aging (Figure 10.7).

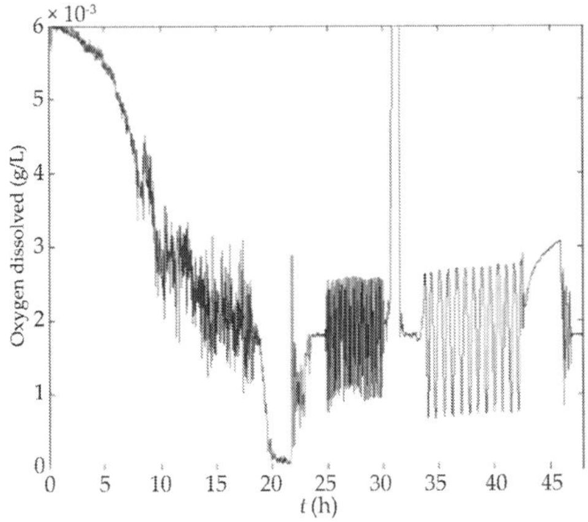

Figure 10.7: PI control of the dissolved oxygen concentration in an industrial fed-batch bioreactor, (*Source: G. Caldwell Darwin (2013), Robotics and Automation in the Food Industry: Current and Future Technologies, Woodhead Publishing*).

This can be clarified by seeing at the shut circle flow of the substrate fixation S. With the PI controller, this is composed as takes after (Figure 10.8):

$$\frac{dS}{dt} = -k_1\mu X + \frac{S_{in} - S}{V}\left[K_p(S^* - S) + \frac{K_P}{\tau_i}\int_0^t (S^* - S)d\tau\right] \tag{10.53}$$

$$= \frac{1}{V}\left(-k_1\mu V_0 X_0 e^{at} + (S_{in} - S)\left[K_p(S^* - S) + \frac{K_P}{\tau_i}\int_0^t (S^* - S)d\tau\right]\right) \tag{10.54}$$

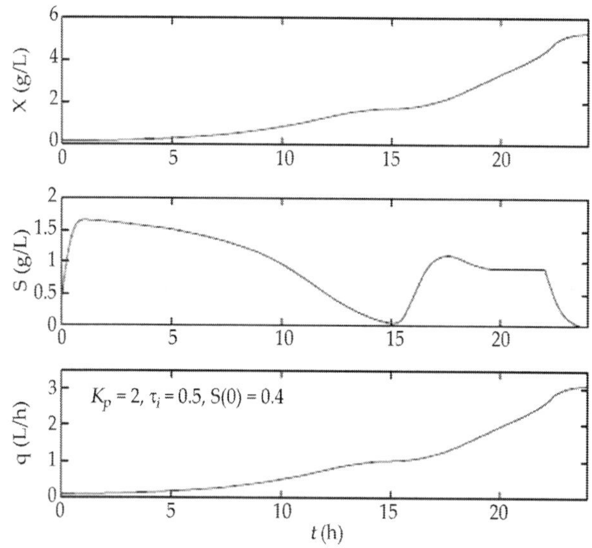

Figure 10.8: PI control of fed-batch bioreactor: constant gain and S(0) = 0.4 g/L, (*Source: G. Caldwell Darwin (2013), Robotics and Automation in the Food Industry: Current and Future Technologies, Woodhead Publishing*).

Under the supposition that S stays near S* (and that in outcome μ remains roughly steady). Note that the main term (exponential) on the correct hand side is just remunerated by terms that don't differ similarly. This conduct is represented (Figure10.5) for our situation ponder for an underlying condition S (0) = 0.4 g/L, not the same as the ideal esteem (disregarding a hostile to windup activity).

An answer comprises then to expressly join the exponential development term into the control input. In a PI controller, this should be possible by means of the relative pick up Kp (and just on the corresponding activity). On the off chance that we signify the basic pick up of the PI controller:

$$K_i = \frac{K_p}{\tau_i} \tag{10.55}$$

Where K_i is kept steady, and on the off chance that we pick the corresponding activity pick up an incentive as takes after:

$$K_p = \bar{K}_p e^{at} \tag{10.56}$$

as proposed in the above condition which is currently modified as takes after:

$$\frac{dS}{dt} = \frac{e^{\mu t}}{V}[\bar{K}_p (S^* - S)(S_{in} - S) - k_1 \mu V_0 X_0] + \frac{S_{in} - S}{V} K_i \int_0^t (S^* - S) d\tau \tag{10.57}$$

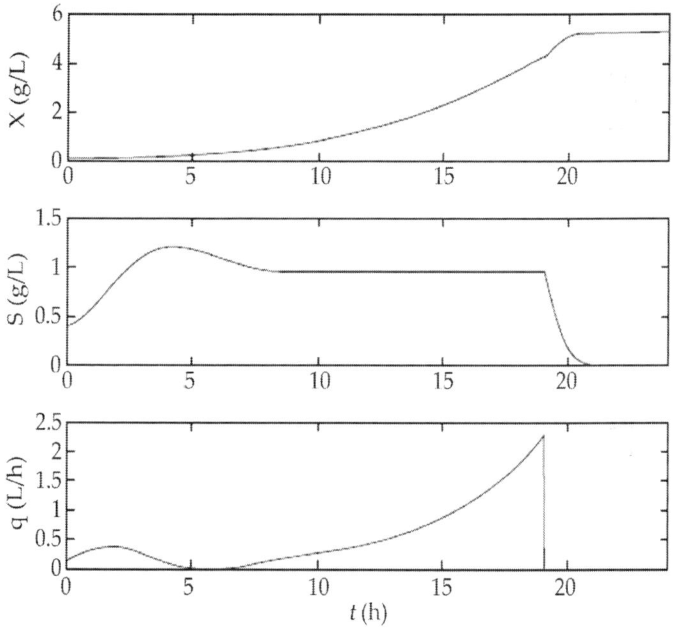

Figure 10.9: PI control of a fed-batch bioreactor; proportional gain with an exponential term.

Source: G. Caldwell Darwin (2013), Robotics and Automation in the Food Industry: Current and Future Technologies, Woodhead Publishing

Note that if the exponential term in the pickup K_p is fittingly picked (i.e. as close as conceivable to the ideal estimation of the particular development rate μ) we present in the control input an exponential activity that enables us to deal with the exponential biomass development. This is shown in Figure

10.6 where an indistinguishable conditions from in Figure 10.5 have been considered.

10.9. REAL-TIME OPTIMIZATION

Most versatile control plans archived in the writing are created for direction to known set-focuses or following known reference directions. In a few applications, in any case, the control target could be to streamline a goal work, which can be a component of obscure parameters, or to choose the coveted estimations of the state factors to keep an execution work at its extremum esteem. Self-advancing control and extremum-chasing control are two strategies to deal with these sorts of enhancement issues. The assignment of extremum looking for is to locate the working set-focuses that expand or limit a goal work. Since the early research deal with extremum control in the 1920s, 26 a few utilizations of extremum control approaches have been accounted for. Krstic et al. displayed a few extremum control plans and security examination for extremum looking for of direct obscure frameworks and a class of general non-straight frameworks.

The suggestions for the biochemical business are clear. In this area, it is perceived that even little execution changes in key process control factors may bring about considerable monetary increases.

The extremum-chasing structure proposed by Guay and Zhang accept that the target work is unequivocally known as a component of the framework states and questionable parameters from the framework dynamic conditions. Parametric vulnerabilities make the on-line remaking of the genuine cost unimaginable, with the end goal that lone an expected esteem in light of parameter gauges is accessible. The control objective is to at the same time distinguish and manage the framework to the least cost working point, which relies upon questionable parameters. The principle preferred standpoint of this approach is that one can promise some level of transient execution while accomplishing the enhancement destinations when a sensible practical estimation of the target work is accessible.

The proposed conspire uses express structure data of the target work that relies upon framework states and obscure plant parameters. This plan depends on Lyapunov's strength hypothesis. Therefore, worldwide security is guaranteed amid the looking for of the extremum of the non-straight blended tank bioreactors. It is additionally demonstrated that once a specific level of industriousness of excitation (PE) condition is fulfilled, the meeting of the extremum-chasing system can be ensured. In this area we focus on the versatile extremum-chasing control of bioreactors working in the fed-cluster mode, however, a few different choices of comparable plans (counting the utilization of general estimation, for example, manufactured neural systems (ANN)) have been proposed in the writing for various bioprocess arrangements, yet in addition for concoction reactors, 43– 46 mash and paper forms and biomedical frameworks.

10.9.1. Adaptive Extremum-Seeking Control of Fed-Batch Bioprocess

Fed group bioreactors speak to a vital class of bioprocesses, basically in the sustenance business and in the pharmaceutical business yet additionally for instance for biopolymer applications (PHB). One of the key issues in the activity of encouraged cluster reactors is to advance the generation of a union item (e.g., penicillin, chemicals, and so on) or biomass (e.g., cook's yeast). They are thusly perfect possibility for ideal control systems. An escalated inquire about movement has been dedicated to ideal control of (bolstered cluster) bioreactors for the most part in the 1970s and in the 1980s. However by and by, in view of the extensive vulnerability identified with the demonstrating of the procedure progression, 15 poor execution might be normal from such control systems, and in spite of the fact that from the earlier alluring, ideal control has not been to a great extent connected to modern bioprocesses. Elective methodologies have been proposed for taking care of the procedure vulnerabilities with a versatile control plot. In the present approach, we propose to go above and beyond by including a static ideal pursuit in the versatile control conspire.

Consider the dynamical model of the basic microbial development process in a fed-batch reactor Equations (10.2)–(10.4) with one vaporous item:

$$y = k_2 \mu X \tag{10.58}$$

where y (g/L/h) is the creation rate of the response item and k 2 is a yield coefficient. An ordinary circumstance in bioprocess applications is the point at which the biomass focus isn't accessible for on-line estimation while the vaporous outpouring rate y (e.g., CO_2) is simpler to quantify online. We consider here that lone S and y are quantifiable while the biomass fixation X isn't accessible for input control. Here we consider the extremum-chasing issue for the bioprocess display Equations (10.2)–(10.4) with a particular development rate μ communicated by the Haldane show (see Figure 10.1).

The control objective is then to outline a controller with D as the control activity to such an extent that the biomass creation VX accomplishes its greatest toward the finish of the fed-cluster task. As clarified in the first area, the augmentation will be finished if the particular development rate is kept at its most extreme esteem:

$$S^* = \sqrt{K_S K_I} \qquad (10.59)$$

From the above contemplations, we realize that if the substrate fixation S can be settled at the set-point S * then the generation of biomass is expanded. Be that as it may, since the correct estimations of the Haldane display parameters K_S, μ_0 and K_I, are generally obscure, the coveted set-point S* isn't accessible. In this work, a versatile extremum-chasing calculation is created to look through this obscure set-point to such an extent that the biomass generation toward the finish of the reactor task, that is, v (tf) X (tf) (with tf the last time of the fed-group activity) is expanded.

REFERENCES

1. Braatz, R. D., & Hasebe, S., (2002). Particle size and shape control in crystallization processes. 6th International Conference on Chemical Process Control, AIChE Symposium. *Series, 98*, 307–327.
2. Bruin, S., & Jongen, Th. R. G. (2003). Food process engineering: the last 25 years and challenges ahead. *Comprehensive Reviews in Food Science and Technology, 2*,pp.42.
3. Caldwell G. Darwin, (2013). Robotics and Automation in the Food Industry: Current and Future Technologies. Woodhead Publishing. Page number: 267-287
4. Henry, C., & Chapman, C., (2002). The Nutrition Handbook for Food Processors. [ebook] Woodhead Publishing. CRC Press Boca Raton, FL. Available at: http://citeseerx.ist.psu.edu/viewdoc/download?doi=10.1.1.453.8190&rep=rep1&type=pdf [Accessed 8 May 2018].
5. James, S., Gigiel, A., & Hudson, W., (1983). The Ultra Rapid Chilling of Pork. *Meat Science 8*, 63–78.
6. Mizrahi, S., (1996). Leaching of Soluble Solids During Blanching of Vegetables by Ohmic Heating. *J. Food Eng. 29*(2), 153–166.
7. Ramaswamy, H., & Marcotte, M., (2006). Food processing principles and applications. CRC, Boca Raton, FL.Page Number: 198-236
8. Ramaswamy, H., Marcotte, M., Sastry, S., & Abdelrahim, K., (2014). Ohmic Heating in Food Processing. CRC Press. 1st edition
9. Taylor, A., (1972). Influence of Carcass Chilling rate on Drip in Meat. In: Meat Research Institute Symposium No. 2. Meat Chilling–Why and How? Bristol: Meat Research

CHAPTER 11

FOOD PACKAGING TECHNOLOGY

"With a little more tweaking, we could make orange juice in the orange without any packaging or processing."

—Homaro Cantu

CONTENTS

Food packaging has evolved from simply a container to hold food to something today that can play an active role in food quality. Many packages are still simply containers, but they have properties that have been developed to protect the food. These include barriers to oxygen, moisture, and flavors. Packaging has allowed access to many foods all through the year that otherwise could not be preserved. It is interesting to note that some packages have actually allowed the creation of new categories in the supermarket. In essence, packaging should provide information about product identity, the mount and the quantity of servings, nutritional content, ingredient contents, consumer information, advertising and other promotional offers. This chapter discusses food packaging requirements in compliance with global food safety guidelines and protection against expected shelf life.

11.1. AN OVERVIEW OF FOOD PACKAGING

Food packaging is defined as a coordinated system of preparing food for transport, distribution, storage, retailing, and end-user to satisfy the ultimate consumer with optimal cost (Coles et al., 2003). Food packaging is an essential part of modern society; commercially processed food could not be handled and distributed safely and efficiently without packaging. The World Packaging Organization (WPO) estimates that more than 25% of food is wasted because of poor packaging (WPO, 2009). Thus, it is clear that optimal packaging can reduce the large amount of food waste. Moreover, the current consumer demand for convenient and high-quality food products has increased the impact of food packaging. The purpose of this chapter is to provide background knowledge for those who are interested in or may become involved in the development of food packaging and/or processing.

Packaging is both strategically and tactically important in the exercise of the marketing function. Distinctive or innovative packaging is often a key to the leverage that most companies seek in the case where there is competition between brands. In the United Kingdom, for example, the development of the famous widget for canned draught beers opened up marketing opportunities and new distribution channels for large breweries. There are various major packaging systems which are described as follows.

11.1.1. Primary Packaging

The first–level package that directly contacts the product is referred to as the "primary package." For example, a beverage can or a jar, a paper envelope for a tea bag, an inner bag in a cereal box, and an individual candy wrap in a pouch are primary packages, and their main function is to contain and preserve the product. Primary packages must be non-toxic and compatible with the food and should not cause any changes in food attributes such as color changes, undesired chemical reactions, flavor, etc.

11.1.2. Secondary Packaging

The secondary package contains two or more primary packages and protects the primary packages from damage during shipment and storage. Secondary packages are also used to prevent dirt and contaminants from soiling the primary packages; they also unitize groups of primary packages. A shrink wrap and a plastic ring connector that bundles two or more cans together to enhance ease of handling are examples of secondary packages.

11.1.3. Tertiary Packaging

The tertiary package is the shipping container, which typically contains a number of the primary or secondary packages. It is also referred to as the "distribution package." A corrugated box is by far the most common form of tertiary package. Its main function is to protect the product during distribution and to provide for efficient handling. In inter-state and international trade, a quaternary package is frequently used to enhance the handling of tertiary packages. This is generally a metal container up to 12 m in length which can hold many pallets and is intermodal in nature. That is, it can be transferred to or from ships, trains, and flatbed trucks, by giant cranes. Certain containers can also have their humidity, gas atmosphere and temperature controlled and this is essential in certain situations such as the transportation of frozen foods (Figure 11.1).

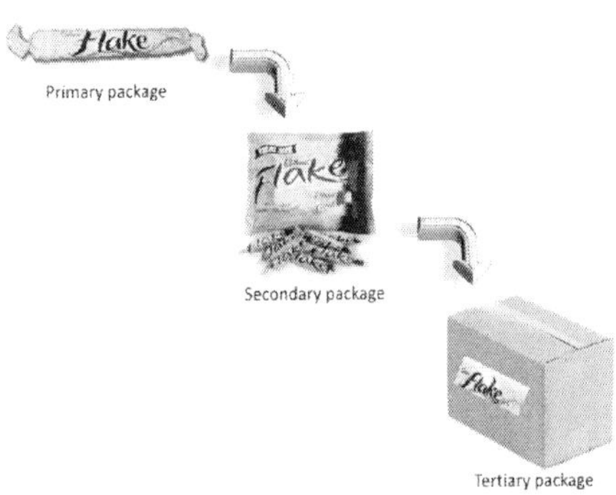

Figure 11.1: Types of packaging in the food industry, (*Source: https://www.slideshare.net/naveenaugust/packaging-labeling-in-food-industries*).

11.1.4. Unit Load

A unit load means a group of tertiary packages assembled into a single unit. If the corrugated boxes are placed on a pallet and stretch wrapped for mechanical handling, shipping, and storage, the single unit is referred to as a unit load. The objective is to aid in the automated handling of larger amounts of product. A fork-lift truck or similar equipment is used to transport the unit load.

11.1.5. Consumer/Industrial Packaging

Packaging systems can also be categorized into consumer and industrial packaging. Consumer packaging means a package that will be delivered to the ultimate consumer in the retail store. Usually, primary and secondary packages fit in this category. Industrial packaging means a package for warehousing and distribution to the retail store. Tertiary packages and unit loads fit in this category. Not all package systems are actually composed of a set of primary, secondary, and tertiary packages. For example, the packaging system for potato chips usually consists only of a flexible barrier bag and a corrugated shipping container before they are palletized, while mayonnaise jars are sold in a club store as a two-pack consisting of plastic bottles, shrink wrap, corrugated boxes, and pallet. Often, the distinction between consumer and industrial packaging is more clear-cut than between primary, secondary, and tertiary packaging.

11.2. METALS

Steel and aluminum are used for metal container and closure construction for food and drink products. Both are relatively low-cost materials that are nontoxic, having adequate strength and are capable of being work hardened.

11.2.1. Metal Requirements for Packaging

Metal packages for food products must perform the following basic functions if the contents are to be delivered to the ultimate consumer in a safe and wholesome manner;

(a) Preserve and protect the product.

(b) Resist chemical actions of product.

(c) Withstand the handling and processing.

(d) Withstand the external environment conditions.

(e) Have the correct dimensions and the ability to be practically interchangeable with similar products. from other supply sources (when necessary).

(f) Have the required shelf display properties at the point of sale.

(g) Give easy opening and simple/safe product removal.

(h) Be constructed from recyclable raw materials.

In addition, these functions must continue to be performed satisfactorily until well after the end of the stated shelf life period. Most filled food and drink containers for ambient shelf storage are subjected to some form of heat process to prolong the shelf life of the product. For food cans, this will normally provide a shelf life of up to 2–3 years or more. The heat process cycles used to achieve this are particularly severe and the containers must be specifically designed to withstand these conditions of temperature and pressure cycles in a steam/water atmosphere. Following heat processing, when the can temperature has returned to ambient, there will normally be a negative pressure in the can, i.e. a vacuum. Under these conditions, the food product itself does not provide any strength to the can to resist external loads.

In the case of carbonated beverage cans, which form the bulk of drink cans filled, once the container is closed, the carbonation pressure continues to provide significant physical support to the container until the moment of opening. In the case of still liquids, such as juices, juice drinks, and wine, nitrogen gas may be used to provide the necessary internal pressure for rigidity and compression strength of thin-walled DWI containers.

11.2.2. Steel

Steel is used in the form of a low–carbon steel which is initially produced as black plate. This is then converted into tinplate or tin-free steel (TFS) for container and closure manufacturer.

Tinplate is created by electrolytically coating black plate with a thin layer of tin. The tin is coated on both sides of the plate in thickness to suit the internally packed product and the external environment. Different thicknesses of tin may be applied to each side of the plate. Tin, plated in sufficient thickness, provides good corrosion-resisting properties to steel and is suitable for direct contact with many products including specific foodstuffs such as white fruits (e.g., peaches, apricots, pineapple, and pears) and certain tomato-based products (e.g., tomatoes in brine and beans in tomato sauce). However, for most foods and drinks it is necessary to apply an organic coating to the inside surfaces of the tinplate container to provide an inert barrier between the metal and the product packed. This barrier acts to prevent chemical action between the product and container and to prevent taint or staining of the product by direct contact with the metal. The tin surface assists in providing good electrical current flow during welding processes. Being a very soft metal, it also acts as a solid lubricant during the wall ironing process of forming two-piece thin wall cans.

TFS, also referred to as electrolytic chromium/chrome oxide coated steel (ECCS), is created by electrolytically coating black plate with a thin layer of chrome/chrome oxide. This must then be coated with an organic material to provide a complete corrosion-resistant surface. The metallic layer of ECCS provides an excellent key for adhesion of liquid coatings or laminates to the surface. ECCS is usually marginally less expensive than tinplate. However, being a matt surface, after coating with clear lacquer it does not provide a reflective surface like tinplate. ECCS in its standard form is not suitable for welding without prior removal of the chrome/chrome oxide layer. The Japanese steelmakers have developed modified tin-free metallic coatings for steel that do permit satisfactory welding of this material.

Figure 11.2: 0.18-0.4 mm TFS for beer caps/ food can packing, (*Source: https://www.alibaba.com/product-detail/0-18-0-4mm-TFS-with_60283812504.html*).

11.2.3. Aluminum

Aluminum for light metal packaging is used in a relatively pure form, with manganese and magnesium added to improve the strength properties. This material cannot be welded by can–making systems and can only be used for seamless (two–piece) containers. The internal surfaces of aluminum containers are always coated with an organic lacquer because of the products normally packed. Packaging applications of aluminum as a single material, or combined with paper and plastics, cover all industrial needs: rigid containers (cans), semi-rigid containers (trays, jars, and compressible tubes), flexible pouches and bags, thin films for wrapping, easy-open closures for aluminum or tinplate cans, capsules and caps for bottles or jars, etc. The use of aluminum as an alternative to tinplate for rigid containers is limited to the packaging of beer and carbonated beverages. For these containers, aluminum alloys with variable concentrations of magnesium (up to 5%), manganese (1.5%), and traces of other elements like Fe, Si, Cu, Zn, etc., have been developed.

11.2.4. Aluminum Foil

Aluminum foil is made by rolling pure aluminum metal into very thin sheets, followed by annealing to achieve dead-folding properties (a crease or fold made in the film will stay in place), which allows it to be folded tightly. Moreover, aluminum foil is available in a wide range of thicknesses, with thinner foils used to wrap food and thicker foils used for trays. Like all aluminum packaging, foil provides an excellent barrier to moisture, air, odors, light, and microorganisms. It is inert to acidic foods and does not require lacquer or other protection. Although aluminum is easily recyclable, foils cannot be made from recycled aluminum without pinhole formation in the thin sheets (Figure 11.3).

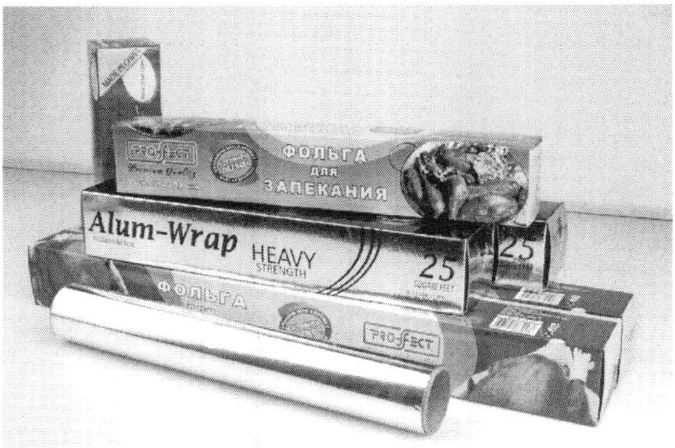

Figure 11.3: Aluminum Foil for food packaging, (*Source: https://cnhuayun.en.made-in-china.com/productimage/AXgnBxJbuUcY-2f1j00fSJtsPeabhcD/China-Aluminium-Foil-for-Food-Packaging.html*).

11.2.5. Laminates and Metallized Films

Lamination of packaging involves the binding of aluminum foil to paper or plastic film to improve barrier properties. Thin gauges facilitate application. Although lamination to plastic enables heat sealability, the seal does not completely bar moisture and air. Because laminated aluminum is relatively expensive, it is typically used to package high-value foods such as dried soups, herbs, and spices. A less expensive alternative to laminated packaging is metalized film. Metalized films are plastics containing a thin layer of aluminum metal (Fellows and Axtell, 2002). These films have improved barrier properties to moisture, oils, air, and odors, and the highly reflective surface of the aluminum is attractive to consumers. More flexible than laminated films, metalized films are mainly used to package snacks. Although the individual components of laminates and metalized films are technically recyclable, the difficulty in sorting and separating the material precludes economically feasible recycling.

11.3. CAN-MAKING PROCESSES

The most common form of packaging for thermally processed products is the can. The strength and non-breakability of metal makes it an ideal material for food packaging. Metal is also an excellent light and gas barrier, is quick and easy to seal and can withstand the temperature extremes of food processing. Cans are made from tinplate, tin-free steel or aluminum. There are two basic styles of cans: three-piece and two-piece. As the name indicates, a three-piece can is made from three pieces (a body blank and two ends), and a two-piece can is made from two pieces (one body and one end).

11.3.1. Three-Piece Cans

Three-piece 'sanitary' cans, consisting of a can body and two end pieces, are used to hermetically seal heat-sterilized foods and also to package powders, syrups and cooking oils. They are made from mild steel that is rolled to a strip 1.8 mm thick, and then dipped into hot dilute sulphuric acid, cold-rolled to 0.15–0.50 mm thick, and then temper-rolled to give required hardness and surface finish. Further cold reduction produces steel with greater stiffness and so thinner sheet can be used (known as 'double cold-reduction' steel). Tin is applied by electroplating to a different thickness of tin coatings on each side of the steel (e.g., 2.8–1.2 gm^2 or 0.1–0.3 mm thick). However, this produces a dull surface, due to the porous finish and it is heated quickly in a bath of hot oil or by electric induction (known as 'flow brightening') to slightly melt the tin and improve surface brightness and resistance to corrosion. It is then treated with chromate solution to stabilize the finish. The tin may be coated with the following lacquers to prevent interactions with foods. Epoxy-phenolic compounds are widely used, are resistant to acids, and have good heat resistance and flexibility. They are used in canned meat, fish, fruit, pasta and vegetable products. They are also coated with zinc oxide or metallic aluminum powder to prevent sulfide staining with meat, fish, and vegetables.

For food packaging, most three-piece cans are made using welded side seams. Soldering was the original method, but has been mostly discontinued due to concerns about lead contamination. Lead solder is no longer permitted for packaging of products sold in the US as well as in many other countries. Next, the body is flanged, and one end of the can is attached using a double seaming process. The double seam forms a hermetic seal by interlocking the end cover and body of the can with a rubbery sealing compound between them. After filling, the second end is attached in the same manner.

Cans for food packaging are often subjected to external and internal pressure during processing and storage. The can body may be rippled or beaded to increase its strength (Figure 11.4).

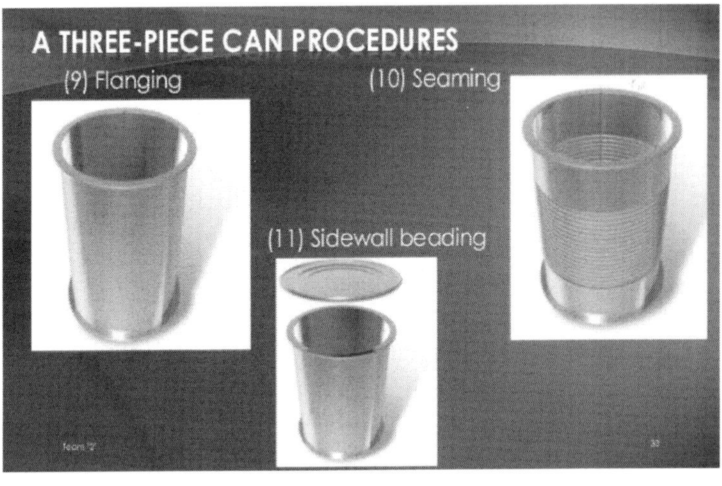

Figure 11.4: Procedures for a Three-piece can, (*Source: https://www.slideshare.net/DaliaElGhayesh/can-manufacturing*).

11.3.2. Two-Piece Single and Multiple Drawn (DRD) Cans

Pre-coated, laminated and printed tinplate or TFS is fed in sheet or coil form in a reciprocating press that may have single or multiple tools. At each tool station, the press cycle cuts a circular disc (blank) from the metal and whilst in the same station draws this into a shallow can (cup). During the drawing process, the metal is reformed from flat metal into a three-dimensional can without changing the metal thickness at any point. After this single draw, the can may be already at its finished dimension. However, bypassing this cup through a similar process with different tooling, it may be redrawn into a can of smaller diameter and greater height to make a draw-redraw can (DRD). This process may be repeated once more to achieve the maximum height can. At each of these steps, the can base and wall thickness remain effectively unchanged from that of the original flat metal. Following this body–forming operation, necking, flanging and beading operations follow according to the end use and height-to-diameter ratio of the can (as for three-piece welded cans). For all two-piece cans, pinhole and crack detection on finished cans is carried out in a light–testing machine. This measures the amount of light passing across the can wall using high levels of external illumination.

One advantage of two-piece cans is that there is only one can end instead of two, meaning that one major critical control hazard point is eliminated. The single drawing process is also used to make aluminum- or steel-tapered shallow trays for eventual heat-sealing with coated metal foil. The container bodies are constructed from metal laminated with organic film. The single drawing process is also used for the manufacture of folded aluminum baking trays and takeaway containers. In this process, the aluminum is allowed to fold, as the metal is converted from a flat sheet into a shaped container (Figure 11.5).

Figure 11.5: Two-piece single and multiple drawn (DRD) cans, (*Source: https.// silgancontainers.com/products/cans/*).

11.3.3. Two-Piece Drawn and Wall Ironed (DWI) Cans

The DWI cans are constructed from uncoated tinplate or aluminum. However, DWI cans for processed food are only made from tinplate as thin wall aluminum cans do not have sufficient strength to withstand the heat process cycles. For this process, the coiled metal, as it is unwound, is covered with a thin film of water-soluble synthetic lubricant before being fed continuously into a cupping press. This machine blanks and draws multiple shallow cups for each stroke. The cups are then fed to parallel body–making machines which convert the cups into tall cans. This is the drawing and ironing process where the cups are first redrawn to the final can diameter and then rammed through a series of rings with tungsten carbide internal surfaces which thin (iron) the can walls whilst at the same time increasing the can height.

During this process, the can body is flooded with the same type of lubricant used in the cupping operation. In addition to assisting the ironing process, the lubricant cools the can body and flushes away any metallic debris. No heat is applied to the can during this process—any heat generated being from the cold working of the metal as it is thinned. After the forming of the can body the uneven top edge of the can is trimmed to leave a clean edge and a can of the correct overall height. Trimmed can bodies are passed through chemical washers and then dried. This process removes all traces of lubricant and prepares the metal surfaces for internal and external coating and ultimately external decoration (drink cans only).

For drink cans, the clean cans are coated externally with a clear or pigmented base coat that forms a good surface for the printing inks. The coating is then dried by passing the cans through a thermally heated oven. The next step is a highspeed printer/decorator which applies the printed design around the outside of the can wall in up to eight colors plus a varnish. A rim-coater coat applies a heavy varnish to the base of each can in order to provide added protection against scuffing during distribution and external corrosion; especially as such products are often kept in the cold humid conditions of chilled refrigerators. The cans now pass through a second oven to dry the ink and varnish.

The inside of each can is coated with lacquer using an airless spray system. The special lacquer is applied to protect the can itself from corrosion and prevent its contents from interacting with the metal. This lacquer is finally dried in an oven at a temperature of about 210 °C. Following this, the can body now passes through a necker/flanger machine where the diameter of the top wall is first reduced (necked–in) before the top edge is flanged outwards to accept the can end. After all the mechanical forming operations have been completed, every can is tested by passing it through a light tester which automatically rejects any cans with pinholes or fractures.

For food cans, which will ultimately receive a paper label, an external coating is applied by passing them under a series of waterfalls of clear lacquer which protects the surface against corrosion. The lacquer is dried by passing the cans through a heated oven. Subsequently, the can body now passes through a flanging machine where the top of the can is flanged outwards to accept the can end, which will be fitted after the can is filled with product. The flanged can is next passed through a beading machine which forms circumferential beads in the can wall, to give added strength to the can. After all the mechanical forming operations have been completed, every can is tested by passing through a light tester which automatically rejects any cans with pinholes or fractures. The inside of each can is then coated with lacquer using an airless spray system. The special lacquer is applied to protect the can itself from corrosion and prevent its contents from interacting with the metal. This lacquer is finally dried in a thermal oven at a temperature of about 210 °C (Figure 11.6).

Figure 11.6: Two-piece drawn and wall ironed (DWI) cans, (*Source: https://silgancontainers.com/products/cans/*).

11.3.4. Shelf Life of Canned Foods

Foods that can be safely stored at room temperature, or "on the shelf," are called "shelf stable." These non-perishable products include jerky, country hams, canned and bottled foods, rice, pasta, flour, sugar, spices, oils, and foods processed in aseptic or retort packages and other products that do not require refrigeration until after opening. Not all canned goods are shelf stable. Some canned food, such as some canned ham and seafood, are not safe at room temperature. These will be labeled "Keep Refrigerated." In order to be shelf stable, perishable food must be treated by heat and/ or dried to destroy foodborne microorganisms that can cause illness or spoil food. Food can be packaged in sterile, airtight containers. All foods eventually spoil if not preserved (Table 11.1).

Table 11.1: Shelf Life of Canned Foods

Food	Storage On Shelf	Storage After Opening
Canned ham (shelf-stable)	2 to 5 years	3 to 4 days in the refrigerator
Low-acid canned goods Examples: canned meat and poultry, stews, soups (except tomato), spaghetti (noodle and pasta) products, potatoes, corn, carrots, spinach, beans, beets, peas, pumpkin.	2 to 5 years	3 to 4 days in the refrigerator
High-acid canned goods. Examples: juices (tomato, orange, lemon, lime, and grapefruit); tomatoes; grapefruit, pineapple, apples and apple products, mixed fruit, peaches, pears, plums, all berries, pickles, sauerkraut and foods treated with vinegar-based sauces or dressings like German potato salad and sauerbraten.	12 to 18 months	5 to 7 days in the refrigerator
Home canned foods	12 months. Before using, boil 10 minutes for high-acid foods; 20 minutes for low-acid foods	3 to 4 days in the refrigerator

Adapted from United States Department of Agriculture Food Safety Information Source: https://www.fsis.usda.gov/wps/wcm/connect/77ffde83-dc51-4fdf-93be 048110fe47d6/Shelf_Stable_Food_Safety.pdf?MOD=AJPERES.

11.4. GLASS

Glass is one of the oldest packaging materials used for food. It is almost completely chemically inert and therefore is suitable for packing many products and all foods. The glass used for food packaging is made by heating a mixture of sand (approximately 70 to 74% silica), broken glass or 'cullet' (15–30% of total weight), and 15% sodium carbonate Na_2CO_3 (added to silica to lower the melting point) by weight. It is known as soda-lime glass. Alumina (Al_2O_3) improves the chemical durability of the glass, and refining agents reduce the temperature and time required for melting, and also helps remove gas bubbles from the glass. Various compounds can also be added to give the glass some color (to block out light from transmitting through the glass), for example, FeO and Cr_2O_3 will produce a green tint, and cobalt (II) oxide-aluminum oxide ($CoAl_2O_4$) will result in blue glass. Amber is usually obtained by melting a composition containing iron oxide under strongly reduced conditions. Carbon is also added. Amber glass has UV protection properties and could well be suited for use with light-sensitive products.

Glass is a super-cooled liquid (i.e. a liquid that solidifies without any significant structural changes). Although generally considered fragile, newly formed glass fiber actually has the strength to support twice the weight of an equivalent steel fiber. Unfortunately, in practice, glass only has about one-thousandth of its theoretical strength because it is a liquid and so any defect will spread unhindered throughout its mass. Glass is a poor conductor of heat and sudden changes in temperature (Δ60–65 °C) may cause breakage due to heat shock. Sudden heating, which puts the surface of the glass under compression, is less dangerous than sudden cooling, which puts the surface in tension. Glass bottles and jars have some of the advantages of metal cans: they are impervious to microorganisms, pests, moisture, oxygen, and odors; they do not react foods or have chemicals that migrate into foods; they can be heat processed; they are recyclable, and (with new lids) they are re-useable; they are rigid, to allow stacking without damage; and unlike metal cans, they are transparent to display the contents.

The main disadvantages of glass are: the higher weight than most other types of packaging, which incurs higher transport costs; containers are easily broken, especially when transported over rough roads; they have more variable dimensions than metal or plastic containers, and there are potentially serious hazards from glass splinters or fragments that can contaminate foods. Glass containers are still widely used for foods such as juices, wines, beers, pickles/chutneys, and jams, especially in countries that have a glass-making factory, but their disadvantages and the high cost for imported containers in other places mean that they are steadily being replaced by plastic containers (Figure 11.7).

Figure 11.7: Grease-proof paper for food-grade packaging, (*Source: http://moneypanorama.com/1022323/global-food-greaseproof-papers-market-2017-nordic-paper-delfortgroup-krpa-holding-vicat-group-pudumjee-group/*).

11.5. PAPERS

Kraft paper is a tough paper which is used for 25–50 kg multi-wall sacks for powders, flour, sugar, fruits, and vegetables. It can be bleached white, printed or used unbleached (brown). It is usually used in several layers or 'plies,' to give the necessary strength. Sack material is described from the outer ply inwards with respect to the number and weight (or 'substance' in g per m^2) of the layers. For example, 2/90, 1/80 means that there are three plies, the outer two having a weight of 90 gm^{-2} and the inner having a weight of 80 gm^{-2}. Vegetable parchment is made from sulfate pulp which is passed through a bath of concentrated sulphuric acid to swell and partially melt the cellulose fibers, to make them plasticized.

This closes the pores and fills spaces in the fiber network to make the surface more intact than kraft paper, and thus makes the paper unaffected by grease and oils and gives greater wet strength properties. It is used to pack butter, cheese, and fresh fish or meat. Both types of paper preserve foods from dust and soils, but they have insignificant water vapor or gas barrier features and are not heat sealable. Sulfite paper is lighter and fragile than sulfate papers. It is used for grocery bags and sweet wrappers, as an inner liner for biscuits and in laminations. It may be glazed to enhance its wet strength and oil resistance, when it is known as MG (machine glazed) sulfite paper. Greaseproof paper is produced from sulfite pulp in which the fibers are more vigorously beaten to produce a tighter structure. It is fats- and oil-resistant, and although this property is lost when the paper becomes wet, it is commonly used for wrapping fish, meat and dairy products. Glassine is akin to greaseproof paper, but is given supplementary calendering to increase the density and produce a close-knit configuration and a high gloss finish. It is more resistant to water when dry but loses the resistance once it becomes wet. Tissue paper is a soft non-resilient paper used, for example, to protect fruits against dust and bruising.

11.5.1. Paperboard

Paperboard is thicker than paper with a higher weight per unit area and often made in multiple layers. It is commonly used to make containers for shipping—such as boxes, cartons, and trays—and seldom used for direct food contact. The various types of paperboard are as follows (Soroka, 1999);

(a) Whiteboard: Made from several thin layers of bleached chemical pulp, whiteboard is usually used as the inner layer of a carton. Whiteboard may be coated with wax or laminated with polyethylene for heat sealability, and it is the only form of paperboard endorsed for direct food contact.

(b) Fiberboard: Fiberboard can be solid or corrugated. The solid type has an inner whiteboard layer and outer kraft layer and provides good protection against impact and compression. When laminated with plastics or aluminum, solid fiberboard can improve barrier properties and is used to pack dry products such as coffee and milk powder. The corrugated type, also known as corrugated board, is made with two layers of kraft paper with a central corrugating (or fluting) material. Fiberboard's resistance to control abrasion and crushing damage makes it widely used for conveying bulk food and case packing of retail food products.

(c) Solid board: Possessing strength and durability, solid board has compound layers of bleached sulfate board. When laminated with polyethylene, it is used to create liquid cartons (known as milk board). Solid board is also used to package fruit juices and soft drinks.

(d) Chipboard: Chipboard is made from recycled paper and often comprises blemishes and impurities from the original paper, which makes it unsuitable for direct contact with food, printing, and folding. It is often lined with whiteboard to improve both appearance and strength. The least costly form of paperboard, chipboard is used to make the outer layers of cartons for foods such as tea and cereals.

11.6. PLASTICS

Plastics are a special group of polymers that can be formed into a wide variety of shapes using controlled heat and pressure at moderately low temperatures, compared to metals and glass. Plastics are actually a subdivision of polymers, but in packaging, the terms tend to be used interchangeably. There are hundreds of identified "species" of synthetic polymers but in practice, only a few polymers are often used for food packaging. The use of plastics has increased more rapidly than any other material, and plastic is now the second most used material for packaging. Each plastic has its own exclusive properties, based on its chemical configuration. The performance and interaction with a variety of foods are different for each material. Thus, the plastic material for the packaging of a specific food is selected to function well within the parameters of the application. This subsection focuses on the properties and applications of the plastics that are most commonly used for food packaging.

11.6.1. Polyethylene (PE)

Polyethylene, polymerized from ethylene, is the plastic most commonly used for food packaging. PE generally has flexibility, good moisture control, oil and chemical resistance, and good impact strength. PE is also an inexpensive plastic, so for applications where its performance is suitable, this plastic is usually the most economical choice. The simplest form of PE is a completely unbranched structure of $-CH_2-$ units. However, some side branches are always formed during polymerization. If the branches are relatively few and short (2–4 carbon atoms), the structure can fold and pack tightly and yields high-density polyethylene (HDPE). On the contrary, if there are many long branches, PE becomes low-density polyethylene (LDPE). Low-density polyethylene is softer and more flexible, and has lower tensile strength than HDPE. Since it has relatively weak intermolecular forces, LDPE has a low melting temperature, 105–115 °C, so it is a useful material for heat sealing.

LDPE also has good impact and tear strength. Common applications for LDPE include stretch wraps, shrink wraps, and many types of bags and pouches. LDPE is also used as an adhesive layer for multilayer composite structures, as a coating on paper to provide water protection (such as in milk cartons), inflexible lids for plastic tubs, in squeezable plastic tubes, in soft squeeze bottles, and in a variety of other applications.

11.6.2. Polypropylene (PP)

Polypropylene is polymerized from propylene gas, which is a relatively low–cost feedstock like ethylene (Soroka, 2008a). As with the PE family, PP has good chemical and grease resistance. Barrier properties of PP are similar to those of HDPE; it is a good water vapor barrier but a poor gas barrier. The polypropylene structure includes methyl groups ($–CH_3$) attached to every other carbon in the polymer main chain; subsequently, PP has a lower density and a higher glass transition temperature (the temperature above which a plastic becomes soft and flexible) and higher melting temperature than PE. At freezing temperatures, there is proximity between unmodified PP and its glass transition temperature, and therefore tends to have serious brittleness issues. On the other hand, PP is suitable for use with products that require moderately high temperatures such as hot filling or reheating (but not cooking) in a microwave oven. One of the main uses of PP in food packaging is enclosures (caps). Particularly for threaded caps, while HDPE is deformed too readily and loses sealing force under stress, PP maintains its original toughness and performs successfully.

11.6.3. Polystyrene (PS)

Polystyrene is a linear addition polymer of styrene resulting in a benzene ring attached to every other carbon in the main polymer chain. It is a material that is brittle and clear and has high surface gloss. The use of PS in food packaging is aesthetically appreciated, but the material cannot generally be used when extended shelf life is required because of its poor water vapor and gas barrier properties. The brittleness of PS restricts its application where good impact resistance is required. In order to reduce the tendency to fracture, oriented polystyrene (OPS) is commonly used. Typical applications include produce and meat trays, lids for drink cups, and low-priced party glasses. High-impact PS (HIPS) is a PS copolymer with polybutadiene (synthetic rubber). Adding the synthetic rubber causes HIPS to become opaque but improves the impact resistance considerably. HIPS is commonly used for disposable cutlery, tubs, and other thermoformed containers. Polystyrene foam incorporates small bubbles within the plastic, which increase the suppressing properties and insulating ability of PS. PS foam is usually called expanded polystyrene (EPS). While it is not rare for this material to be called styrofoam, that name is copyrighted for Dow Chemical Company's EPS building insulation and should not be used to describe a packaging material. Foamed PS is generally used for disposable coffee cups, meat and produce trays, egg cartons etc. (Figure 11.8 and Figure 11.9)

Figure 11.8: A Polystyrene tray containing raw chicken breast, (*Source: https://www.canadianpackaging.com/features/game-chicken-april-2017-canadian-packaging/*).

Figure 11.9: Plastics recycling facility (PRF) in Rushden, Northamptonshire, England, (*Source: http://zerowastezone.blogspot.com.ng/2016/02/commercial-plastic-film-recycling-zero.html*).

11.7. LOGISTICS ACTIVITY-SPECIFIC AND INTEGRATION ISSUES

Although they have been discussed separately, the three packaging functions, protection, utility, and communication directly influence each other. The importance of each function varies between logistical operations. Sometimes there are trade-offs between these functions. For example, unitization increases handling protection and productivity, but may decrease in-transit cube utilization and require different markings from packages shipped break-bulk. From the packages' point of view, there are many variations in logistical systems. In typical food and consumer product supply chains, unit loads are shipped by full truckload (TL) to the wholesale level. Here orders are picked, in one of several common methods, and then distributed in mixed loads on various types of pallets, carts or platforms to a variety of retail store types. On the other hand, some food products are floor loaded in trailers and delivered directly to a retail store without the use of pallets. Foodservice supply chains likewise include both direct and wholesale channels. And in e-commerce, each package may travel alone or at least with strangers, using a small parcel transport system.

In each type of system, packaging must not only provide protection, utility, and communication for specific operations, but it must also facilitate transitions between operations. In logistical systems, products change ownership and location, and packaging must be designed to satisfy a number of functional needs and users. For example, a slip sheet packaging system should not be implemented to improve trailer cube utilization without considering how it will increase the consignee's unloading time if he/she does not have the appropriate handling equipment. The more complex the system, the greater the need to study its handling methods, transport modes, facility dimensions, damage sources and communication needs, in order to design packages that will optimize the system. The next sections review the general functional requirements for packaging during each link in the supply chain: food processor, transport, warehousing and customer service issues.

11.7.1. Packaging Issues in Food Processing and Retailing

Food processors are concerned primarily with the packing operation. Their emphasis is on the cost and quality of incoming packaging materials, controlling the quality and productivity of the packing/filling process, hygiene, and mechanization. In instances where food safety is threatened, packaging can be a critical control point in a product manufacturer's hazard analysis critical control (HACCP) plan. These are such important packaging issues that many food processors never find the time to look beyond their receiving and shipping docks. But there are other opportunities for packaging to improve both inbound and outbound logistics.

11.7.2. Retail Customer Service Issues

Once a supply chain customer, such as a retailer, manufacturing plant, or contractor, receives a shipment, logistical packaging has to perform a new set of functions. The package needs to be opened easily without damaging the contents. Handling and the unpackaging operation should be quick and efficient, and reclosure may be desired. The product should be easy-to-identify, and the package may be required to display, provide special instructions, for installing and using the product. The package should minimize the customer's cost of disposal. Traditionally, retailers did not have much control over the packaging that they received. Most large manufacturers planned packaging to best suit their own operations, and small retail customers had little power to request improvements. Increasingly, however, food marketing channels are dominated by large retailers who demand specific types of packaging. For example, grocery chains increasingly specify the durability of shipping containers and the information needed, prohibit some box styles, and help their suppliers to redesign packages to better maximize cube utilization in transit. Some retailers even control the consumer packaging and sell their own brands.

11.7.3. Warehousing Issues

Traditional warehouse order picking can be regarded as a packaging operation. In a normal warehouse for consumer goods, large packages such as. full TLs or full pallet loads are received. These are disassembled into smaller components, e.g., shipping containers or consumer packages as orders are picked. The orders are then repacked into mixed loads as carriage to logistical customers such as retailers. Sometimes, the orders are so small that individual items are picked from shipping containers and re-boxed into a delivery package. Sometimes, the delivery package is simply an original box or pallet repacked with a mixed load. In other cases, returnable pallets or totes are used to supply to the retailer. Contemporary warehouses are called distribution centers, to accentuate the fact that they only generate income by moving goods. To a distribution center, storage signifies unproductive assets. The output of warehousing is very significant because order picking and materials handling are commonly very labor-intensive. There is a great deal of importance on tracking and handling productivity. Cases or orders picked per hour, trucks loaded or unloaded per hour, and pallets received and put away per hour are examples of warehousing productivity measures.

Packaging can augment productivity to order picking operations when products are packed in order sizes and cases that do not need additional labor to open or split. For example, if five is the standard order quantity, then products should be bundled into fives, rather than the usual case quantities of 12, 24 or 48. The steadiness of packing old-fashioned dozens is probing, given more logical reasons to specify other counts. The drift in distribution is to boost the order filling process, and many warehouses are now no more than cross-dock tasks. They quickly assemble the mixed load orders as shipments are received, or simply convey already mixed loads from a single manufacturer to delivery trailers. In cross-dock situations, there is an increased demand for packaging to be segmental and spontaneously recognizable to facilitate computerized sorting.

REFERENCES

1. Coles, R., McDowell, D., & Kirwan, M., (2003). Food Packaging Technology. Blackwell Publishing, Oxford [ebook] Available at: [Accessed 12 May, 2018].
2. Environmental Protection Agency., (EPA) (2010). Municipal Solid Waste Generation, Recycling, and Disposal in the United States: Facts and Figures for 2009. [online]Available at: www.epa.gov/epawaste/ nonhaz/ municipal/pubs/msw_2010_rev_factsheet.pdf [Accessed 16 May, 2018].
3. Fischer, C., & Davisen, C., (2010). Europe as a Recycling Society: The European Recycling Map. ETC/SCP Working Paper. Denmark: European Topic Center on Sustainable Consumption and Production (ETC/SCP).
4. Food and Drug Administration (FDA), (2002). Guidance For Industry, Preparation of Food Contact Notifications and Food Additive Petitions for Food Contact Substances: Chemistry Recommendations Final Guidance. Center for Food Safety and Applied Nutrition, Office of Food Additive Safety. www.fda.gov/food/guidanceregulation/guidancedocumentsregulatoryinformation/ingredientsadditivesgraspackaging/ucm081825.htm, accessed 20 November 2013.
5. Robertson, G., (2013). Food Packaging Principles and Practice. 3rd edn. Boca Raton, Florida: CRC Press, pp.728.
6. Soroka, (2002). Fundamentals of Packaging Technology, Institute of Packaging Professionals.3rd edition

INDEX

D

E

F

Q

R

S

T

U

V

W